D1362348

Structural Change in the American Economy

Harvard Studies in Technology and Society

The volumes in this series present the results of studies by the Harvard University Program on Technology and Society. The Program was established in 1964 by a grant from the International Business Machines Corporation to undertake an inquiry in depth into the effects of technological change on the economy, on public policies, and on the character of the society, as well as into the reciprocal effects of social progress on the nature, dimensions, and directions of scientific and technological developments.

Books in the series

Anthony G. Oettinger with Sema Marks, *Run, Computer, Run: The Mythology of Educational Innovation*, 1969

Emmanuel G. Mesthene, *Technological Change: Its Impact on Man and Society*, 1970

Anne P. Carter, *Structural Change in the American Economy*, 1970

Anne P. Carter

Structural Change in the American Economy

Harvard University Press, Cambridge, Massachusetts 1970

Distributed in Great Britain by Oxford University Press, London
Library of Congress Catalog Card Number 73–95516
SBN 674–84370–3
Printed in the United States of America

To Sarah and both Franklins,
without whom this book
might have been written earlier
—or not at all

Preface

This book is a mere drop in an enormous—and barely damp—intellectual bucket. Nevertheless, it represents a substantial investment of human and financial resources. The ideas and much of the data are products of a long-term and continuing program of research at the Harvard Economic Research Project, going back more than fifteen years. My debt to the Project is very great. The final phases of the research were supported by the Harvard University Program on Technology and Society under a long-term grant from the International Business Machines Corporation.

I have drawn on the work of many colleagues. Charlotte Taskier, Alfred Conrad, Alan Strout, Barbara Sundquist, Alex Korns, Ann Wyman, Samuel Rea, Frances Mesher, and Darlene Butler, among others, contributed essential parts of the basic data and analysis. Mrs. Butler also coordinated much of the initial spadework. Computations were rendered moderate by a matrix manipulation program, PASSION, written by C. William Benz. During the last two years, Brookes Byrd, Linda Saltman, and Daniel Ford assisted in computations and research with skill, ingenuity, and superb humor. Mrs. Byrd worked closely with me on problems of analysis and presentation.

None of this work would have been possible without the informational groundwork laid by the Office of Business Economics. Albert Walderhaug, Beatrice Vaccara, Jean Frazier, and Nancy Simon answered endless questions about the 1947 and 1958 input-output studies; Mrs. Vaccara and her staff undertook an official reconciliation of the 1947 and 1958 input-output tables and thus revised and improved the data. Much of the work reported here is inseparable from research performed under contract with the Interagency Growth Project. The cooperation and guidance supplied by Jack Alterman was essential, as was his continuing moral support.

Millicent Milanovich, Virginia Rosenthal, and Charlene Long deciphered my handwriting, corrected my grammar, and typed the manuscript. The figures were drawn by Franklin Fancher.

Darlene Butler, Karen Polenske, Andrew Brody, Wassily Leontief, and Brookes Byrd read the manuscript at various stages and offered useful criticisms and suggestions. My analytical and philosophical debt to Professor Leontief will be obvious to any reader.

I am more than grateful for my husband's unique understanding and optimism and for the flexibility of our children.

Contents

Figures

Tables

Part I Structural Change and Industrial Specialization

Chapter 1 Structural and Technological Change

1.1 Intermediate Inputs and the Economics of Technological Change

In our era of population explosion, thoughtful individuals are beginning to weigh their social responsibilities before adding yet another human baby to the crowded world. In the world of ideas, the book explosion poses an analogous threat, and it behooves prospective authors to compare the benefits against the social costs of adding yet another volume to a professional literature that threatens to crowd and choke new ideas with its own profusion. But books, like babies, are not often conceived in a spirit of social dedication and soul searching, and once started, they grow with autonomous persistence. An author feels called upon to furnish a rationale for foisting his intellectual offspring on the world. In all fairness, it is more likely to be an excuse than a justification.

While the economics of technological change has only recently become fashionable, the literature is already proliferating at a rate that makes it difficult for the individual scholar to assimilate and, at the same time, to add to it. Still, it continues to grow. Lave (1966:207–220) has compiled a thirteen-page bibliography on the mainstream of modern economic thought in this area, and to that must be added a much greater amount of material accumulating in the related, almost indistinguishable, fields of production analysis, growth, and economic development. *A priori*, the marginal value of any contribution is likely to be small. A *raison d'être*, or at least a rationale, for another book on technological change is called for at the outset: What new areas will it explore; what new insights will it contribute in order to earn its space on crowded bookshelves?

The principal novelty of the present study lies in its explicit concern with intermediate inputs, as well as primary factors, in the analysis of changing economic structure. A distinguished group of economists presented contributions in the area of production functions at a 1964 conference on income and wealth. At that meeting, Evsey Domar pointed out that the papers (recently published in a 515-page book: Brown, ed. 1967) contained not a single reference to inputs of materials. In studies ranging over many different levels of detail—from three-digit Standard Industrial Classifications (SIC) to an aggregated national economy—and representing different time periods and nations, it was tacitly accepted that modern analysis deals with only two factors of production: labor and capital. In the interests of reality, economists concede that more detail is desirable—that labor should be specified by types of skill and education, or by how much it has "learned by doing"; that capital may be stratified by vintage; that perhaps the classical third factor, natural resources, should be introduced if and when there is information. However, it is very much in the current tradition to deal only in primary factors: the coal and ore and steel and chemicals and fibers and aluminum foil; sausage casings, wire products, wood pulp, electronic components, trucking, and business services that establishments furnish to each other are netted out. These remain enclosed in the economic black box that converts primary inputs into final output—value added to gross national product.

Modern practice has much to recommend it. It spotlights what many of us care most about: people as workers and as consumers. Furthermore, this traditional approach limits itself to the variables for which there is relatively full information, often enough of it to permit fairly sophisticated statistical manipulation. Information on intermediate inputs—the transactions within the black box—is harder to find, often fragmentary, clumsier to process. Perhaps only confusion, or at best awe, will result from attempts to analyze the contents.

The present study is rooted in the premise that an explicit analysis of changing intermediate input requirements adds more to insight than it does to confusion—particularly in the understanding of technological change. The study of changes in intermediate input requirements is easily justified in its own right. Many practical problems of business and government require an understanding of how, and at what rate, use of plastics or truck transportation or producers' services is changing. Indeed, it is difficult to conceive of studying some central aspects of technical change—such as invention or diffusion of new techniques—without introducing specific intermediate inputs. New technology involves new products and new ways of combining old products. Many of these new products are sold industrially, and some never reach the ultimate consumer. These are an essential part of the picture. They turn up as variables in questions and answers about our economic system that cannot be discussed at a highly aggregative level. Most important, they are indispensable in bridging the gap between engineering and technical information, on the one hand, and economic description, on the other. As detailed planning and forecasting become standard practice for business and government, the integration of ideas on new techniques, trends, and commonsensible business judgment in a consistent economic framework is required (Carter 1968).

Blending of these disparate elements is not possible in a highly aggregative system or in terms of net factor inputs alone. Working with the details of intermediate, as well as primary, input requirements has the advantage that it permits the absorption of a great deal of fragmentary information from extra-economic sources.

Explicit introduction of intermediate inputs also contributes to the understanding of the conventional labor-capital-output relationships.

Superficially, it is clear that long-run changes in labor productivity are rooted in changes in the organization of production, that new materials, components, communications, as well as new types of capital goods, have been prerequisite to continued rises in primary factor efficiency. The major methodological presupposition of modern economics seems to be, nevertheless, that these organizational changes are uninteresting—and certainly not essential for understanding change. But they are definitely part of change, and they may even affect the net relationships. If they are to be netted out, in our most sophisticated work, then it is at least worthwhile to investigate how this otherwise neglected aspect of technology acts on the more familiar elements.

Some of the present findings should be reassuring. This study shows that intermediate input requirements are characterized by relative stability, as compared with primary inputs; the study explains why this is so. Thus, it should contribute some welcome theoretical and empirical basis for the established practice of ignoring them (Eisner 1967). Even more important is the provision of an analytical and informational basis for studying intermediate input requirements in relation to those questions where they cannot reasonably be overlooked.

This study sets out to show how technological change has affected industrial specialization, as well as direct primary input requirements, and how these components of change are interrelated. It is conducted at an intermediate level of detail. At various stages, the economy is divided into from 38 to 76 industries. Certainly, it would not be wise to undertake such a large-scale study of economically interdependent developments without placing it explicitly in a general-equilibrium framework. Much of the theoretical structure is standard, although some of the formulations used in later phases of the analysis are new. In any case, the primary emphasis is empirical rather than methodological: the principal findings are not indifferent to the specific data used. The conclusions are factual; because the data are subject to error, the findings, specific and general, could possibly be wrong. It seems high time that proponents of economic methodologies tasted their own puddings.

1.2 Input-Output and Structural Description

Input-output structure

Input-output tables and coordinated information on labor and capital coefficients provide the central informational foundation of the study, while input-output theory provides the basic theoretical framework. Excellent introductions to input-output analysis, such as Miernyk (1965); Chenery and Clark (1959); as well as the Leontief (1951) classic, are available; and no comprehensive survey need be offered here.

An input-output table gives a detailed accounting of the amounts of goods and services that individual industries buy from and sell to each other in a particular year. Table 1.1 is a 1958 input-output table for the United States, aggregated to only eight sectors for illustrative purposes. Reading across each row, we see the amounts that a particular industry sold to all sectors, including itself, and to the "final demand" or final users categories: households, government, foreign trade, net inventory change, and gross private capital formation. Transactions are measured in dollars, although they could be measured in physical units (kilowatt hours, tons, number of automobiles, and so on). The materials industry, for example, sold $276 million worth of its output to the mining industry, $8,565 million worth to itself, and $3,994 million worth to final demand. Reading down individual columns, we see how much each sector purchased as inputs from other sectors. Column 1 shows that the materials industry purchased $8,565 million worth of materials, $1,505 million worth of metalworking products, and $506 million worth of agricultural products to produce 1958 output. The bottom row of the table gives "value added" for each sector, which is the sum of its payment to labor and of its capital charges, profits, direct taxes, and miscellaneous disbursements. Additional information on the value added components, such as data on labor requirements in physical units and on the stock of capital goods, is usually specified separately.

Input-output coefficients are obtained by dividing the entries in a column, which are an industry's inputs, by that industry's output. Input-output coefficients enumerate the amounts an industry purchased from all other industries and from value added, per unit of its own output. Thus, each input coefficient shows the requirement for a particular input, per unit of a particular output. A column of coefficients, then, gives a detailed quantitative description of the technique of production used by a sector, a sort of recipe for its output, with specifically enumerated inputs as ingredients. As an input-output coefficient table includes a column of input-output coefficients for

Table 1.1 Eight Sector Input-Output Table[a] for 1958 (millions of 1947 dollars)

Sector	Sector[b]								Final demand	Gross domestic output
	1	2	3	4	5	6	7	8		
1. Materials	8,565	8,069	8,843	3,045	1,124	276	230	3,464	3,994	37,608
2. Metalworking	1,505	6,996	6,895	3,530	3,383	365	219	2,946	19,269	45,100
3. Construction	98	39	5	429	5,694	7	376	327	39,348	46,322
4. Transportation eq. & utilities	999	1,048	120	9,143	4,460	228	210	2,226	22,625	41,059
5. Services & trans.	4,373	4,488	8,325	2,729	29,671	1,733	5,757	14,756	137,571	209,404
6. Mining	2,150	36	640	1,234	165	821	90	6,717	−653	11,199
7. Agriculture, etc.	506	7	180		2,352		18,091	26,529	8,327	55,992
8. All other	5,315	1,895	2,993	1,071	13,941	434	6,096	46,338	82,996	161,080
Value added	14,097	22,522	18,320	19,877	148,614	7,344	24,923	57,777		313,475
Total inputs	37,608	45,100	46,322	41,059	209,404	11,199	55,992	161,080	313,475	921,240

[a] Each entry tells the volume of sales by the sector named at the left to the sector numbered at the top.
[b] The sector numbers across the top correspond to the sectors numbered and named in the table.

every sector, it gives a comprehensive structural description of the entire economy for a particular year.

Product and process mix: the sector as an aggregate

While a column of input-output coefficients gives a fairly detailed description of how a particular product or group of products was made, it is itself only a very compressed summary of a much greater body of information. Such a column is obtained by summing reports on values of inputs and outputs for all of the various establishments included in a particular industry. No matter how detailed the classification, each output category will subsume a mix of products: large cars, small cars, powerful cars, economy cars, automobile bodies, automobile chassis, automobile repair parts. Similarly, the input to motor vehicles from the iron and steel industry is the sum of many technically distinguishable inputs: steel sheet, castings, bars, carbon steel, alloy steel. Even within a given establishment, many different products may be produced, each from a different combination of inputs.

In general, no matter how disaggregated the sectoring, there will be structural changes due to changes in "technology," and those due to changes in "product mix." This distinction rests on a model of the sector as an aggregate. The coefficient in any given cell will be a weighted average of subcoefficients characterizing subproducts or components. For example, the coefficient for electric and gas utilities into "nonferrous metals" is an output-weighted average of coefficients for electricity and gas inputs into aluminum, copper, lead, zinc, and so on. The coefficient in this cell will change if (1) electricity requirements per unit of output for any one of these nonferrous metals changes or (2) if the relative importance of, say, aluminum increases relative to that of the other metals. We might want to label the first "technological change" and the second "change due to shifting product mix." Change of the second type can sometimes be reduced or eliminated through disaggregation, although it is clear that product mix problems persist at the establishment level.

From the point of view of the system as a whole, however, the distinction is not at all clear cut. Consider the underlying sources of variation in product mix: Changes in the relative importance of different subsectors' outputs result either from changes in the composition of final demand or from structural changes in other sectors. Aluminum grows in relative importance because metalworking and construction sectors are substituting aluminum for steel or are making new aluminum products. Thus, product-mix changes in nonferrous metals coefficients are very often induced by structural changes elsewhere in the system.

Since information for constructing input-output tables is usually collected on an establishment basis in the United States and since establishments generally produce more than one product, it is difficult to discern a "pure" input structure for a given product. Gigantes (1969), Almon (1969), and others have devised methods of estimating pure product-to-product input structures from the usual types of economic census information. No attempt was made to introduce such techniques for the present study, although the advantage of doing so for future work is apparent.

Additional variety among input structures of establishments and industries making the same product arises from the use of different processes. Steels, which are identical from the users' point of view, may be produced by the basic oxygen process, characterized by one input structure; or by the open hearth process, characterized by another; or by the electric furnace process, with still a third input structure. Even when a single process seems to enjoy a unique advantage for producing a given product at a particular time, several processes may be used simultaneously for historical reasons. New processes—generally requiring new equipment—are introduced gradually, so that industries contain many strata of "older" and "newer" technologies.

In addition to the great variety of input patterns coexisting at any given time in an industry, there are usually product or process alternatives that are known but not actually in use. These will often include obsolete technologies superseded by new ways that are clearly more efficient under any conceivable economic circumstances. They may also include technological alternatives that are likely to be introduced at any time. The extraction of oil from shales in the United States seems a possible example of such a technical alternative. Landsberg and Schurr (1968:130–137) discuss the possibility that it will be economic to begin to produce oil from shales within the next few years.

Structural change and technological change

Because input-output coefficients specify more detail, it is easier to relate particular developments, such as oxygen steelmaking or atomic power generation, to input-output structures than it is to span the gap between such particular developments and, say, national productivity. However, it is clearly nonsense to claim that input-output structures depend only on technology or that they are a comprehensive compendium of technological information. Chenery and other contributors to Leontief et al. (1953); Manne and Markowitz (1963); and Carter (1967a) treat various phases of the relation between technology and economic production functions. Each of them makes it clear that even the most detailed industrial and commodity statistics give only a very incomplete description of the production process and the technology underlying it.

Technology consists of a vast and complex body of scientific, technical, and social knowledge that dictates how goods and services are produced. In most areas, however, its scope is broad and ill defined—from chemical formulae to skills and hunches; from agricultural traditions to microbiology. In this age of specialization, no individual "knows" even an appreciable part of technology and there will always be some essential pieces (the chemistry of the blast furnace or supersonic flight in the 1950s) that are only partially understood. In addition, the "state of the arts" is changing constantly. Thus, operationally, it is virtually impossible to codify this rich and nebulous substructure of technology; it should be kept conceptually distinct from the more limited notion of the Schumpeterian "production function" and from the much more definite and specific kind of production function, "input-output structure."

If we could map the underlying technical substructure, how could it be transformed into an input-output structure? The translation would involve two major stages: (1) establishment of a two-way classification of inputs and outputs and aggregation to a workable number of industry variables and (2) restriction of the range of description to those input and output combinations actually used in the given time interval. The technical substructure is unmapped, and detailed census statistics are the basic data for input-output tables. Only actual inputs and outputs are reported in census tabulations. Thus, alternative combinations that were not used are automatically excluded. Aggregation of the innumerable subcomponents of each input and output is achieved, initially, by the specification of reporting categories and units and, later, by further combinations to achieve a workable sectoring scheme.

The chapters that follow describe changes in the input-output structure of the American economy and often relate these changes to known changes in the underlying detailed productive structure. However, knowledge of the actual underlying technologies is only fragmentary and often qualitative or directional, rather than quantitative. It may therefore be possible to attribute falling steel coefficients in construction to the rise of prestressed concrete but not to specify a mathematical formula (a general production function) relating concrete, steel, and the output of, say, apartment houses.

In traditional economic theory, a clear-cut distinction is made between "substitution" and "technological change," that is, between choice within the context of a given production function and changes in the production function itself. A single input-output structure describes only a single combination of inputs and, hence, includes no alternatives. Both substitution and technological change lead to changes in input-output structure. Inevitably, once a specific mathematical form is chosen for the production function, "structural change" is confined to changes in *its* parameters. In actual practice, it is often difficult to draw the line between bona fide changes in technological possibilities and shifts among "known" alternatives. What is known in the laboratory may not be known on the production line; a process that has been tried on a small scale in one country may have a long way to go before it is economic on a large scale in another. There is a broad spectrum of possible interpretations to the term "known techniques." For most of the analysis that follows, the distinction is not very important. When a material or component becomes relatively cheap, ways are found to use it more. Whether this involves reaching into an old file of instructions or making up new ones may not be an essential question.

Because "technological change" has a common-sense meaning with important social connotations and associations, it is tempting to use this colorful term, instead of the more neutral but unambiguous "structural change." It is standard practice for economists to do this: to define as technological change any change in the parameters of the particular form of production function they have chosen (see, for example, Solow 1957). To relieve the monotony of vocabulary, this liberty will be taken occasionally. The meaning should be clear from the context.

Qualitative change and industrial classification

The crucial role of industrial classification in the description of input-output structure and of structural change should be apparent from all that has been said thus far. Industrial classification is the lens through which all change is observed and measured. Since we cannot see at all without the lens, we cannot say whether it "distorts" the "true" picture. Change will look greater or smaller, depending on whether growing inputs are grouped together, to reinforce each others' growth, or are grouped with declining inputs, so that the changes tend to cancel each other; and on whether greater detail is devoted to static or to changing elements. The plastics sector will grow more or less, depending on whether it includes rubber; the aircraft sector, depending on whether it includes guided missiles.

Although there is a considerable literature on classification and aggregation (see, for example, Balderston and Whitin 1954; Fisher 1958; May 1947; and Kossov 1969), there is no general standard against which distortion imposed by classification can be gauged. To know the results of a computation based on two alternative classification schemes, when one is not a simple aggregate of the other, it is necessary to have the data in terms of both of them, and such information is rarely available. Industrial outputs, employment levels, capital requirements, and interindustrial transactions are the variables of this study; all but the most general findings depend on the definitions of the industries, that is, on the particular classification scheme. SIC categories form the core of the sectoring plans, and there is every reason to believe that they were not "rigged" to favor any specific conclusions. Indeed, the task of rigging conclusions in terms of a two-way classification grid would not be a simple one.

Difficulties in allowing for qualitative change in a fixed classification framework go a long way toward explaining the central paradox of this study: that input-output structure changes slowly in an era noted for rapid technological advance. This paradox turns out to be the unifying theme of many different aspects of the analysis, and major economic reasons, besides this classification problem, are offered to explain it. However, it is important, throughout, to be aware of a general and fundamental "classification bias" that limits our perspective on structural change. The object of any classification scheme is to group like objects into categories. As new items are produced, the job of classification is to group the new with old ones. Are guided missiles "ordnance" or "aircraft"? Is the molding of fiberglass "plastic products" or "boat building"? Whatever the new item, it is bound to start out as a small one and so be tucked into an old category until it becomes large enough—old enough—to warrant new category space of its own (see U. S. Bureau of the Census 1968). Thus, what might, in a very detailed scrutiny, appear to be very rapid growth of a new activity, appears as slower growth in a larger category that subsumes older activities as well. Furthermore, since each category is an aggregate of several items, at a minimum, the path of the most rapidly growing component items is bound to appear as damped by less rapid growth, or decline, in others. Finally, a fixed classification scheme tells nothing of the changing qualitative composition of each category. However today's clothing or chemicals or office equipment differ from that of twenty years ago, the industries delivering them have the same fixed labels. To the extent that "different" cannot be expressed as "more" or "less," it is not measured at all.

The longer the time span covered, the more difficult it is to make meaningful comparisons of qualitatively changing variables. Over short periods, the problem is more serious for some sectors than for others: more serious in machinery, for example, than in coal—but sooner or later it affects them all. Unfortunately, this problem is not peculiar to our particular analytical approach, but appears inevitably in all economic time-series analysis.

There are ways of avoiding the problem of quality change altogether. Leontief (1965) gets around the problem by calling each year's output of any given industry a different product: this year's automobile is not assumed identical with last year's, and so on. In a dynamic context, he describes successive input structures, with each year's outputs being produced from inputs produced in preceding years and without requiring that the inputs or outputs be identical or comparable over time. While this approach avoids the unrealistic assumption that products remain qualitatively uniform, it is not helpful in pointing up regularities in how structures change. Hence, it gives no clue to the prediction of future structures from past.

1.3 Plan of the Book

This study is divided into two parts. Part I is primarily descriptive of changing intermediate input structure, that is, the shifting industrial division of labor. Chapters 1, 2, and 3 describe the data and methods of the study. Chapter 4 is a broad survey of structural change between 1939 and 1961. Chapters 5, 6, and 7 are each concerned with the details of specific developments in three major areas of the economy: (1) general sectors, consisting primarily of services, energy, and transportation; (2) materials producers; and (3) metalworkers. As a survey of structural change, Part I is self-sufficient. It can be read and understood without the more complex analytical material of Part II.

Several general conclusions emerge from Part I of the study. Over time, the economic system requires less and less labor and capital to produce a fixed final demand, and labor and capital productivity definitely improve. Total intermediate output required to produce a fixed final demand remains fairly stable and even increases slightly. This finding implies that somewhat greater indirectness or specialization has come with technological change. Contrary to the popular "golden age" proportional-growth assumption (see Phelps 1966), there are decided changes in the relative importance of individual intermediate sectors in the total.

Input structures of individual industries are apparently sensitive to changes in relative prices of the inputs. This observation implies that the underlying technologies in each sector do allow for substitution. Alternatives either exist or are developed over a ten-year period, so that factor combinations can change in response to price changes. There is some tendency for rows of coefficients to move in parallel across the board. For instance, almost all input coefficients for steel and coal fall while coefficients for aluminum and gas and electric utilities rise. This parallelism suggests that substitution may explain some aspects of changing structure, a facet of technological change that is considered further in Part II.

Part II evaluates and analyzes structural change in terms of total primary factor requirements. Labor, capital, and replacement coefficients are introduced in Chapter 8. Both labor and capital coefficients are falling in most sectors, with capital falling generally more slowly. There is no evidence that labor productivity has improved in proportion to changing capital intensity. Data on capital requirements are the weakest link in our informational chain. Probably increasing labor productivity, along with other structural changes, is made possible by major qualitative changes in capital goods that are not readily measured. Changes over time in direct and indirect labor and capital requirements are computed for alternative final demands and individual end items. Some implications for the study of labor-capital substitution in terms of aggregate and other production functions of the constant elasticity of substitution (CES) family are cited. Chapter 9 distinguishes direct primary factor saving from "adaptive change." The latter refers to economies in primary factors achieved through reshuffling of intermediate input requirements to take advantage of differential rates of improvement in other sectors. Adaptive change depends on the price mechanism. In Chapter 10, linear programming techniques are used to compute the optimal mix of new and old input structures and to study the sensitivity of the choice of technologies to changes in interest rates and the wage structure.

Chapter 11 synthesizes much of the material presented in all the preceding chapters. Using hybrid matrices—composed of 1947 structures for some sectors and 1958 structures for others—it surveys the extent to which structural changes in different sectors were coupled; that is, the extent to which the advantage of a new structure in one sector depends on introduction of new structures in others. By and large, each new structure is better than its older counterpart, even ignoring developments elsewhere in the system. Adaptive change adds to its advantage but does not change its direction. This generalization rests on the relative importance of direct labor saving in the structural change of most sectors.

Chapter 12 introduces real gross investment as a limit to the rate of introduction of given new techniques in the economy. This approach explains "layering," the existence of new and old structures side by side. A linear programming system is formulated, with all structural change embodied in new investment goods. Each sector has two alternative technologies, that of 1947 and an estimated new one. A sector can expand capacity for producing with old or new technology by purchasing old or new types of capital goods. Given final demand, the system determines investment for expansion and replacement and factor costs. Computed investment and primary factor costs are compared with actual investment and costs for several intervals between 1947 and 1961. Chapter 13 summarizes the general conclusions of the study. Detailed data and sectoring plans are given in the appendices.

Chapter 2 The Data

2.1 Input-Output Information

This is certainly not the first study of changing input-output structure; indeed, one hesitates to cite any in particular, for fear of neglecting others. A thorough bibliography of such studies is given by Taskier (1961) and United Nations (1964 and 1967). The bulk of the work in this field concerns countries other than the United States, such as the Netherlands, Norway, and Japan, from whom sequences of comparable input-output tables have been available for some years. In general, the emphasis of these studies of structural change has been more statistical and methodological than the present one and less concerned with specific, concrete descriptions of developments in industry (see, for example, Tilanus 1966; Theil 1966; Sevaldson 1969; and Middelhoek 1969). The difference in orientation may be explained partly in terms of the research interests of the analysts and partly in terms of the materials available. In contrast to the United States, tables of some twelve or more consecutive years furnish bases for bona fide statistical analysis for both the Netherlands and Norway. These tables are more aggregated than their American counterparts and, in this respect, furnish less satisfactory data for the analysis of specific technological developments.

Leontief himself pioneered in the analysis of structural change in the United States. In Leontief (1953), he analyzed 1919–1939 changes for this country. Although its data are much less detailed, that study supplies major methodological roots for the present one. The time lag between Leontief's pilot study and the more ambitious empirical implementation of its ideas is explained by bottlenecks in computing techniques and in the supply of data. The essential role of the large-scale computer in the analysis of this type of information will be self-evident throughout this volume. Ten years ago, we should have found the computations too expensive and too slow and cumbersome to warrant this kind of detailed analysis, nor would the information have been available for processing. Indeed, at this point the limiting factor is data supply, not computing power. Until a few years ago, there was only one United States input-output table, the Bureau of Labor Statistics 1947 study (Evans and Hoffenberg 1952), that was sufficiently reliable and afforded adequate detail for the sort of structural analysis to be undertaken here. The publication of the 83-order 1958 table by the Office of Business Economics in 1964–1965 (Goldman et al. 1964 and Office of Business Economics 1965) provided a second full-scale input-output table, permitting a detailed study of structural change in the United States. With these two official tables as a firm basis for comparison, it seemed worthwhile to undertake further revisions of the cruder 1939 table of the Bureau of Labor Statistics, which had been elaborated and reworked, over a long period, at the Harvard Economic Research Project.*

The data core of the present study consists, then, of three input-output tables for the years 1939, 1947, and 1958, with coordinated auxiliary information on labor and capital requirements to go with each. In addition, the Office of Business Economics made available unpublished 1961 vectors of final demand and total output compatible with the 1958 study. Each of the input-output tables originated at a different time and was constructed by a different group of research workers. The 1958 table and the vectors of 1961 final demand and outputs use identical conventions and are fully comparable. They were constructed and published by the U. S. Department of Commerce, Office of Business Economics, as a regular part of the United States national accounts.† The 1958 recession may well have introduced some abnormalities into that year's input-output structure, but there is no satisfactory method of gauging the quantitative impact of cyclical factors on intermediate input structure. Now that input-output tables are being assembled for full census years on a regular basis, we can look forward to steadily improving information and a wider basis for evaluating cyclical effects.

Our 1947 table is based on a much more detailed 450-order study released in the early 1950s. Although the detailed materials are unpublished, the mimeographed documentation is excellent and deserves special mention. To render the 1947 material comparable with that available for 1958, it was necessary to construct a detailed alignment of the two classification schemes and to adjust for price changes (see Section 2.3) and for the many differences in accounting conventions underlying the two tables—differences in the treatment of taxes, receipts for contract work, research and development activities, rents and royalties receipts by establishments primarily engaged in other industrial activities, and numerous other changes in conventions of measuring outputs of individual sectors. This reconciliation was a major research task in itself. The initial work on alignment, reconciliation, aggregation, and price adjustment took the better part of a year's work of a team of economists and programmers at the Harvard Economic Research Project. After the initial alignment, deflation, and aggregation work, the Office of

* A partial description of the revisions of the 1939 table is given in Strout (1967).

† The published 1958 table contains 83 endogenous sectors. Information for the disaggregation of the food, nonferrous metals, and utilities sectors was later made available in Office of Business Economics (1966); portions of it are used in the present analysis.

Business Economics undertook many adjustments and revisions of the 1947 study necessary to improve comparability and to refine and correct the price deflators.* Even though the task of aligning two such large-scale input-output tables is never really completed, their general comparability seems clear.

Available 1939 input-output materials are much cruder than those for the later years. The study was undertaken cooperatively by the Bureau of Labor Statistics and Harvard in 1941. It provided input-output data for important government analyses, including the pioneering study by Cornfield et al. (1947); but the 96-order table was never completed by the Bureau of Labor Statistics. Revision and completion of the study at 96 order was undertaken at the Harvard Economic Research Project in 1952, first by the author and later and more intensively by Alan Strout. Further adjustments required to align the 1939 study with the 1947 and 1958 materials were undertaken in 1965–1966. Despite the skilled and conscientious efforts of those who constructed and reworked the 1939 matrix, there is really no firm basis for judging the accuracy of the estimated input-output coefficients. Perhaps more fastidious economists would have chosen to ignore this material altogether, but the temptation to include an input-output table for a third time was irresistible. Because of quality considerations, 1939 materials are used less intensively than those of 1947 and 1958 in this study.

The classification bases for the three original input-output studies were very different, and it was necessary to seek a reasonably detailed sectoring scheme that would span them all. A 38-order classification was devised to permit comparison of the 1939, 1947, and 1958 materials. Whenever the 38-order classification scheme is used, all data are adjusted to a common 1947 price basis.

The classification basis for 1947–1958 comparisons, not including 1939, can be much more detailed than 38 order, and two point 1947–1958 comparisons are made at 76 order. The 76-order classification is essentially the 1958 Office of Business Economics 83-order scheme with a few sectors transferred to final demand and some others aggregated to eliminate major problems of alignment arising out of changes in the SIC. The 1947–1958 comparisons at 76 order are also made in 1947 prices. However, some comparisons of 1947–1958 changes with 1970 projections for materials are made in 1958 prices in Chapter 6. The 38- and the 76-order classification schemes and other sectoring information are given in Appendix B.

Sources of auxiliary information on labor and capital requirements, investment, and replacement are cited and discussed in Chapter 8. For the most

* The officially comparable tables are available on request from the Office of Business Economics.

part, this information is unpublished, and there is no firm basis for evaluating its accuracy. At the outset, then, it must be acknowledged that the data for this study are of uneven quality. This fact opens the analysis to challenge at many points. Perhaps this very criticism can contribute to the vital task of correcting and improving the scanty information thus far available.

2.2 Scope of Structural Description

The present analysis deals with changing structure of the so-called intermediate or endogenous industries only. These include a very wide range of sectors in agriculture, mining, manufacturing, transportation, and services (see Appendix B for a detailed listing). Final demand or bill-of-goods sectors are treated as exogenous. Their composition is taken as given and is not analyzed. Final demand sectors are commonly distinguished from intermediate in that the former are, in some sense, ultimate consumers: their consumption of goods and services is the basic objective of the system. Sectors classified in final demand are generally expected to be more sensitive to changes in taste and in public policy than are intermediate sectors; hence, economists are sometimes reluctant to attribute a "structure" to them. Actually, there is no clear empirical basis for assuming that the input structures of households or of government sectors are more (or less) stable over time than are the input structures of manufacturing or mining sectors, whose structures are presumed to be technologically conditioned. The scope of government sectors varies greatly over time and among countries.

The distinction between final demand and intermediate sectors is quite arbitrary, and economists generally recognize that the line must be drawn differently to meet the specific needs of each investigation. Classical economists—Ricardo, Marx, Malthus—viewed households as an industry that supplied primary factors of production and saw consumer goods as inputs required by households. This practice is continued in many modern "closed" growth models (see Morishima 1964). Similarly, government and foreign trade activities are subject to a wide range of possible analytical treatments, depending on the country and on the analytical goals.

For the present study, the Office of Business Economics distinction between final demand and intermediate activities is followed fairly closely. They have distinguished seven exogenous sectors including household consumption, gross private capital formation, net inventory change, exports, competitive imports, state and local government, and federal government. Our analysis required two major adjustments to this classification scheme. First, in the

1958 Office of Business Economics treatment, there are both exogenous government-industry sectors and endogenous government-enterprise sectors. The latter includes the postal service and some government activities performed by the private sector as well, principally power generation and local transportation. To eliminate inconsistencies in the treatment of these government activities for 1939, 1947, and 1958, government enterprises with private counterparts were reassigned to the corresponding private sectors, and all other government activities were treated as exogenous. Second, the ordnance sector, which is endogenous in the Office of Business Economics table, was transferred to final demand.

Competitive imports, that is, imports of products that are also produced domestically, are treated as a negative vector in final demand. This treatment excludes goods produced outside of the United States from industry outputs. Since the coefficients are on a domestic output base, fluctuations in the imports of goods that have domestic counterparts do not in themselves lead to coefficient change.*

Exogenous, as well as endogenous, sectors change in structure over time, and both are subject to the influence of new technological developments. Just as the relative importance of steel as a material for metalworking and construction operations decreases, so its importance as an element in foreign trade and government purchases also declines; nor is this tendency surprising. The competition of other materials induces reductions in steel use both domestically and abroad and both privately and in government activities. After all, metalworking industries in Japan are subject to many of the same innovations as those in the United States, and government construction of, say, military bases abroad has many technological features in common with domestic construction activities. Thus, changes in the structures of foreign trade, government, and even households often resemble changes in the structures of intermediate sectors.

The inclusion of a gross capital formation vector in final demand signifies another limit to the scope of the present study of technological change. Transactions in an input-output table measure interindustry sales on current account, including only those items that are consumed as intermediate inputs. Sales of new capital goods, which are investments, are lumped as sales to a single gross capital formation column in final demand; they are not identified by industry of destination. Thus, changes in the amounts and kinds of different equipment and construction items are not covered in a survey of changing

* The published Office of Business Economics 1958 Study (1965) is on a total output base, but the Office of Business Economics also distributes a computer tape containing a domestic base matrix (see Vaccara and Simon 1968).

input-output structure per se. The analytical distinction between capital and current account items is not clear cut. In fact, all inputs are both "flows" and "stocks" as they proceed through the productive pipeline. In prevailing accounting practice, inputs are classified as current or as capital depending on whether they stay in the pipeline for a relatively short time—materials and fuels—or a relatively long time—buildings and machinery. Thus capital and current inputs are in practice distinguished arbitrarily by dividing a continuous spectrum of longevities.

New types of capital goods are at the core of technological change and are of prime importance in explaining current account structural changes as well as stock requirements. A capital flow table for 1958—showing interindustry transactions on capital account—and capital stock coefficients for 1947 and 1958—showing total value of fixed capital per unit of capacity for each sector—throw some light on the capital aspect of structural change. These data will be discussed later, in Chapters 8 and 9, as they are used. For the optimizing computations presented in Part II, both the flow and the stock aspects of capital goods were taken into account. Replacement matrices were estimated and added to current flow matrices for each year to allow for current requirements for "maintaining capital intact" in addition to the intermediate inputs usually reported. Capital coefficients were used to measure the stock aspect of plant and equipment requirements. Inventories representing the stock aspect of conventional intermediate flows had to be neglected altogether for lack of basic information.

2.3 Value Units, Physical Units, and Price Deflation

Input-output tables used for our structural analysis are all in value terms, that is, they record dollar volumes of transactions among sectors. As raw data, the tables for different years reflected the prices prevailing when the transactions actually took place. For a given year, coefficients can be interpreted as value ratios—value of the output of industry i required per dollar of output of industry j—or as ratios of physical input requirements to outputs, with the units of inputs and outputs taken to be "a dollar's worth" of each particular commodity. As long as prices remain constant, the "dollar's worth" is a meaningful physical unit; that is, the amount one can buy for one dollar (Leontief 1951). Leontief points out that the basic properties of his economic system are uniquely determined by the (relative) value figures of all the different kinds of inputs and outputs: "Two systems with identical value patterns will have also the same price and output reactions. Even if the prices and quantities taken separately were quite different (because of real or nominal differences), the identity of the value figures shows that each of the two setups could be transformed into the other by changes of purely dimensional kind" (1951:65). The "basic properties of an economic system" he cites are, of course, value relations—the reaction of prices or outputs, expressed in value terms, to a given structural or exogenous change, also expressed in value terms.

A number of scholars, including Tilanus (1966:37) and Sevaldson (1963), note that coefficients expressed in given-year prices are sometimes more stable over time than those expressed in constant prices. This phenomenon is not too difficult to explain: the price of each sector depends on the prices of its cost elements; hence, input and output prices tend to move in parallel. Furthermore, inputs whose relative prices are declining tend to be substituted for those whose relative prices are rising. Thus, substitution works toward stabilizing the value of inputs registered in any given cell. In order to relate structural change to technological change, however, it is still necessary to disentangle the price and quantity components. To discuss whether less steel or more aluminum was used per unit of automobile output in two different years requires that the units in which steel, aluminum, and automobiles are measured be fixed over the period of comparison. With changing prices, of course, the "dollar's worth" is a changing unit. Therefore, it was essential to deflate all of the basic input-output materials to a common price basis.

Actual price deflation was done at a very detailed level. To derive 1947–1958 deflators, separate price indices were applied for many 450-order industries in the 1947 input-output table, and 1939–1947 price indices were based, wherever possible, on the equivalent of four-digit SIC detail. Price deflation, however, is always a difficult and sensitive task, and price deflation of input-output matrices is no exception. Underlying the difficulty is the problem of quality change discussed in Section 1.2. When the qualitative character of a product is changing, choice of common units for comparing the new and the old is necessarily arbitrary. Unavoidably, this choice of units is central to the construction of price deflators. Assume that the cost per Buick to the consumer remains fixed. Does an increase in its horsepower mean a cut in price? Reliance on published price indices shifts the burden of choice to other shoulders, but it does not eliminate the central problem.

How will an error in a single price deflator affect the analysis? Suppose that the 1947–1958 price index, used for deflating the 1958 output of the steel industry to 1947 prices, is "too large." This means, in effect, that all purchases of steel and their sum, that is, total steel output, will be understated as compared with corresponding 1947 transactions. All coefficients along the steel *row* will be understated by a fixed proportion. Since their denominator will be understated, all coefficients in the steel *column* will be overstated by the same fixed proportion. It would be as if the Census of Manufactures had shifted from short to long tons between 1947 and 1958 without anyone noticing. Such dimensional "distortions" of the steel row and column are exactly compensating in their effects on other sectors: steel requirements are understated, but inputs required to produce steel are overstated. Thus, an error in the steel price does not bias estimates of requirements from mining or castings. Errors in price deflators distort output estimates only for those sectors in which the errors actually occur (Leontief 1953:41).

This finding is analytically convenient, but it must be qualified. In most instances, 450-order price relatives for 1947–1958 were weighted with 450-order 1947 transactions cell by cell and then combined into 83-order deflators. This produced matrices for converting the 1958 tables to 1947 prices. The relative importance of a given 450-order sector varied from cell to cell along any given 76- or 38-order row; in effect, different deflators were used to deflate each element in the row of the aggregated table. Furthermore, distinct sets of deflators were used to deflate final demand and primary and secondary transactions. Similar deflation procedures were also applied to the 1939 table. These deflators, as opposed to row deflators, make it more difficult to localize the impact of possible errors in prices. The fact that the 1947–1958 price deflators prove fully consistent with price information in the national accounts provides some reassurance.

Chapter 3 Basic Methodology and Notation

3.1 Graphs

Changes in the overall contours of industrial interdependence rest, ultimately, on changes in individual coefficients and groups of coefficients. The graphs and computations introduced in Part I will serve primarily to digest this massive volume of detail. Theoretically, the computations and graphs involve no radically new departures. All are based on static input-output theory as originally set forth in Leontief (1951). Within this context, the analysis in Part I relies on only a few of the many methods that have been, or can be, devised to summarize the various aspects of structural change. Other interesting approaches to this type of problem are employed by Bureau of Labor Statistics (1966); Vaccara (1969); Domar (1967); Tilanus (1966); and Theil (1966:168–255).

Because the prime objective of Part I is to convey a clear impression of the specific patterns of structural change that actually took place, graphs are employed liberally. They are particularly necessary in describing changes in direct coefficients, as there are so many of them. Direct coefficients are presented in more detail than other information.

Three-way scatter diagrams, such as figure 5.3, are used to show, at a glance, which are the large and which are the small coefficients in a given row and how they change over time. Numbers next to each point identify consuming sectors. Each axis measures direct input coefficient values for a particular year, and the 45-degree line in each half guides the reader in judging whether coefficients were larger in one year than in another. Coefficients that were larger in 1958 than in 1947 appear below the 45-degree line in the right section; and those larger in 1947 than in 1939 appear above the line in the left section. Clustering on one side of the line means that many coefficients tended to move (increase or decrease) in the same direction. Logarithmic scales are used, and distance from the 45-degree line measures (relative) rates of coefficient change. Bar graphs are used to present the more compact computed summary measures. These graphs will be introduced in the course of the substantive exposition and should involve no particular difficulty. Graphs, as opposed to the more customary tables, serve to stress the directional and order-of-magnitude aspects of the material. This seems appropriate in view of the statistical shortcomings of basic data discussed in Chapter 2.

3.2 Descriptions of Structural Change Based on the Leontief Inverse Matrix

Changes in individual coefficients do not generally occur independently of one another. Most often, each coefficient change is part of a complex of interrelated shifts in which the specialized roles of individual supplying sectors are realigned. In studying this process of mutual adjustment, indirect linkages among sectors must not be overlooked. The construction industry buys aluminum directly from the nonferrous metals sector and indirectly through suppliers of fabricated metal products, who, in turn, use aluminum as one of their principal inputs. Every sector purchases electricity directly from the utilities industry and indirectly from many others, since electricity is an input to all their inputs and to all their inputs' inputs, and so on.

A measure of total (direct-plus-indirect) interdependence between each pair of sectors in an economy can be derived from a direct coefficient table by computing the Leontief inverse matrix. Each element of the inverse matrix measures the total amount of the product of a given industry required (directly and indirectly) to deliver a unit of a particular sector's product to final demand. The inverse coefficient for chemicals into textiles measures the total amount of chemicals required directly and indirectly to deliver one unit of textiles to final demand. Chemicals will be required directly in making textiles; indirectly in making synthetic fibers, agricultural products, and other inputs used by textiles; and indirectly again in making inputs into synthetic fibers, into agricultural products, and into other textile inputs, and so on. Almon (1967:15–30) gives a thoughtful exposition of the meaning of the Leontief inverse matrix and how to compute it.

Since direct requirements often constitute a major portion of total requirements, direct and inverse coefficients for a given cell generally move in the same direction over time. However, this is not always so. Yan and Ames (1965) point out that pairs of sectors may be related by first-order, second-order, third-order, or higher order linkages, depending on whether one furnishes a direct input or an input-of-an-input or an input-of-an-input-of-an-input and so on, of another sector. As sectoral division of labor changes, some of the indirect linkages may be weakened, others strengthened; hence, in general, changes in inverse coefficients will not necessarily be proportional to changes in corresponding direct coefficients.

For example, the increased use of synthetic fibers in the textile industry between 1947 and 1958 resulted in a decrease in direct purchases from the chemicals industry. Decreased use of natural fibers led to a cut in requirements for the chemicals used to clean them and in chemicals used in finishing. While the direct coefficient for chemicals decreased, that for synthetic fibers rose. Production of synthetic fibers requires a large chemical input. Thus, while the direct linkage of textiles with chemicals was weakened between 1947 and 1958, the total (direct-plus-indirect) dependence of textiles on chemicals increased. Over the same period, more aluminum was used in buildings, but direct coefficients for aluminum into construction went down. The major substitution of aluminum for steel into construction took place within the sector supplying heating, plumbing, and structural metal products to the construction industry; but this substitution is not apparent in coefficients for direct aluminum purchases by the construction sector itself.

Measures of structural change based on inverse coefficients have some important advantages over direct coefficient comparisons. Inverse coefficients are insensitive to certain troublesome changes in industrial division of labor and in accounting practice. Steel for refrigerator doors, for example, may be purchased directly from the iron and steel sector in one year and indirectly, through the stampings sector, in another year. Furthermore, one suspects that some apparent changes in industrial specialization come from changes in the methods of estimating basic input-output flows or from the inevitable vagaries of census tabulation, in which multiproduct establishments are classified according to their principal products and may shift from industry to industry with small changes in product mix. While differences in direct coefficients might reflect changes in specialization or in classification procedures, in addition to changes in the technical structure of an industry, the first type of change tends to be "washed out" in comparisons of inverse coefficients.

Measures based on the inverse matrix, however, have their own disadvantages. In particular, they tend to obscure the primary locus of change. Increased total direct-plus-indirect electricity requirements for producing household equipment may mean that more electricity is used directly in making household equipment, that more is used in making components of household equipment, that electricity-intensive aluminum has been substituted for steel, or all of these simultaneously. Clearly, it is necessary to present both types of information in analyzing changing input-output structures.

3.3 Summary Measures

In some ways it would be ideal to present tabulations of changing individual inverse coefficients along with cell-by-cell changes in direct coefficients. The volume of detail, however, quickly becomes unmanageable, and it is necessary to limit it. Direct coefficients will be presented at the most detailed level possible while measures based on the inverse matrix will be used for broader summaries or to describe changes at an intermediate level of detail. Many of the summary measures of structural changes used in Part I share the same central logic: they ask (that is, compute) what would be the amounts of individual inputs required to deliver a fixed final demand with the input-output coefficients of successive years. Each year's inverse matrix measures total requirements with that year's input structure in all sectors, per unit delivery to final demand. Multiplying the inverse coefficients along each row by a final demand vector of given level and composition yields estimates of total output requirements to deliver that particular bill of final demand. Total intermediate output requirements from each sector are obtained simply by subtracting the specified final demand vector from the estimated vector of total output requirements. Thus it is possible to postulate any given historical or hypothetical bill of final demand and study how intermediate requirements for any sector's output changes with changing input-output structures. Such computations show, for example, that more chemicals, plastics, and electronic components and less steel, wood, and nuts and bolts are required to deliver the same final demand with 1958 input structures than with 1947 ones. Many variants of this type of measure are introduced, and more precise and detailed definitions accompany them.

Sometimes input requirements are studied at the level of detail of the individual supplying sector: iron and steel, synthetic materials, electronic components. Sometimes inputs are combined into broader functional groups: services, materials, general metalworkers. Similarly, study of changing requirements for delivering a complete bill of final demand is elaborated by comparisons of requirements for delivering individual segments of final demand. Final demand is partitioned into eight, and sometimes nine, "subvectors": food and tobacco, textiles and clothing, paper and chemicals, construction, machinery, electric and service equipment, transportation equipment, services and transportation, and furniture. Changes in requirements of various inputs to deliver each of these types of final products separately are analyzed and compared.

3.4 Notation Used in Part I

Conventional matrix notation is used to describe the computations that were performed. The following discussion introduces most of the symbols that will be used.

Matrices are denoted by uppercase letters, printed in boldface type. The term $\mathbf{A}^t$ represents a matrix of input-output coefficients for a particular year t. In the matrix $\mathbf{A}^t$, each element a_{ij}^t measures the amount of the product of industry i purchased per dollar of output of industry j in year t.

The term $\mathbf{Q}^t$ stands for the Leontief inverse matrix $(\mathbf{I} - \mathbf{A}^t)^{-1}$ for year t. In the matrix $\mathbf{Q}^t$, each element q_{ij}^t measures the amount of the product of industry i required directly and indirectly to deliver a dollar's worth of output of industry j to final demand.

The term $\mathbf{B}^t$ is a capital coefficient matrix for year t. Each element b_{ij}^t measures the value of the stock of the product of industry i required per dollar of capacity output of industry j. The term $\mathbf{B}_i^t$ denotes row i of the capital coefficient matrix.

Vectors are represented by lowercase letters in boldface type. The terms $\mathbf{l}^t$ and $\mathbf{b}^t$ stand for vectors of labor coefficients (man-years per dollar of output) and capital coefficients (value of capital stock per dollar of capacity output per year), respectively. Each element l_j^t (or b_j^t) of these vectors is the labor (or capital) requirement per dollar of output of a particular sector j. The term l_j^t is converted to value units by means of a wage vector, $\mathbf{w}^t$. Each element w_j^t of $\mathbf{w}^t$ is the average wage paid for year t in industry j. Each element b_j^t of the capital coefficient *vector* is the sum of the elements of column j of the capital coefficient matrix; that is, $b_j^t = \sum_{i=1}^n b_{ij}^t$, where n is the number of sectors.

The terms $\mathbf{m}^t$ and $\mathbf{k}^t$ are vectors of total (direct-plus-indirect) labor (measured in man-years) and capital (measured in dollar values of fixed capital stock) requirements, respectively, per unit delivery to final demand. Thus,

$$\mathbf{m}^t = \mathbf{l}^t\mathbf{Q}^t$$

and

$$\mathbf{k}^t = \mathbf{b}^t\mathbf{Q}^t$$

The terms $\mathbf{l}^t$, $\mathbf{b}^t$, $\mathbf{m}^t$, and $\mathbf{k}^t$ enter our computations as row vectors $(1 \times n)$.

The term $\mathbf{x}^t$ is a total output vector for year t. Each element x_j^t measures the total dollar value of output of a particular sector j in year t.

The term $\mathbf{y}^t$ is a final demand vector for year t. Each element y_j^t measures the total dollar value of the product of sector j that is delivered to final users—to government, households, foreign trade, gross private capital formation, and the inventory account in year t.

Similarly, $\mathbf{z}^t = \mathbf{x}^t - \mathbf{y}^t$ is a vector of intermediate output levels in year t, measured in dollar value of transactions.

The terms $\mathbf{x}^t$, $\mathbf{y}^t$, and $\mathbf{z}^t$ enter our computations as column vectors $(n \times 1)$.

Multiple superscripts are used to designate the results of computations in which elements of more than one year are combined. For example, in Chapter

4, a vector of total intermediate output requirements to produce 1961 final demand with 1939 input structures would be written:

$$\mathbf{z}^{39,61} = \mathbf{Q}^{39}\mathbf{y}^{61} - \mathbf{y}^{61}$$

A circumflex over a vector symbol transforms it into a diagonal matrix, with the elements of the original vector arranged along the principal diagonal:

$$\hat{\mathbf{y}}^t = \begin{pmatrix} y_1^t & & & \\ & y_2^t & & \\ & & \ddots & \\ & & & y_n^t \end{pmatrix}$$

Scalar products are represented by Greek letters to distinguish them from the vectors of their component elements. The equation

$$\mu^t = \mathbf{l}^t\mathbf{Q}^t\mathbf{y}^t = \mathbf{m}^t\mathbf{y}^t$$

represents total employment (labor requirements) for the economy as a whole, whereas $\mathbf{m}^t$ and $\mathbf{y}^t$ are vectors of employment coefficients and final demand in specific sectors. More specifically,

$$\mu^{39,61} = \mathbf{l}^{39}\mathbf{Q}^{39}\mathbf{y}^{61} = \mathbf{m}^{39}\mathbf{y}^{61}$$

is a scalar that represents the sum of employment required in all industries of an economy with 1939 structure to deliver the 1961 final demand.

"Subtotals" are also analyzed. How, for example, do employment requirements in each of four specific groups of sectors change with changing input-output structure? To compute employment requirements in the rth group of sectors, for a given input-output structure and final demand, we partition the row vector of labor coefficients into four parts:

$$\mathbf{l}^t = ({}_1l_1^t, \ldots, {}_1l_{n_1}^t|, \ldots, |{}_rl_1^t, \ldots, {}_rl_{n_r}^t|{}_4l_1^t, \ldots, {}_4l_{n_4}^t)$$

where r identifies the subvector of the labor coefficients and n_r is the number of sectors in subvector r.

Let ${}_r\mathbf{l}^t$ be a $(1 \times n)$ vector consisting of zeros except for the n_r elements of the rth subvector of the partitioned labor row:

$${}_r\mathbf{l}^t = (0, 0, \ldots|0, 0, \ldots|{}_rl_1^t, \ldots, {}_rl_{n_r}^t|0, 0, \ldots)$$

Then employment required in the rth industry group with 1961 final demand and 1939 input-output structure becomes:

$${}_r\mu^{39,61} = {}_r\mathbf{l}^{39}\mathbf{Q}^{39}\mathbf{y}^{61} \quad (r = 1, 2, 3, 4)$$

Note that ${}_{r}\mu^{39,61}$ is a scalar representing the total labor requirement in one of the four industry groups.

Later on we compute the total amounts of specific inputs required to deliver particular subvectors of final demand—to deliver a given year's final demand for, say, food items only or for transportation equipment only. Eight subvectors of final demand, designated ${}_{g}\mathbf{y}^{t}$, are usually distinguished, and the final demand vector is partitioned into eight parts. Total labor required to deliver one of the eight subvectors of 1961 final demand with 1939 input-output structures is written:

$${}_{g}\mu^{39,61} = \mathbf{l}^{39}\mathbf{Q}^{39}{}_{g}\mathbf{y}^{61} \quad (g = 1, 2, \ldots, 8)$$

where

$${}_{g}\mathbf{y}^{61} = \begin{bmatrix} 0 \\ 0 \\ \vdots \\ 0 \\ {}_{g}y_{1}^{61} \\ \vdots \\ {}_{g}y_{n_g}^{61} \\ 0 \\ \vdots \\ 0 \end{bmatrix}$$

The term n_g is the number of elements in the gth subvector of final demand, and ${}_{g}\mu^{39,61}$ is a scalar representing employment required to deliver the gth subvector.

Note the distinction between ${}_{r}\mu$, which represents total employment in a given industry group, and ${}_{g}\mu$, which specifies the sum of employment in all sectors required to deliver a particular subvector of final demand. In general, a Greek letter with an r pre-script will denote a quantity such as employment, total capital stock, or total intermediate inputs in a given broad industry-of-origin group; a Greek letter with a g pre-script denotes such a quantity required to deliver a specific subvector of final demand.

3.5 Notation Used in Part II

Part II concentrates on evaluating changing structure in terms of total factor content. While linear programming and other relatively complex types of computations are introduced, most of them can be discussed in terms of the standard notation just described. Primary factor requirements, measured in dollars per dollar of output in year t, are represented by $\mathbf{f}^t$, with labor and capital combined in fixed proportions within each sector. Wage and interest rates are used as weights for labor and capital charges in total factor content:

$$\mathbf{f}^t = \mathbf{l}^t\hat{\mathbf{w}}^t + \alpha^t\mathbf{b}^t$$

where $\mathbf{w}^t$ is a vector of wage rates and α^t is an estimated interest rate for year t. The merits and pitfalls of this mode of combining primary factors are discussed in Chapters 9 and 10.

Hybrid input-output matrices, combining 1947 column structures for certain sectors with 1958 structures for others, will also be used in Part II. Compound superscripts serve to describe them. The first term in the superscript denotes the year whose structures prevail in sectors *not* specially singled out; that is, the first term is the "base year" of the matrix. The second term in the superscript identifies the year of structure for particular sectors whose base-year structures have been replaced. Subscripts on the second superscript tell which sectors do not have base-year structure. Thus, $\mathbf{A}^{47,58_{j,k}}$ is a 1947 coefficient matrix but with 1958 column structures for industries j and k. Similarly, $\mathbf{x}^{47,58_{j,k}}$ and $\nu^{47,58_{j,k}}$ denote outputs and total factor requirements computed on the basis of the specified hybrid matrix $\mathbf{A}^{47,58_{j,k}}$.

3.6 Index-Number Problem

In comparing intermediate output requirements to deliver a fixed bill of goods with changing input-output structure, one inevitably encounters an index-number problem. Different final demands will yield different changes in intermediate output requirements because the inverse coefficients will be weighted differently depending on the compositions of the final demands. For example, suppose that aluminum requirements for making automobiles have been increasing more rapidly than aluminum requirements for making other kinds of output. Under these conditions, a bill of goods with a large automobile component will bring out a larger rate of increase in aluminum requirements than one with less emphasis on automobiles. The problem is analogous to that of choosing between base- and given-year weights in the computation of a conventional price or production index—base-year final demand and given-year final demand weights yield different "indices" of changing structure.

In order to judge the importance of the index-number problem for the findings of this study, we performed some of the basic computations with several alternative fixed final demands and compared the results. While choice of final demand must, of course, affect the detailed rates of change computed, the directions and general orders of magnitude of the findings do not seem to be very sensitive to shifts from one year's bill of final demand to another. Appendix A reports the numerical findings in detail.

For practical reasons, some single bill of final demand must be chosen as a basis of comparison. The 1961 bill of goods was most often used. Use of the 1961 final demand made it possible to add a fourth observation to many of the comparisons, even in the absence of a complete 1961 matrix (which had not been released at the time the computations were performed).

The introduction of 1961 as a fourth point of comparison in the absence of a full input-output matrix for that year is explained in Section 4.1. It is well to bear in mind the type of bias inherent in using the bill of goods for a later, rather than an earlier, year. Some tendency for final demand structure to change in the same directions as intermediate input structures has already been explained. Because of this tendency, final demands for later years will tend to give greater weight to elements that are growing and smaller weight to those that are declining in relative importance than will base-year final demands. No attempt has been made to assess the importance of this bias as compared with other types of changes in final demand over time.

In philosophical terms, the index-number problem arises when we try to generalize from quite specific to broader findings. Narrowly, the meaning of the measures computed and presented in succeeding chapters should be unambiguous. Index-number questions arise when we interpret these specific measures as indicators of some "deeper" generalizations; that is, when we wonder whether the contours of structural change are in fact independent of the specific methods used to describe them. We hope that the weights "do not matter." At some level they always do. The problem pervades all empirical research in economics and, indeed, all empirical research of any kind where measurement is involved. Fortunately, it is not a difficult matter to estimate alternative variants of the summary measures.

If the reader has particular interest in using alternative weights, he can perform the computations using the information given in the data appendices. Experiments with changing final demand weights (Appendix A) and sensitivity tests with regard to weights for the value added components (Chapter 10) suggest that plausible changes in weighting afford little challenge to the broader conclusions of the analysis. However, specific numerical results will always depend on the specific weighting systems used in computing them.

Chapter 4 Outlines of Changing Industrial Specialization

Without guidelines, it is difficult to inspect a disaggregated structure and get a clear impression of where and how much it has changed. In this chapter, we look at the combined effects of all industries' coefficient changes at once and consider the effects on the relative rates of growth of individual sectors, and on the broad outlines of industrial specialization. This survey provides the frame and perspective for the more detailed analysis in Chapters 5, 6, and 7.

4.1 Changing Total Intermediate Input Requirements

Figure 4.1 shows the changing gross output level summed for all industries in the American economy over the period 1939–1947–1958–1961. The width of each bar measures the total level of production (in 1947 dollars), and the bars are spaced proportionally to the time interval between them. The part of output constituting the gross national product is shown by the top segment of each bar. The remainder consists of intermediate output, that is, production required to supply the input requirements of all sectors (including requirements from establishments classified in the same sector) in order to deliver this gross national product, or final demand, to ultimate consumers—to government, households, foreign trade (net exports), inventory changes, and gross private capital formation. Note that intermediate output happens to be roughly equal to the value of the gross national product each year.

In earlier days of national income accounting, intermediate production was eliminated to avoid "double counting." This is reasonable if one is primarily concerned with measuring an economy's "success"—the net amount the nation has managed to produce, whatever its methods. Yet this duplicative portion of economic activity is precisely the focus of our present analysis. For it is the composition of interindustry sales that mirrors most directly the effects of changing technology and the organization of production. Intermediate inputs are the specific goods and services used to produce the gross national product. As methods of production change, more of one kind of input will be required and less of another—more chemicals, less steel, and so on—and the interdependence of individual supplying sectors will be changed accordingly.

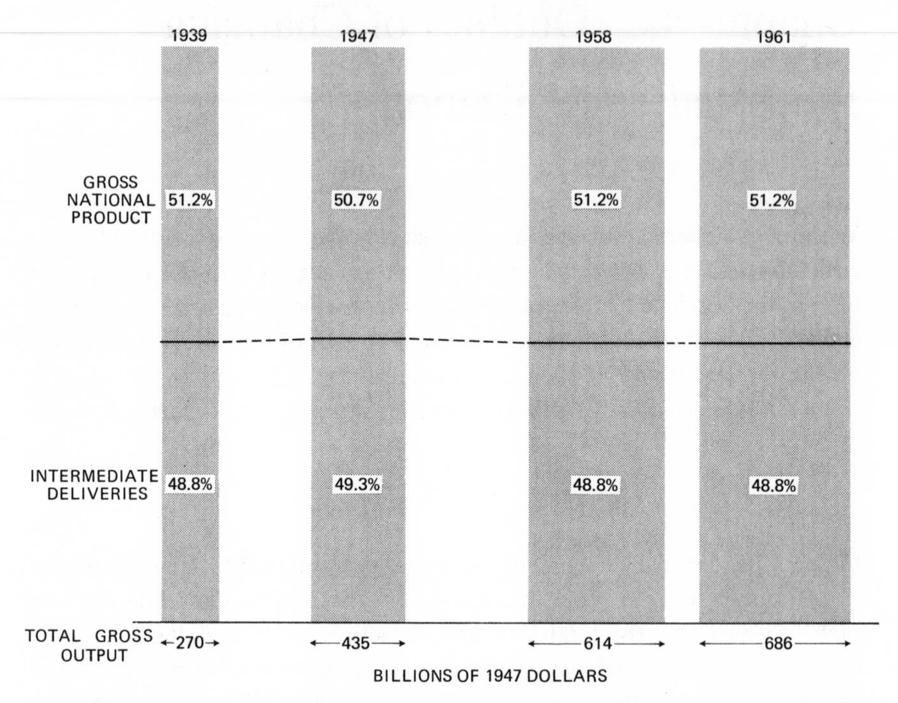

4.1 Total Gross Output, Gross National Product, and Intermediate Deliveries for 1939, 1947, 1958, and 1961.

We begin to survey change by performing the following "mental experiment." We fix the gross national product at a given level and industrial composition, and then we examine the intermediate output requirements to produce this same final demand with the input-output structures of different years. By comparing the indermediate outputs required to do the same job, we can separate the effects of changes in the structure of industry from changes in the final demands made on the system. Table 4.1 shows the intermediate output levels that would have been required to produce final demand of 1961 level and composition with the input structures of 1939, 1947, 1958, and 1961 in all sectors. The 1961 intermediate outputs consistent with this bill of final demand are simply the difference between actual total output and final demand for that year. For the other years, the intermediate outputs were computed on the basis of the three input-output coefficient matrices, which were standardized at 38 order and expressed in 1947 prices:

$$\mathbf{z}^t = \mathbf{Q}^t\mathbf{y}^{61} - \mathbf{y}^{61} \qquad (4.1)$$

where $\mathbf{z}$ is a vector of intermediate output levels, $\mathbf{y}^{61}$ is the 1961 final demand

Table 4.1 Intermediate Output Requirements for Delivering Total 1961 Final Demand with 1939, 1947, 1958, and 1961 Technology; 38 Order (millions of 1947 dollars)

Sector	1939	1947	1958	1961	Annual rate of growth (percent)
I. General industries					
(31) Utilities	5,341	6,576	10,105	10,931	3.4
(30) Communications	2,857	3,143	5,250	5,287	3.1
(32) Trade	14,242	20,381	22,886	23,594	2.1
(36) Auto. repair	1,339	3,286	2,401	2,460	2.0
(35) Business serv., etc.	11,011	12,925	15,886	15,792	1.7
(14) Petroleum refinining	5,549	7,830	7,710	7,523	1.2
(37) Institutions, etc.	1,854	3,045	2,604	2,622	1.2
(33) Finance & insurance	7,742	10,028	9,367	8,789	0.5
(34) Real estate & rental	15,046	14,734	15,332	16,227	0.3
(29) Transportation	13,900	16,237	12,827	13,448	−0.5
(4) Coal mining	4,426	4,008	2,068	1,743	−4.5
Total[a]	83,306	102,192	106,436	108,416	
II. Chemicals					
(13) Chemicals	12,474	12,580	17,309	18,191	1.9
III. Materials					
(17) Stone, clay, & glass	4,810	5,966	6,623	6,675	1.4
(15) Rubber prod., etc.	3,301	3,564	4,110	4,549	1.4
(19) Nonferrous metals	8,820	9,353	8,098	8,057	−0.6
(11) Wood & products	9,695	9,558	7,924	7,845	−1.1
(1) Agriculture, etc.	69,745	59,438	52,289	49,790	−1.5
(18) Iron & steel	15,862	16,159	11,412	10,896	−1.9
Total[a]	112,232	104,038	90,457	87,812	
IV. Metalworking					
(25) Aircraft	919	1,494	3,120	2,775	5.5
(27) Instruments, etc.	1,093	1,469	2,062	2,098	3.0
(23) Electrical eq., etc.	6,223	6,736	9,907	10,644	2.6
(21) Nonelectrical eq.	3,013	5,367	5,507	5,404	2.4
(22) Engines & turbines	390	834	739	615	1.8
(20) Metal forming	10,580	13,048	12,905	12,313	0.6
(24) Motor vehicles	6,866	7,539	7,580	7,601	0.4
(26) Trains, ships, etc.	705	734	655	630	−0.6
Total[a]	29,790	37,220	42,475	42,080	
V. All other					
(8) Food	16,502	17,077	18,814	18,439	0.6
(6) Nonmetallic mining	1,936	1,547	1,856	1,959	0.3

Table 4.1 *continued*

Sector	1939	1947	1958	1961	Annual rate of growth (percent)
(10) Textiles & products	18,672	15,628	18,449	18,124	0.1
(12) Paper & publishing	17,907	17,924	18,109	17,606	
(5) Petroleum mining	7,811	8,035	7,642	7,435	−0.2
(28) Misc. manufactures	2,554	1,814	2,084	2,228	−0.4
(38) Scrap	1,896	2,758	2,194	1,543	−0.8
(16) Leather & shoes	2,082	1,692	1,518	1,386	−1.7
(2) Iron mining	939	674	640	620	−1.7
(3) Nonferrous mining	1,790	1,369	1,201	1,076	−2.1
(7) Construction	14,396	11,749	7,757	7,244	−3.2
Total[a]	86,485	80,267	80,265	77,662	
Total intermediate requirements	$324,288	$336,296	$336,941	$334,160	

[a] All totals have been rounded to the nearest million.

vector, and $\mathbf{Q}^t$ is the inverse matrix for the given year. To facilitate interpretation, the entries in table 4.1 are grouped by type of industry. Within each group, sectors are arranged in descending order of their average relative rates of change between 1939 and 1961. Annual rates of growth in intermediate output from each sector, over the entire period, are shown in the last column of the table. These growth rates are the slopes of linear time trends, fitted to the logarithms of computed requirements from each sector. Each of the changes listed in table 4.1 can be explained in terms of the specific technological developments they reflect, and many will be analyzed more fully in later chapters. Certain general features, which are apparent at the outset, are discussed in the following sections.

Specialization or division of labor among industries

The column sums in table 4.1 measure the gross volume of intermediate inputs from all industries (identical to the gross volume of intermediate outputs) required to produce 1961 deliveries to final demand with 1939, 1947, 1958, and 1961 input structures, respectively. Note that the dollar volume of intermediate inputs (in constant prices) is quite stable, growing slightly over time—the total volume of inputs required to produce the same final product tends, if anything, to be a little greater with newer, than with older, techniques of production over this time interval. At first glance, this may appear paradoxical. If technological change is to be considered technological progress, how can more inputs have been required to produce the same deliveries to final demand at a later date? Actually, an increased volume of intermediate inputs means an increase in specialization. It represents a change in the division of labor among establishments, but it does not in itself imply a deterioration of technology. The later technologies use slightly more intermediate inputs but less primary inputs, labor, and capital (see Sections 4.2 and 4.3). As an economy develops, it sometimes becomes advantageous for individual establishments to become more specialized, that is, to cover a shorter vertical sequence or a narrower horizontal assortment of activities. Each establishment may fabricate a particular kind of component in volume instead of a more varied assortment of parts, or it may perform a specialized service function. If, as Adam Smith suggests, division of labor depends on the extent of the market, then this tendency is to be expected as the total volume of production expands. Perhaps one should ask why apparent division of labor has not increased more. One answer is suggested later, in Chapter 11.

Changes in the relative importance of broad industrial groups

While the total level of intermediate input requirements remains relatively stable, requirements from some groups of industries expand and from others, contract. Thus, inputs from the general industries—producers of fuels, transportation, trade, communications, and other services—have been expanding steadily, as have requirements from the chemicals sectors. Metalworking inputs rise as the complexity of these products increases and as specialization within the metalworking block grows. However, materials inputs tend to decline as the size and weight of many different equipment items decrease, as waste of materials is reduced, and as cheaper materials are substituted for more expensive ones.

Figure 4.2 summarizes some of these tendencies. It shows total output requirements to produce the 1961 gross national product with 1939, 1947, 1958, and 1961 technology. Total outputs (intermediate outputs *plus* the constant final demand) are shown to facilitate comparisons with figure 4.1. Note that figure 4.2 shows computed output levels required to deliver a *fixed* (the 1961) final demand, while figure 4.1 shows output levels required to deliver the *actual* bills of final demand of the respective years. The last bar in figure 4.2 is identical in width with the one in figure 4.1. The other bars are wider than those in figure 4.1, since they represent requirements to deliver a larger final demand. Each bar is divided according to the shares contributed by the major industry groups distinguished in table 4.1. The subdivisions illustrate the changing relative importance of general industries, materials producers, and other groups of sectors.

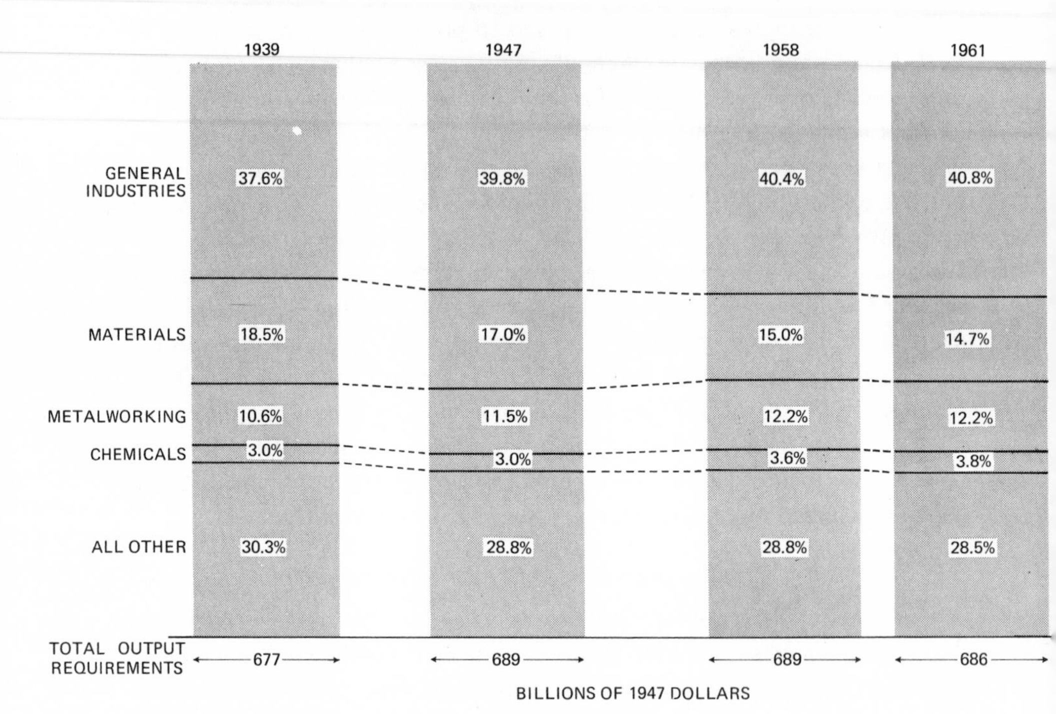

4.2 Total Gross Output Levels Required to Deliver 1961 Final Demand with 1939, 1947, 1958, and 1961 Technologies.

Marked trends in requirements from individual sectors

Within each of the broad industrial groups, requirements from certain individual sectors change persistently. While most such shifts occur very gradually, it is easy to single out some more dramatic instances. On the one hand, to deliver the same bill of goods, the American economy required less than half as much coal with 1961 as with 1939 structure. On the other hand, inputs from the utilities sector roughly doubled, and inputs from several of the service sectors and from chemicals increased by as much as 50 percent.

Among materials, there are rapid percentage increases in relatively small sectors such as rubber and plastic products* and in stone, clay, and glass. In the larger sectors—in iron and steel and in wood and products—there are similar percentage rates of decrease. In the metalworking group, there is spectacular growth in aircraft, instruments and cameras, electrical and service equipment, and nonelectrical equipment. Table 4.2 singles out those sectors

* In the 38-order classification, the major portion of rubber and plastic materials production is included in chemicals, and hence the growth rate shown for rubber and plastic products is modest. A clearer impression emerges in Chapter 6.

Table 4.2 38-Order Sectors whose Intermediate Deliveries are Most Affected by Structural Change (percent)

Sector	Annual rate of change in requirements
Gaining	
(25) Aircraft	5.5
(31) Utilities	3.4
(30) Communications	3.1
(27) Instruments, etc.	3.0
(23) Electrical eq., etc.	2.6
(21) Nonelectrical eq.	2.4
(32) Trade	2.1
(36) Auto. repair	2.0
(13) Chemicals	1.9
(22) Engines & turbines	1.8
Declining	
(4) Coal mining	−4.5
(7) Construction	−3.2
(3) Nonferrous mining	−2.1
(18) Iron & steel	−1.9
(2) Iron mining	−1.7
(16) Leather & shoes	−1.7
(1) Agriculture, etc.	−1.5
(11) Wood & products	−1.1
(38) Scrap	−0.8
(26) Trains, ships, etc.	−0.9

that are gaining and those that are losing importance most rapidly as a result of structural change. Rates of increase with fixed final demand, but changing technology, are shown for twenty sectors: the ten for which total intermediate requirements increase most and the ten for which they decrease most rapidly.

4.2 Changing Total Labor Requirements

While the volume of transactions required to deliver a fixed bill of final demand has tended to increase a little, total labor requirements have been declining rapidly. Figure 4.3 shows, for the fixed bill of 1961 final demand, how the total volume of employment required in the different industry groups changes over the period 1939–1961. These employment estimates for 1961 were obtained by summing the products of the given 1961 output levels and

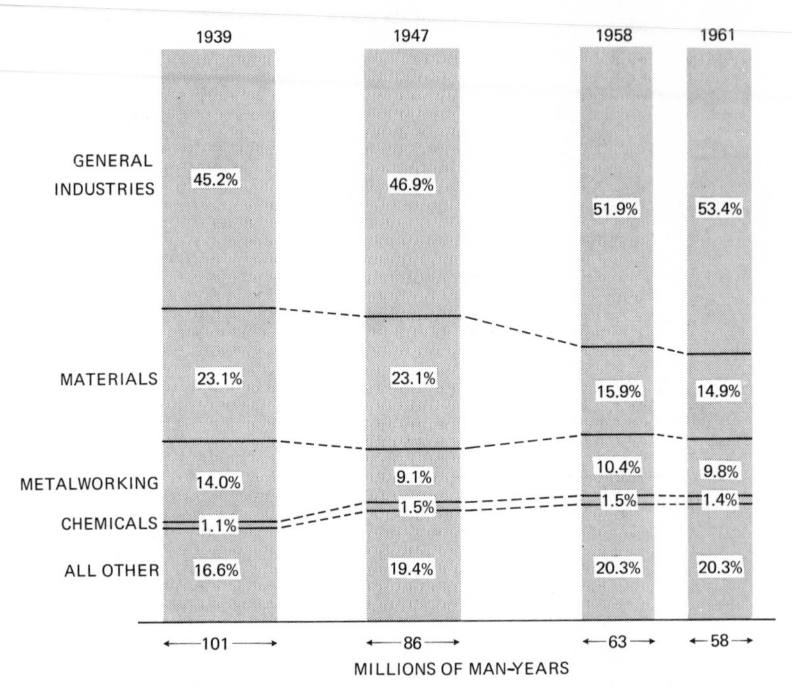

4.3 Employment Required to Deliver 1961 Final Demand with 1939, 1947, 1958, and 1961 Technologies.

their respective labor coefficients—employment (in man-years) per dollar of output—for each sector. Employment estimates for the other years were computed analogously:

$$\mu^t = \mathbf{l}^t \mathbf{Q}^t \mathbf{y}^{61} \tag{4.2}$$

where

μ^t = employment estimate (in man-years) in year t
$\mathbf{Q}^t$ = inverse matrix for year t
$\mathbf{l}^t$ = (row) vector of labor coefficients for year t
$\mathbf{y}^{61}$ = 1961 final demand vector

Subtotals of employment in each of five major industrial groupings, ${}_r\mu^t$, are also shown:

$${}_r\mu^t = {}_r\mathbf{l}^t \mathbf{Q}^t \mathbf{y}^{61} \quad r = \begin{array}{l}\text{general industries,}\\ \text{metalworking,}\\ \text{materials, chemicals,}\\ \text{all other}\end{array} \tag{4.3}$$

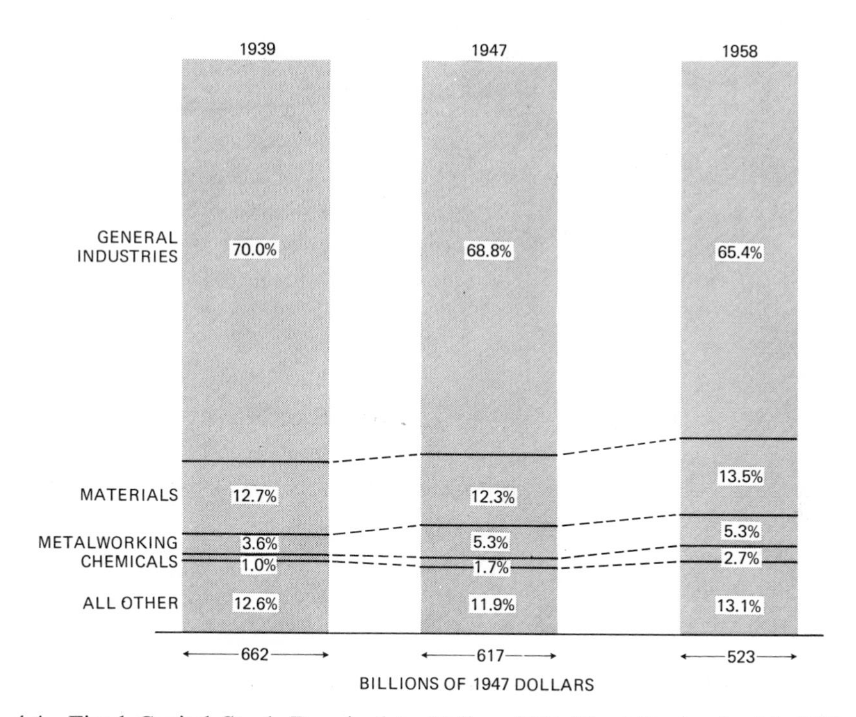

4.4 Fixed Capital Stock Required to Deliver 1961 Final Demand with 1939, 1947, and 1958 Technologies.

Total labor requirements decline rapidly because labor coefficients fall markedly in almost all industries and at rates that easily offset any increases in the volume of total transactions. Productivity increases and changes in specialization often go hand-in-hand with the reorganization of industry for greater efficiency. For example, increases in power use underlie some decreases in labor requirements through mechanization; increases in purchased communications services reduce manpower requirements in sales, administration, and the maintenance of large inventories.

By 1958–1961, more than half of the employment generated by fixed final demand was in the general industries group, while employment in material-producing sectors had fallen to the level of 15 percent.

4.3 Changing Total Capital Requirements

Figure 4.4 shows the third aspect of the general picture. It gives changes in the amounts of fixed investment in each industry group required to deliver 1961 final demand with the input structures and total capital stock coefficients of 1939, 1947, and 1958. Capital requirements were computed by the same method as the labor requirements. The 1961 capital coefficients, necessary for making a 1961 estimate of investment requirements, were not available. Totals and subtotals of capital requirements for 1939, 1947, and 1958 were computed as follows:

$$\kappa^t = \mathbf{b}^t\mathbf{Q}^t\mathbf{y}^{61} \qquad {}_r\kappa^t = {}_r\mathbf{b}^t\mathbf{Q}^t\mathbf{y}^{61} \qquad r = \begin{array}{l}\text{general industries,}\\ \text{metalworking,}\\ \text{materials, chemicals,}\\ \text{all other}\end{array} \tag{4.4}$$

where

κ^t and ${}_r\kappa^t$ = dollar value of the stock of total fixed investment and of fixed investment in each group required to deliver 1961 final demand with the technology of each of the years 1939, 1947, and 1958

$\mathbf{b}^t$ = row vector of total dollar value of fixed capital required to support a unit (dollar's worth per year) of output in each sector in year t

It is important to bear in mind that the capital stock coefficients available for this analysis are crude and serve only to establish orders of magnitude for the quantities under discussion. It is particularly difficult to secure capital coefficients that are comparable over time. The difficulty lies not merely in a dearth of information but also, fundamentally, in the changing qualitative character of the capital goods composing the stock. Qualitatively different capital goods of several vintages coexist in the capital stock at any given time, and awkward fictions are required to convert the aggregate capital stock for any sector or group of sectors into common dollar units. Figure 4.4 shows that the order of magnitude of the value (in constant dollars) of capital stock required to deliver a given bill of final demand decreases over the interval 1939–1947–1958, particularly after 1947. Hickman (1965:151–153) reports similar tendencies for the postwar period.

The relative contributions of the five broad sectoral groups in terms of labor, capital, and total intermediate requirements are very different. General industries, in 1958, contribute roughly 40 percent of intermediate inputs, but they account for more than 50 percent of employment and 65 percent of capital stock requirements. Rates of change in the relative importance of the broad groups are not the same when judged in terms of labor, capital, and intermediate requirements. Thus, chemicals contributes a growing proportion of the total value of intermediate inputs. Because gains in the productivity of

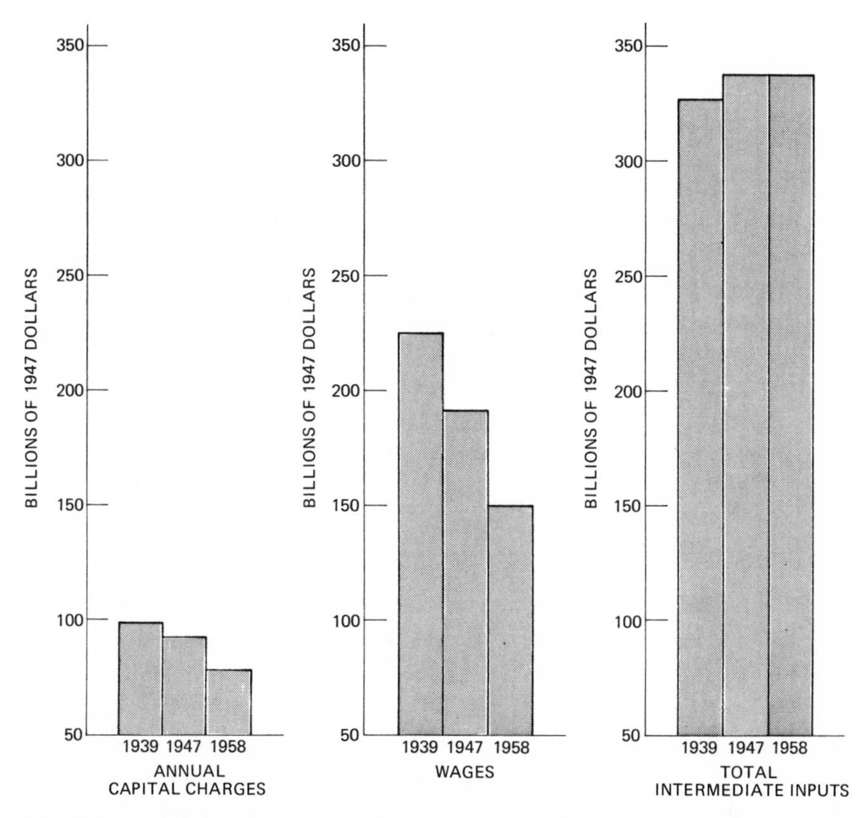

4.5 Value of Intermediate Inputs, Wages, and Annual Capital Charges to Deliver 1961 Final Demand with 1939, 1947, and 1958 Technologies.

both labor and capital have been comparatively large in chemicals, this group does not account for an increasing proportion of labor or of capital requirements to deliver a fixed final demand. General industries account for the great bulk of the nation's capital stock, but their *relative* importance in terms of capital requirements is declining. The relative decline comes from decreases in the capital coefficients for transportation and public utilities.

In summary, it appears that technological change in 1939–1947–1958–1961 has made it possible for the American economy to produce a given bill of final demand with appreciably less labor and somewhat smaller quantities of fixed capital stock. The net decrease in labor requirements was achieved by means of decreasing direct labor coefficients, along with changes in the relative importance of specific intermediate inputs required to deliver the given final demand. Changes in the skill mix undoubtedly account for some of the apparent gains in productivity. While total fixed capital requirements have decreased too, much of the labor saving would have been impossible without improvements in the performance characteristics of capital goods. Changes in industrial specialization and qualitative change—in intermediate inputs, in labor, and in capital stock—provide the basis for quantitative increases in labor efficiency.

Figure 4.5 is designed to integrate the information on changing input structures presented thus far. It compares labor—measured in constant dollar wage units—and current capital charges—capital stock requirements times a fixed rate to cover depreciation and interest*—for each sector needed to produce 1961 final demand with 1939, 1947, and 1958 technology. In addition, total intermediate input requirements with the technology of each year are shown.

4.4 Changes in Intermediate Input Requirements to Deliver Specific Subvectors of Final Demand

Let us now begin to inquire into the underlying details. Changes in total intermediate requirements for each sector to deliver a given bill of goods reflect many coefficient changes within the input-output matrix. Some of these changes reinforce each other; others tend to cancel each other out in their combined effects. Aluminum, for example, is of growing importance in automobile engines and construction but of declining importance in pots and pans. Some industries use more energy as the level of their mechanization or automation increases, while energy consumption in other sectors shows a net decrease as growing efficiency of energy use outpaces increases in mechanization. Paper is gaining in some types of packaging uses and being displaced by plastics in others. To understand the trends depicted in table 4.1 and the discussion that follows, one would want to know to what extent the net shifts described there represent parallel changes in many different sectors and to what extent they represent the domination of marked changes in a few basic areas. In the last analysis, this brings us back to cell-by-cell changes within a detailed classification framework.

In view of some of the difficulties of interpreting direct coefficient changes, it seems best to approach the details gradually, beginning with a survey of changing requirements for delivering different subvectors of final demand. The 1961 final demand vector is divided into eight major subvectors or "end-

* Fifteen percent was chosen as a crude order-of-magnitude rate for converting capital stock to annual capital charge estimates.

product" groups, such as food and tobacco, construction, and transportation equipment (figure 4.6 contains a complete list). We compute a vector of intermediate output requirements for each subvector of final demand separately, as if final demands for products in all other groups were zero, using first the 1939, then the 1947, and finally the 1958 input-output coefficients:

$$ {}_g\mathbf{z}^t = \mathbf{Q}^t{}_g\mathbf{y}^{61} - {}_g\mathbf{y}^{61} \quad (g = 1, 2, \ldots, 8) \tag{4.5} $$

where

${}_g\mathbf{z}^t$ = vector of intermediate outputs required to deliver the final demand for subvector g (say, food and tobacco, only) with the technology of time t prevailing in all industries

${}_g\mathbf{y}^{61}$ = one of the eight subvectors of 1961 final demand

It is as if we turn first to the 1939 economy, then to the 1947, and then to the 1958, asking each to produce the 1961 bill of goods for food and tobacco only, pretending that all other elements of final demand are zero. We list the resulting intermediate output levels generated for each year. Then we go back and ask the 1939, 1947, and 1958 economies to produce the next final demand subvector alone, list the second set of intermediate output levels, and so on. The result is a set of eight 38-order intermediate output vectors for each year. At the time these computations were made, no input-output matrix—only total outputs and final demand—was available for 1961. Thus, the hypothetical requirements to deliver specific subvectors could not be computed for the terminal year.

Comparison of the corresponding vectors of intermediate outputs for the three years tells how requirements to deliver 1961 final demand for, say, construction were affected by all changes in the input-output matrix over each interval. Such comparisons show the extent to which various broad tendencies noted for the economy as a whole characterize the production of specific types of end products. The right side of figure 4.6 shows the total value of intermediate inputs required to produce each of the eight subvectors of final demand with 1939, 1947, and 1958 technologies:

$$ {}_g\zeta^t = (1, 1, \ldots, 1)\mathbf{Q}^t{}_g\mathbf{y}^{61} - {}_g\mathbf{y}^{61} \quad (g = 1, 2, \ldots, 8) \tag{4.6} $$

To make it easier to compare requirements for subvectors of very different absolute sizes, the input requirements were put on a per-unit-of-final-delivery basis. This was done by dividing each side of equation 4.6 by ${}_g\phi^{61} = (1, 1, \ldots, 1)_g\mathbf{y}^{61}$, that is, by the total value of final demand for subvector g in 1961. Essentially, total intermediate input requirements to produce a dollar's

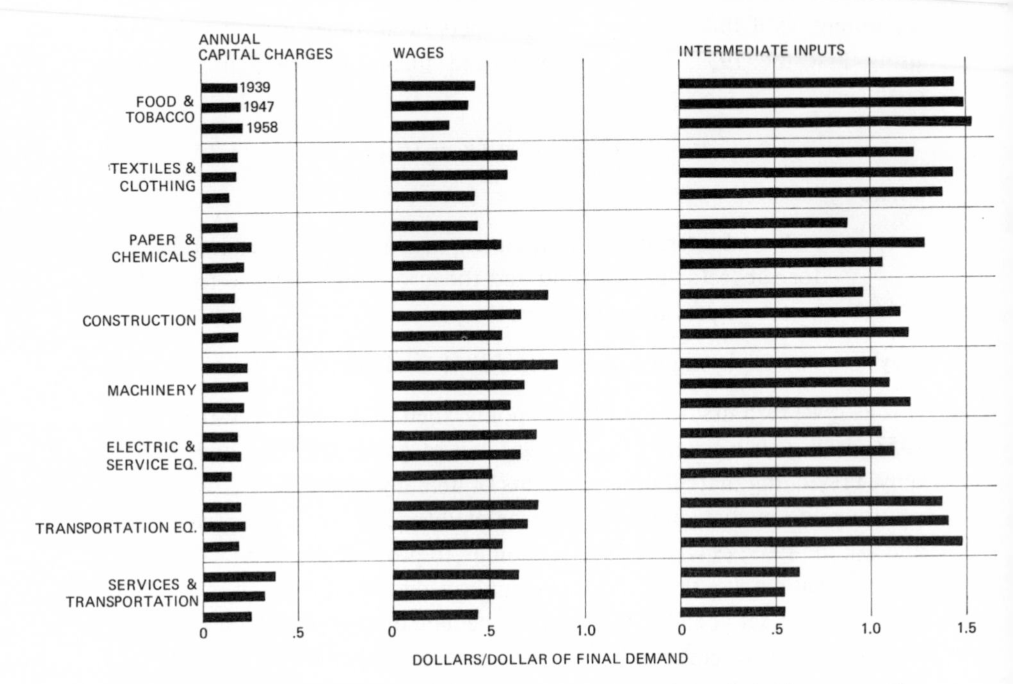

4.6 Value of Intermediate Inputs, Wages, and Annual Capital Charges to Deliver Eight Subvectors of 1961 Final Demand with 1939, 1947, and 1958 Technologies.

worth of each of eight different assortments of final goods are being compared.

The total value of intermediate inputs is slightly higher for delivering some types of final goods than others. Total intermediate requirements are lowest for service and transportation deliveries, highest for food and for transportation equipment. It tends to increase over time for most of the individual subvectors of final demand but to decrease for services and transportation. In paper and chemicals, electrical and service equipment and, to a lesser extent, in textiles and clothing, roundaboutness tends to increase between 1939 and 1947, and fall off between 1947 and 1958.

Indices of gross changes in intermediate input structure: amount of reshuffling

Changes in total intermediate deliveries required to produce each subvector of final demand are the sum of increases in requirements for some inputs,

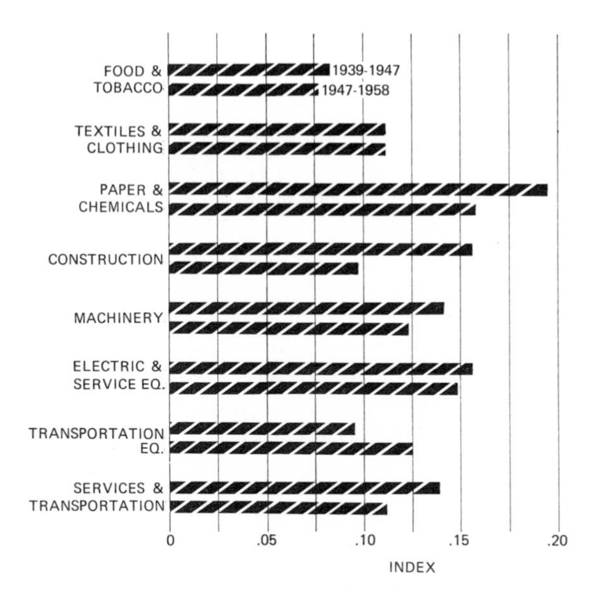

4.7 Amount of Reshuffling: Indices of Gross Volume of Change in Intermediate Requirements from Specific Sectors to Deliver Eight Subvectors of 1961 Final Demand for 1939–1947 and 1947–1958.

decreases in requirements for others. Thus, a great deal of substitution of one input for another can take place with little or no change in total intermediate input requirements. The gross, as well as the net, amount of change is of interest. The sum of absolute differences in total requirements for each individual input between two years can be interpreted as an index of the amount of reshuffling—of the gross amount of changes in specific requirements affecting the production of each subvector of final demand.

$$ {}_g\sigma^t = \frac{(1, 1, \ldots, 1)|{}_g\mathbf{z}^t - {}_g\mathbf{z}^{t-1}|}{(1, 1, \ldots, 1)({}_g\mathbf{z}^t + {}_g\mathbf{z}^{t-1})} \tag{4.7} $$

where

${}_g\sigma^t$ = index of the amount of change in specific intermediate input requirements for a particular subvector of final demand g

The term ${}_g\sigma^t$ measures how much the direct and indirect dependence on particular industries shifted over a period (relative to the total requirements). The more substitution among intermediate inputs and the more transfer of function from one sector to another, the larger ${}_g\sigma^t$ will be.

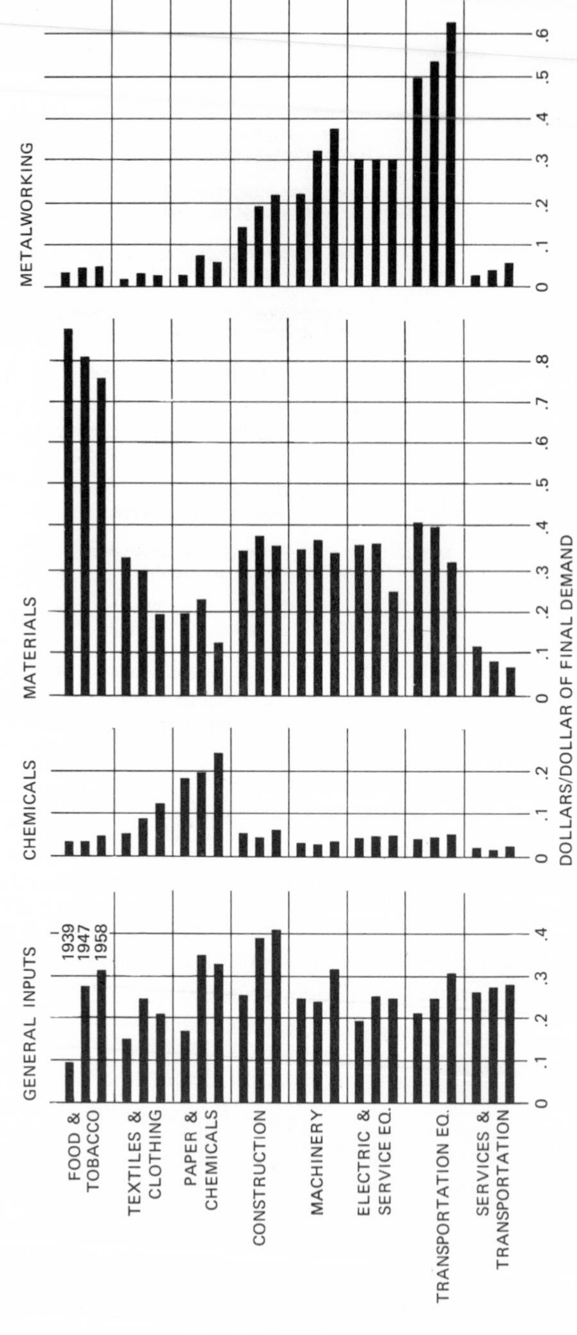

4.8 Intermediate Requirements from Four Supplier Groups to Deliver Eight Subvectors of 1961 Final Demand with 1939, 1947, and 1958 Technologies.

Figure 4.7 gives $_g\sigma^t$ for 1939–1947 and 1947–1958 for each of the eight specified final demand subvectors. It shows that the gross volume of change in intermediate input requirements was greater for delivering chemicals and machinery than for food and textiles. Except in transportation equipment and textiles, the relative volume of "reshuffling" of intermediate inputs was slightly greater between 1939 and 1947 than between 1947 and 1958, but the orders of magnitude are similar. Reshuffling among intermediate inputs and labor and capital is studied in Chapters 9 to 11.

Changing relative importance of supplier groups for each subvector of final demand

Figure 4.8 shows how changes in input structures between 1939 and 1947 and between 1947 and 1958 affected input requirements from the four broad groups of suppliers—general inputs, chemicals, producers of materials, and metalworkers—in delivering the different types of end products. For the most part, over the 1939–1947–1958 interval, total requirements of general inputs and of chemicals tend to increase while materials requirements tend to decrease for each subvector of final demand. Further details of these changes in specialization will be considered in Chapters 5 and 6.

4.5 Changing Labor and Capital Requirements to Deliver Specific Subvectors of Final Demand

Changes in the total labor and capital charges for each subvector of final demand were also computed:

$$\begin{aligned} {}_g\bar{\mu}^t &= \mathbf{l}^t\hat{\mathbf{w}}^{47}\mathbf{Q}^t{}_g\mathbf{y}^{61} \\ {}_g\bar{\kappa}^t &= .15\mathbf{b}^t\mathbf{Q}^t{}_g\mathbf{y}^{61} \end{aligned} \qquad (4.8)$$

where

$_g\bar{\kappa}^t$ = annual capital charges

$_g\bar{\mu}^t$ = wage costs at 1947 wage rates

$\mathbf{w}^{47}$ = vector of 1947 wage rates

0.15 = rough estimate of proportion of capital stock charged to current account each year

Changes in labor costs are shown, next to changes in intermediate requirements already discussed, in figure 4.6. Each bar measures total cost of labor at constant wage rates required to deliver a dollar's worth of each subvector of final demand with the technologies of 1939, 1947, and 1958. While the right side of the figure shows changing degrees of roundaboutness for producing many of the different types of end products, the middle section shows reductions in total labor requirements to deliver each of them. Note that reductions in labor requirements are much greater for some types of final products than for others. Reductions for food and tobacco are relatively modest, and for chemicals and paper they even show a small rise from 1939 to 1947. As compared to changes in intermediate and capital inputs, however, decreases in labor requirements are large and, with the one minor exception in chemicals, persistent.

The left side of figure 4.6 shows changes in annual capital charges per dollar of delivery to each of the eight subvectors of final demand (annual capital charges are derived from estimated stock requirements, as in Section 4.3). The different types of final products vary in the amount of capital required (directly and indirectly) to deliver them. Somewhat less capital is required to deliver a dollar's worth of textiles or construction than a dollar's worth of, say, machinery. Services and transportation remains the most capital-intensive of all subvectors, although its capital requirements decline over the period. For all but the food subvector, capital requirements are lower with 1958 than with 1947 technology. The order of magnitude of capital charges is relatively small compared to labor and intermediate inputs.

Given these general contours of changing structure, we may proceed to explore the ins and outs of particular regions.

Chapter 5 General Inputs

5.1 Economic Role of General Inputs

General industries produce the kinds of inputs—energy, transportation, trade, communications, and other services—required by virtually all establishments in the economy and used in the production of a very broad range of goods and services. Over the period 1939–1961, growing demand for general inputs accounts for a major portion of any observed increases in roundaboutness in the American economy. This, plus the fact that general industries absorb more than half the nation's total employment, and an even larger proportion of fixed capital assets, justifies more detailed scrutiny of this aspect of the structure of production. Compared to requirements for materials, general input requirements are strikingly similar for the different subvectors of final demand (figure 4.8). General input requirements also tend to increase over time for all types of end products. Figure 5.1 supports these conclusions in direct coefficient terms. It shows the sum of direct input coefficients* for all general inputs into each of the 38-order producing sectors for 1939, 1947, and 1958. Most of these coefficient sums range between 0.05 and 0.15, getting slightly higher with time. Exceptionally large consumers of general inputs are the general industries themselves—service sectors buy a lot of services themselves and transportation, a lot of transportation and energy. The significance of such "near-diagonal" purchases is discussed in Section 5.3.

It is easy to explain the relative uniformity of general input requirements among varied consuming sectors. They represent the operations common to almost all types of productive activity: record keeping, finance, communications, advertising, transportation, and trade are required; and energy is used for space heating and cooling and for driving equipment in virtually every industry. In the analysis of change, each industry's structure can be partitioned into a specialized part, whose structural characteristics depend primarily on the specific kinds of goods and services that the industry produces; and into a nonspecialized part, representing the elements common to all sectors in a given economy. Coefficients along a row of general inputs are subject to common influences, and it is particularly worthwhile to consider these across the board in studying or projecting general input coefficients. "Process service" modules, as suggested in early studies by Chenery (1953) and Holzman (1953) might some day be used to handle changing general input requirements in input-output systems. While changes in specialized inputs depend

* In this and all subsequent scatter diagrams, the direct coefficients that are shown do not include fictitious transfers, that is, they are "primary" coefficients.

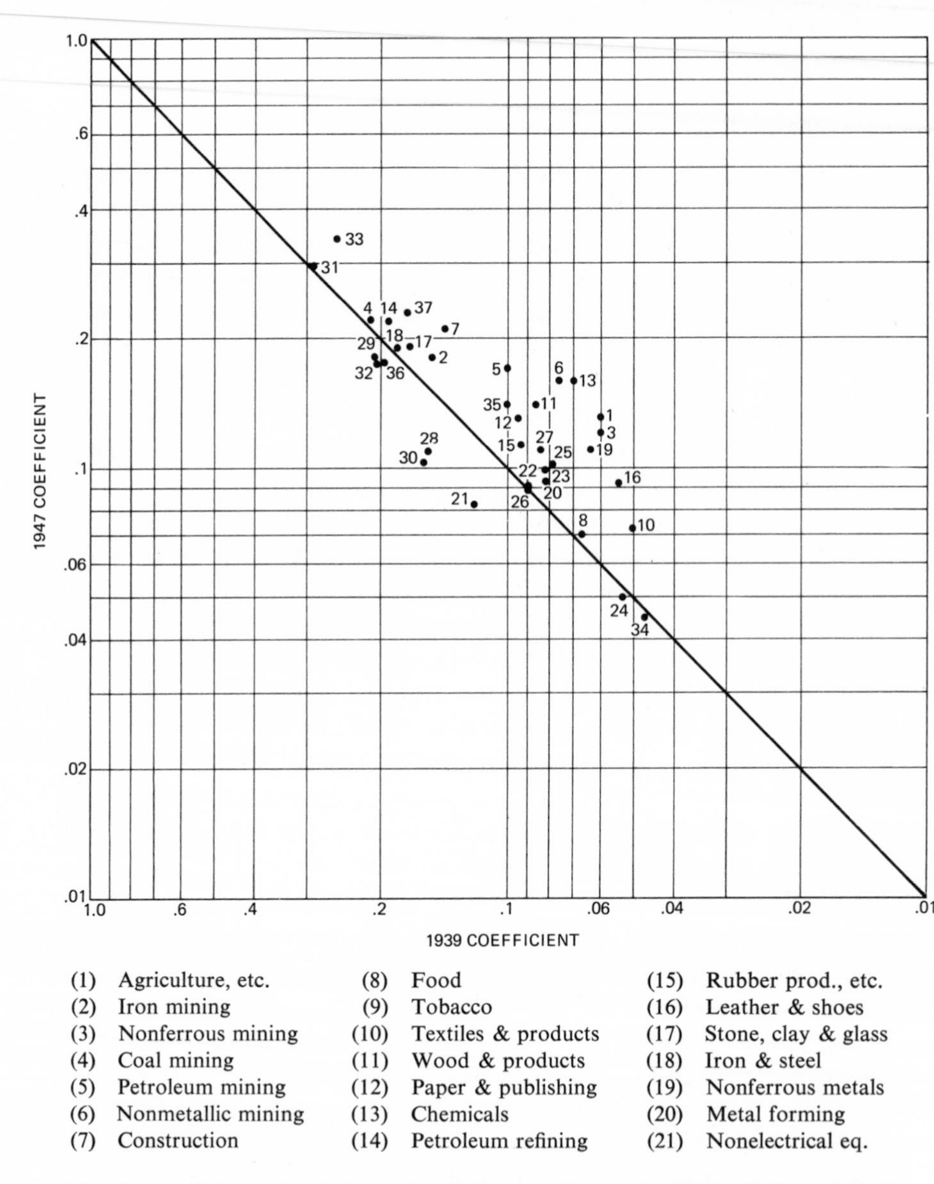

5.1 Sums of Direct Coefficients for all General Inputs for 1939, 1947, and 1958 (dollars per dollar in 1947 prices). Each point indicates the value of the sum of coefficients for a single 38-order consuming sector in each of two years.

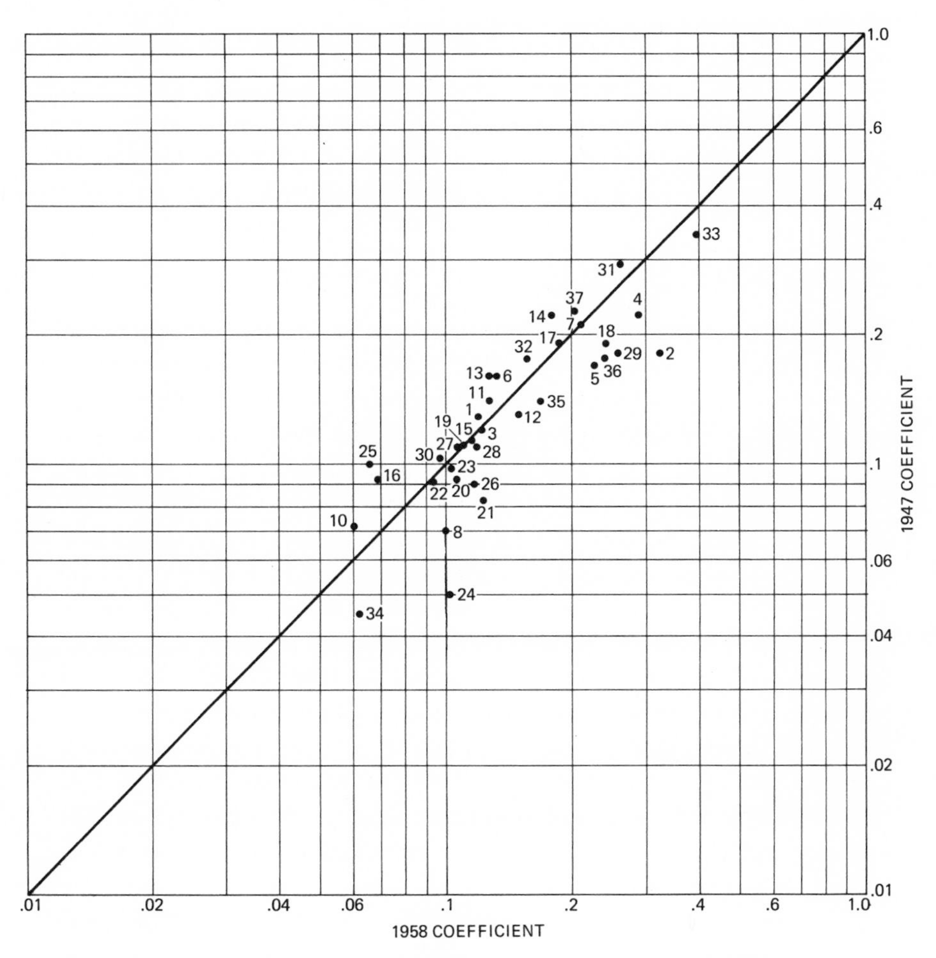

(22) Engines & turbines
(23) Electrical eq., etc.
(24) Motor vehicles & eq.
(25) Aircraft
(26) Trains, ships, etc.
(27) Instruments, etc.
(28) Misc. manufactures
(29) Transportation
(30) Communications
(31) Utilities
(32) Trade
(33) Finance & insurance
(34) Real estate & rental
(35) Business serv., etc.
(36) Auto. repair
(37) Institutions, etc.
(38) Scrap

primarily on the particular technology and circumstances of the consuming industry, general inputs are more directly tied to broad aspects of the economy at large—its size, geography, industrial location patterns, and institutional environment. Thus, general inputs can be expected to vary more from country to country than specialized inputs. Comparisons of United States and Japanese input-output tables confirm this impression. Changing requirements for the various subgroups of general inputs are discussed below.

5.2 Changing Inputs from Energy Sectors

The principal energy-supplying sectors are coal mining (4), petroleum refining (14), and electric and gas utilities (31). Over the past few decades, industry has been consuming a rapidly growing proportion of its energy requirements in the form of electricity rather than through direct fuel consumption (see Schurr and Netschert 1960:180–189). Figure 5.2 shows changes in intermediate requirements from the energy sectors—measured in 1947 dollars per dollar delivery to final demand—for each subvector of final demand. Increase in total value of energy requirements is shown for all types* of end products, and movement is steadily upward for all but two subvectors. The relative importance of different energy suppliers is also changing. Direct-plus-indirect coal requirements decline substantially, while those for petroleum and for electric and gas utilities tend to increase for every final demand category, with a minor exception in petroleum requirements for transportation equipment. Directions of change are the same for 1939–1947 and 1947–1958, with some apparent acceleration in the latter period. The summary figures in table 4.1 suggest a continuation of these tendencies beyond 1958 up to 1961.

Two features of energy use emerge from figure 5.2: (1) similarities in energy intensities for all subvectors in any given year and (2) similarities in their directions and rates of change. These across-the-board similarities represent important common elements in the technology of fuel use in all sectors. Beyond the ubiquitous temperature control and motive power, a few sectors, such as aluminum and steel, require significant inputs of process heat; and petroleum, coal, and gas are raw materials, as well as energy sources, for chemicals. Transportation sectors are relatively fuel intensive. However, exceptional demands of these few sectors are diffused in reckoning with broad end-product categories.

Increases in total energy use are the resultant of two clear, but opposing,

* The general formula for computing sectoral output requirements to deliver individual subvectors of final demand is given by equation 4.5. See also Section 3.4.

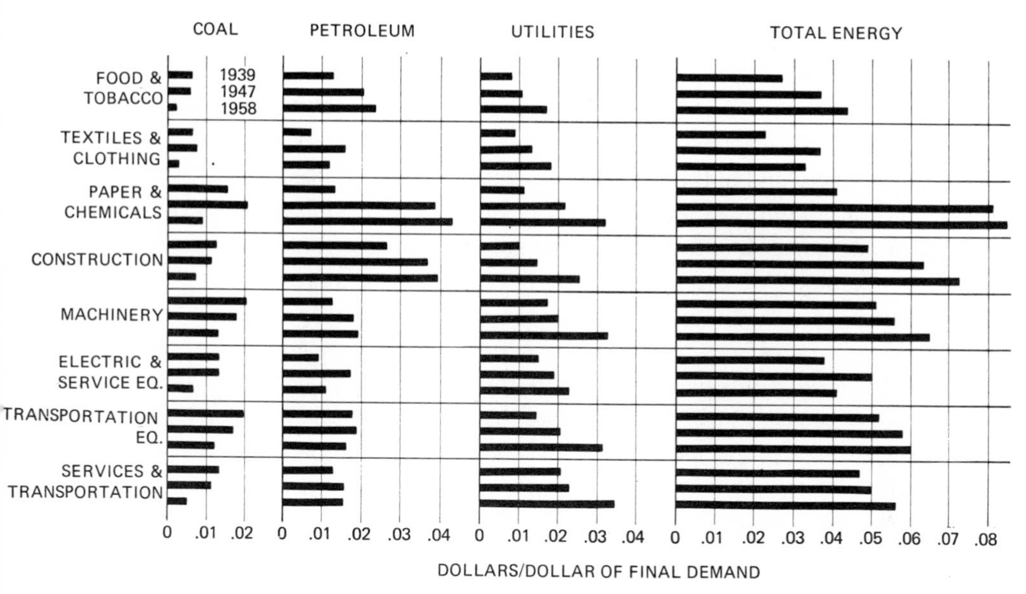

5.2 Intermediate Requirements from Energy Sectors to Deliver Eight Subvectors of 1961 Final Demand with 1939, 1947, and 1958 Technologies.

tendencies. On the one hand, the spheres of atmospheric control and of mechanization or automation are expanding. Offices and factories are being air-conditioned because precision equipment requires it. Air conditioning increases the efficiency of scarce and expensive labor and the competitive advantage of the establishment in bidding for it; and many productive operations are being transferred from man to energy-using equipment. On the other hand, efficiency of energy use has been increasing steadily (see Schurr and Netschert 1960:171–174).

Figure 5.2 bears out other reports (Schurr and Netschert 1960:33–42; U. S. Bureau of the Census 1960:35 and 1965:52) on the changing importance of individual energy-supplying sectors—coal, petroleum, and utilities. Coal has been declining in relative importance with the shift from the use of steam to internal combustion engines in transportation and other uses and to electric power in all sectors. Natural gas and petroleum have been substituted for coal in space heating and electric power generation. The rise in petroleum requirements per unit delivery to final demand represents the reverse side of this coin for the period 1939–1947. Between 1947 and 1958, petroleum levels off with growing competition from natural gas. Without improvements in the efficiency of energy use, less of a decline in coal requirements, or even a rise, might have taken place; and the rise in petroleum requirements might have been much greater. The relation of the utilities sector to other energy suppliers is complex in that the sector subsumes both the distribution of natural gas and the production and distribution of electric power. Thus, the growth of utilities reflects substantial increases both in the relative importance of natural gas and in the proportion of fuel converted to electric energy prior to its consumption by industry. Part of the growth of utilities represents a shift among competing fuels and part represents an increased amount of "preprocessing" of energy before it is consumed by industry, which implies increased "double counting" in the outputs of energy sectors. Increased reliance on electric power has been encouraged by steady improvements in the efficiency of fuel use in electric power generation,* combined with progress in economical transmission of power over long distances. The development of cheap natural gas resources has led to important inroads on other fuels, at first in the South and Southwest and, later, with the spread of major pipeline facilities, throughout the country.

The rise in total energy requirements, measured in figure 5.2, stands in apparent contradiction to findings of Schurr and Netschert (1960:16), who note a decline in energy consumption, measured in British thermal units (btu) per dollar of gross national product, in recent years. However, the contradiction is only superficial and has to do with the difference of units—constant dollars versus btu's—in which energy inputs to the economy are measured. Energy consumption in btu's is estimated in terms of the total energy "content" of each of the fuels. For converting to btu's, the btu content of a ton of coal is taken to be a constant, equal to the amount of energy that can be released under theoretical conditions of one hundred percent efficiency. Estimates for petroleum and natural gas are made, similarly, on the basis of total btu content. Increased thermal efficiency in the use of fuel results in a major decline in total btu requirements per unit of final demand (gross national product). Although efficiency of use of all fuels has been increasing, a major part of the overall improvement has actually resulted from the shift from coal, where thermal efficiency is lowest, to petroleum and natural gas, where it is higher. The picture looks different in (constant dollar) value terms: petroleum and natural gas are priced higher per thermal unit theoretically available, and higher prices are economically justified because a larger portion of their energy content can actually be extracted from the fuels in industrial use. The increased proportion of energy consumed in the form of electric power also increases constant-dollar input from the energy sectors, at the same time that it decreases btu requirements. Indeed, the conversion of a larger proportion of fuels to electric energy has been instrumental in lowering btu requirements at the same time that it raises the total value contribution

* Nuclear power was of negligible economic significance during this period.

of the utilities sector for the delivery of a unit of final demand. Thus, a decline in energy requirements, as measured in btu's per unit of gross national product, may well be consistent with an increase, as measured in deflated dollars.

The tabulation by major subvectors of final demand gives only a summary picture. Figures 5.3, 5.4, and 5.5 supply the underlying direct coefficient detail at 38 order. In these scatter diagrams (figures 5.3 to 5.5) direct coefficients tend to cluster on one side of the 45-degree line or the other. For coal, most of the direct coefficients are larger in 1947 than in 1958, larger in 1939 than in 1947. For utilities, just the opposite tendency appears. The tendencies in petroleum are not so clear cut. Between 1947 and 1958, more direct petroleum coefficients fall than rise. A few fuel-intensive sectors, particularly stone, clay, and glass (17), iron and steel (18), transportation and storage (29), and near-diagonal elements lie well beyond the general range of concentration for all energy sources.

5.3 Increasing Importance of Producer Services

The service sectors to be analyzed here include finance and insurance, real estate and rental, business and personal services and hotels, automobile repair, institutions, research, and entertainment. Trade is considered in a separate section below. Figure 5.6 presents total requirements from the combined service sectors. As in the case of energy, total service requirements do not differ much among the various subvectors of final demand, and they increase between 1939 and 1947 and between 1947 and 1958. The decrease in service requirements for machinery between 1939 and 1947 is unexplained but may well be attributed to crudeness of the data.

Relatively large amounts of services are required to deliver the final demand subvector for services and transportation itself. That is because establishments delivering services to final demand purchase appreciable amounts of services from establishments in their own and other service industries. These near-diagonal purchases are not really general inputs but are more in the nature of semifinished goods or "parts and components"—real estate brokers pay legal fees, advertising agencies rent or purchase the services of more specialized agencies or billboards or broadcast media. Thus, somewhat larger inputs of services are required to deliver a dollar's worth of the service than of most other subvectors of final demand. *In general, direct-plus-indirect requirements for a particular input tend to be large whenever that input is also one of the elements of final demand to be delivered.* This tendency was already apparent for energy sectors and affects many inputs to be discussed below.

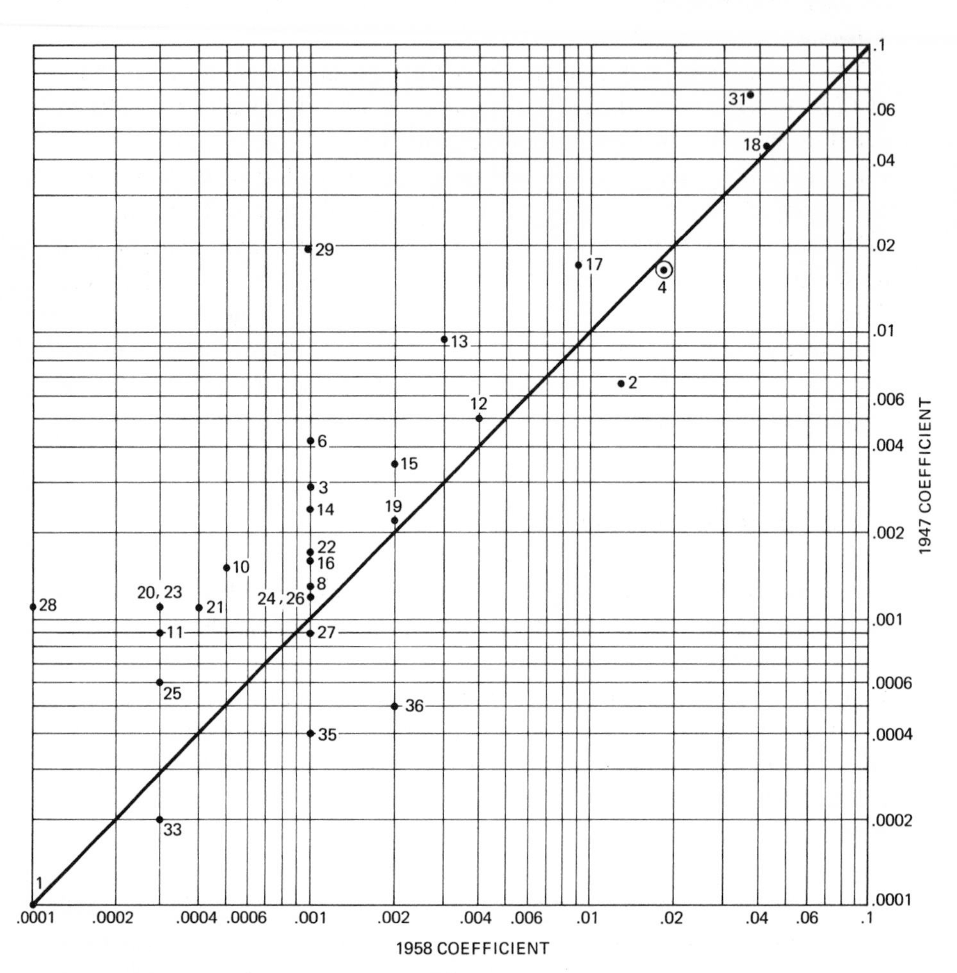

(1)	Agriculture, etc.	(8)	Food	(15)	Rubber prod., etc.
(2)	Iron mining	(9)	Tobacco	(16)	Leather & shoes
(3)	Nonferrous mining	(10)	Textiles & products	(17)	Stone, clay & glass
(4)	Coal mining	(11)	Wood & products	(18)	Iron & steel
(5)	Petroleum mining	(12)	Paper & publishing	(19)	Nonferrous metals
(6)	Nonmetallic mining	(13)	Chemicals	(20)	Metal forming
(7)	Construction	(14)	Petroleum refining	(21)	Nonelectrical eq.

5.3 Direct Coal Mining Coefficients for 1939, 1947, and 1958 (dollars per dollar in 1947 prices). Each point indicates the value of the coefficient for a single 38-order consuming sector in each of two years (for circled points, multiply scales by 10).

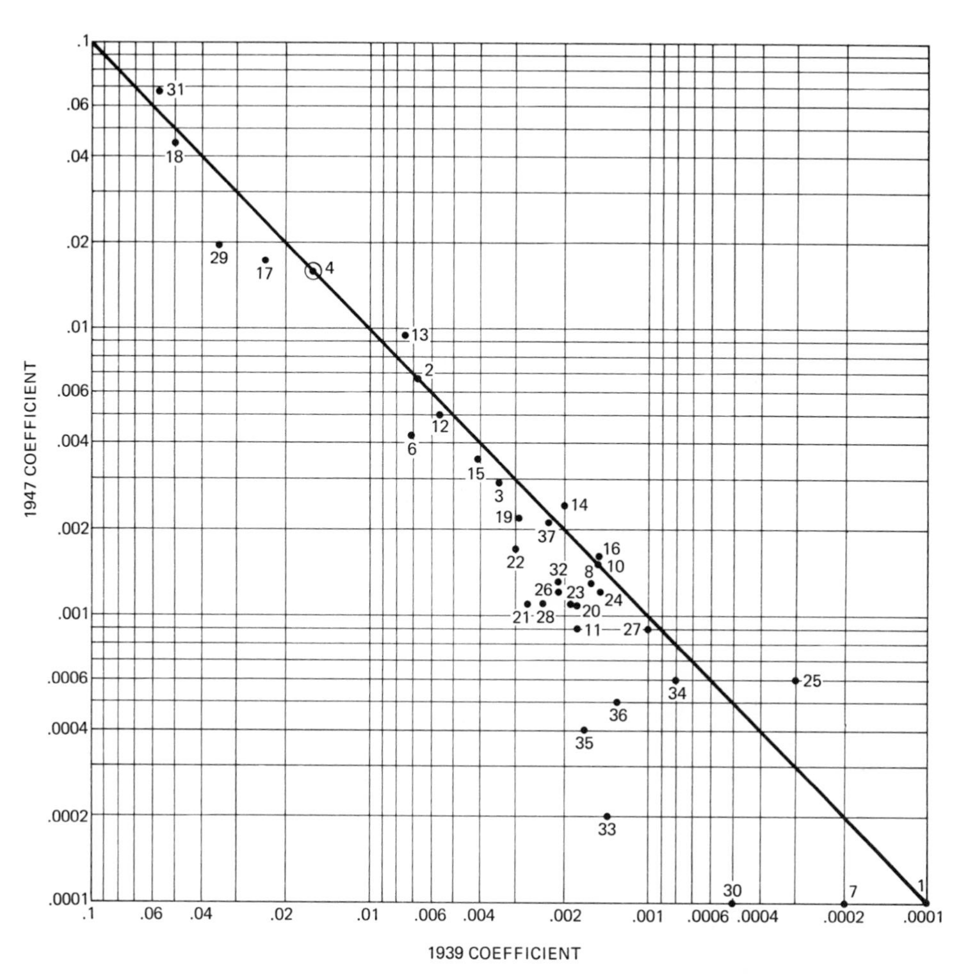

(22) Engines & turbines
(23) Electrical eq., etc.
(24) Motor vehicles & eq.
(25) Aircraft
(26) Trains, ships, etc.
(27) Instruments, etc.
(28) Misc. manufactures
(29) Transportation
(30) Communications
(31) Utilities
(32) Trade
(33) Finance & insurance
(34) Real estate & rental
(35) Business serv., etc.
(36) Auto. repair
(37) Institutions, etc.
(38) Scrap

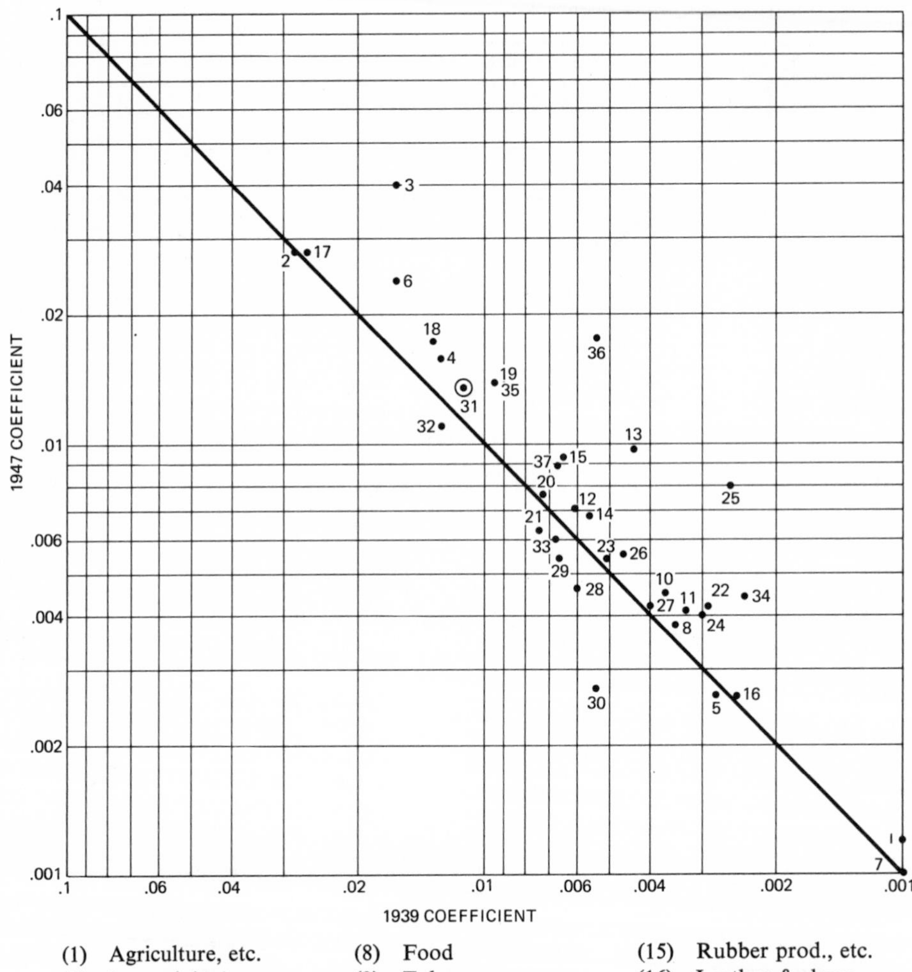

(1)	Agriculture, etc.	(8)	Food	(15)	Rubber prod., etc.
(2)	Iron mining	(9)	Tobacco	(16)	Leather & shoes
(3)	Nonferrous mining	(10)	Textiles & products	(17)	Stone, clay & glass
(4)	Coal mining	(11)	Wood & products	(18)	Iron & steel
(5)	Petroleum mining	(12)	Paper & publishing	(19)	Nonferrous metals
(6)	Nonmetallic mining	(13)	Chemicals	(20)	Metal forming
(7)	Construction	(14)	Petroleum refining	(21)	Nonelectrical eq.

5.4 Direct Gas and Electric Utilities Coefficients for 1939, 1947, and 1958 (dollars per dollar in 1947 prices). Each point indicates the value of the coefficient for a single 38-order consuming sector in each of two years (for circled points, multiply scales by 10).

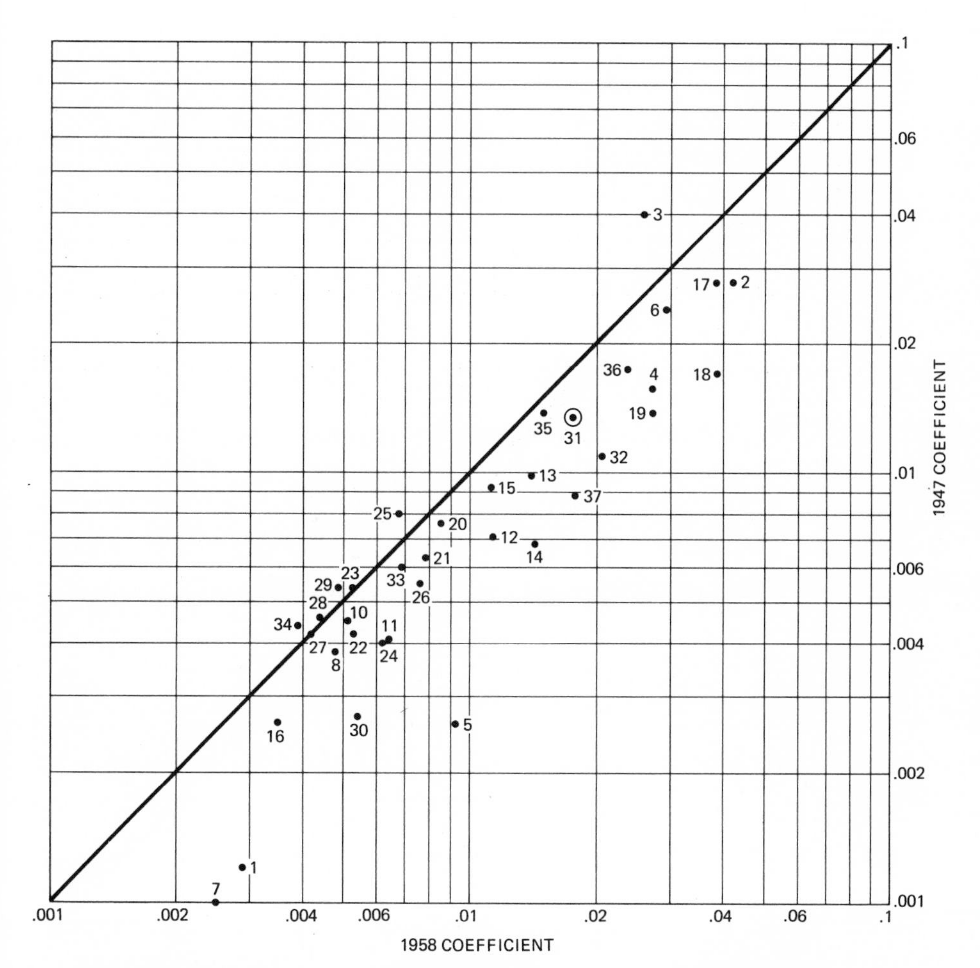

(22) Engines & turbines
(23) Electrical eq., etc.
(24) Motor vehicles & eq.
(25) Aircraft
(26) Trains, ships, etc.
(27) Instruments, etc.
(28) Misc. manufactures
(29) Transportation
(30) Communications
(31) Utilities
(32) Trade
(33) Finance & insurance
(34) Real estate & rental
(35) Business serv., etc.
(36) Auto. repair
(37) Institutions, etc.
(38) Scrap

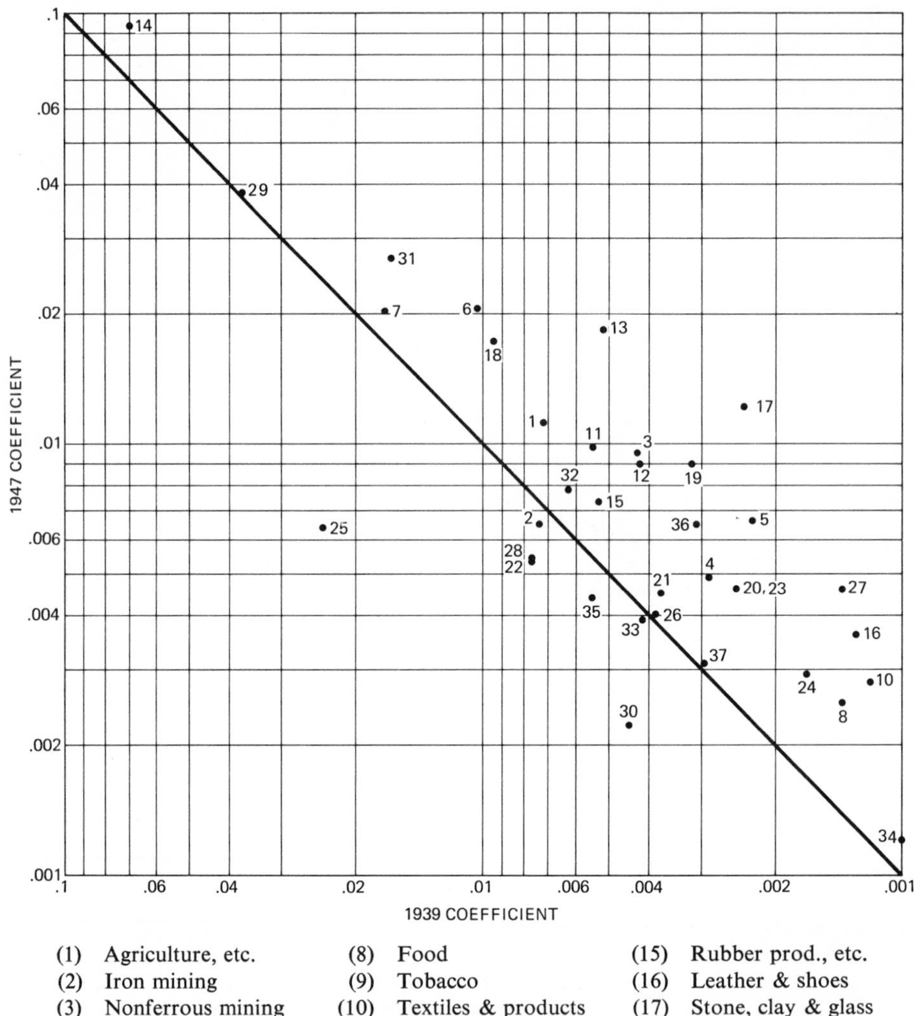

(1) Agriculture, etc.
(2) Iron mining
(3) Nonferrous mining
(4) Coal mining
(5) Petroleum mining
(6) Nonmetallic mining
(7) Construction
(8) Food
(9) Tobacco
(10) Textiles & products
(11) Wood & products
(12) Paper & publishing
(13) Chemicals
(14) Petroleum refining
(15) Rubber prod., etc.
(16) Leather & shoes
(17) Stone, clay & glass
(18) Iron & steel
(19) Nonferrous metals
(20) Metal forming
(21) Nonelectrical eq.

5.5 Direct Petroleum Refining Coefficients for 1939, 1947, and 1958 (dollars per dollar in 1947 prices). Each point indicates the value of the coefficient for a single 38-order consuming sector in each of two years.

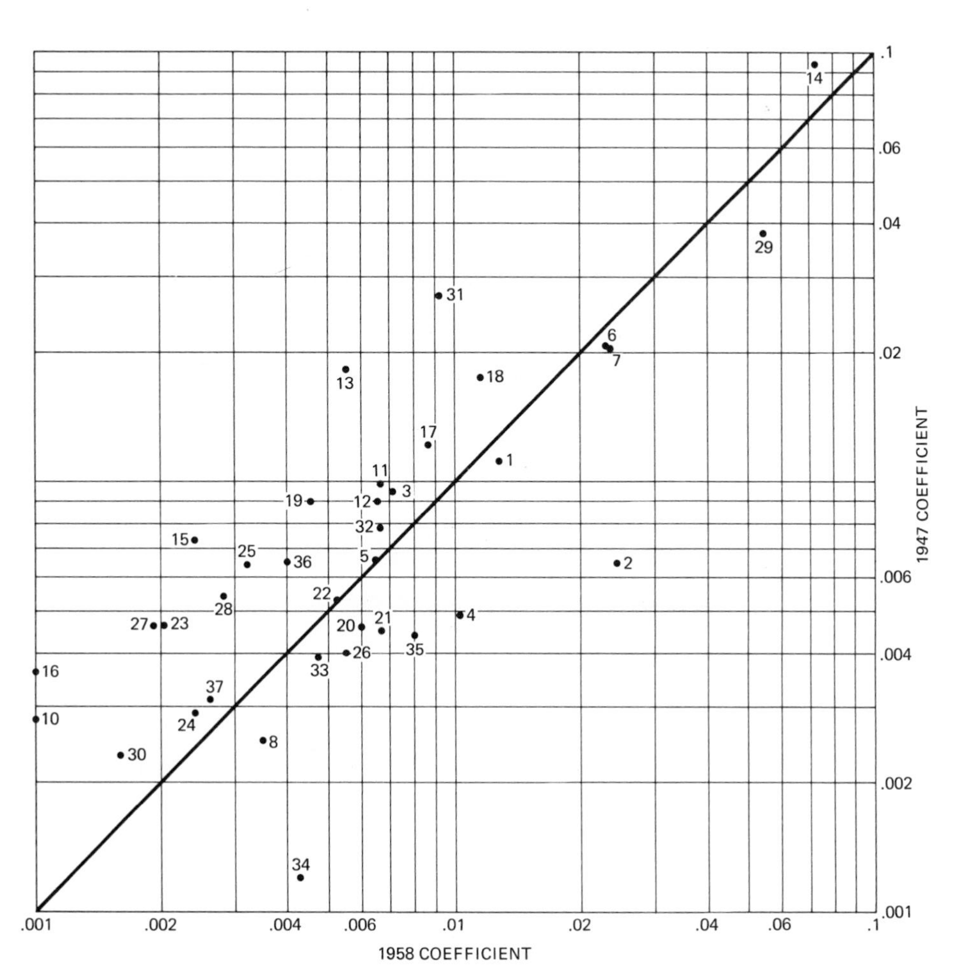

(22)	Engines & turbines	(28)	Misc. manufactures	(33)	Finance & insurance
(23)	Electrical eq., etc.	(29)	Transportation	(34)	Real estate & rental
(24)	Motor vehicles & eq.	(30)	Communications	(35)	Business serv., etc.
(25)	Aircraft	(31)	Utilities	(36)	Auto. repair
(26)	Trains, ships, etc.	(32)	Trade	(37)	Institutions, etc.
(27)	Instruments, etc.			(38)	Scrap

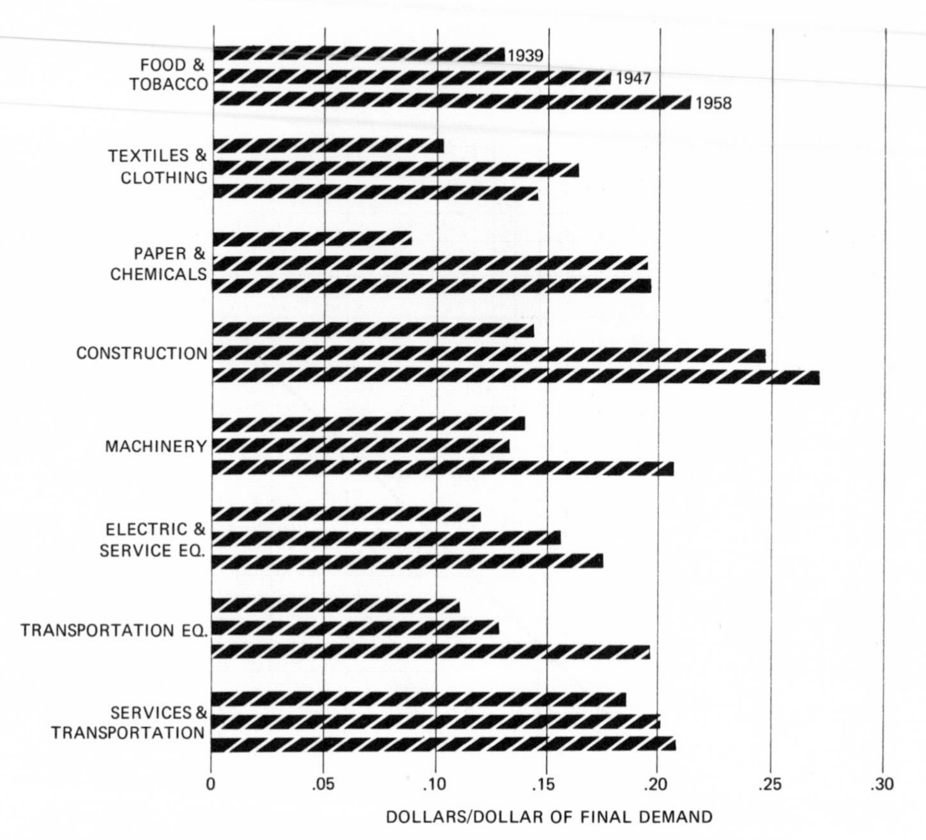

5.6 Intermediate Requirements from All Service Sectors to Deliver Eight Subvectors of 1961 Final Demand with 1939, 1947, and 1958 Technologies.

Figure 5.7 is a scatter diagram for the total of all service coefficients into each consuming sector. Of course, the largest service coefficients are for inputs of services into other services. Increases in services occur for most sectors, particularly in the postwar period. While all service inputs tend to increase, communications, business services, and automobile repair grow most markedly (table 4.1). Advertising is by far the most rapidly growing portion of the business services sector.

Specialization, transfer of function, and rise of producer services

Bear in mind that we are dealing with changing requirements for services to deliver a fixed final demand, so that we observe changes in intermediate service requirements only. The dramatic growth cited here, then, is that of the so-called producer services, that is, services purchased by business establishments rather than directly by consumers. Services provided by government are treated as exogenous and are not included in the totals for rising service requirements.

While some economists (for example, Kuznets 1946; Foote and Hatt 1953; Regan 1963; and Fuchs 1965) have concerned themselves with the growing demand for consumer services in recent years, the growth in producer services has received comparatively little attention. Greenfield (1966) provides a thoughtful exception, but data are very sparse. Certainly, the rise in service inputs represents a significant trend in the organization of production and an important shift in the division of labor among sectors. Unlike changes in energy consumption, changes in service requirements are not generally associated with technological change in the usual sense of the term. However, their explanation is hardly independent of technology. In one sense, they represent social diseconomies of scale, that is, the costs of coordinating a growing volume of more specialized transactions. With rising national income and concomitant increases in intermediate transactions, it is reasonable to expect more than proportional increases in the cost of informational services necessary to coordinate far-flung activities in the system. Perhaps it is not an accident that communications and advertising grow most rapidly of all service inputs. Elements of qualitative change are also present. Consumers, intermediate and final, count on a more reliably distributed product with a larger informational component.

Only part of the growth of service sectors represents actual increases in the total volume of services performed. The rest is the transfer of service functions from firms and establishments primarily engaged in manufacturing and other product-oriented activities to specialized service sectors. Thus far, evidence on the extent of transfer of function is largely impressionistic, but fragmentary statistical evidence is becoming available for a few particular areas. For example, reports of the National Science Foundation (see Bureau of Labor Statistics 1960:13; and National Science Foundation 1961:9–10) show that increasing proportions of research and development were contracted out during the 1950s. Particularly over the period 1947–1958, banks began to perform record-keeping and information-storage functions that were formerly the province of the bookkeeping departments of customer establishments. Similarly, insurance companies assumed an increasing proportion of risks formerly taken by the insured companies themselves. Some of these services, such as workmen's compensation, are mandatory. Specialized record-keeping and computing services have been growing rapidly. Such business services range from very sophisticated data retrieval schemes to assisting small businesses in coping with tax regulations. To a greater and greater extent,

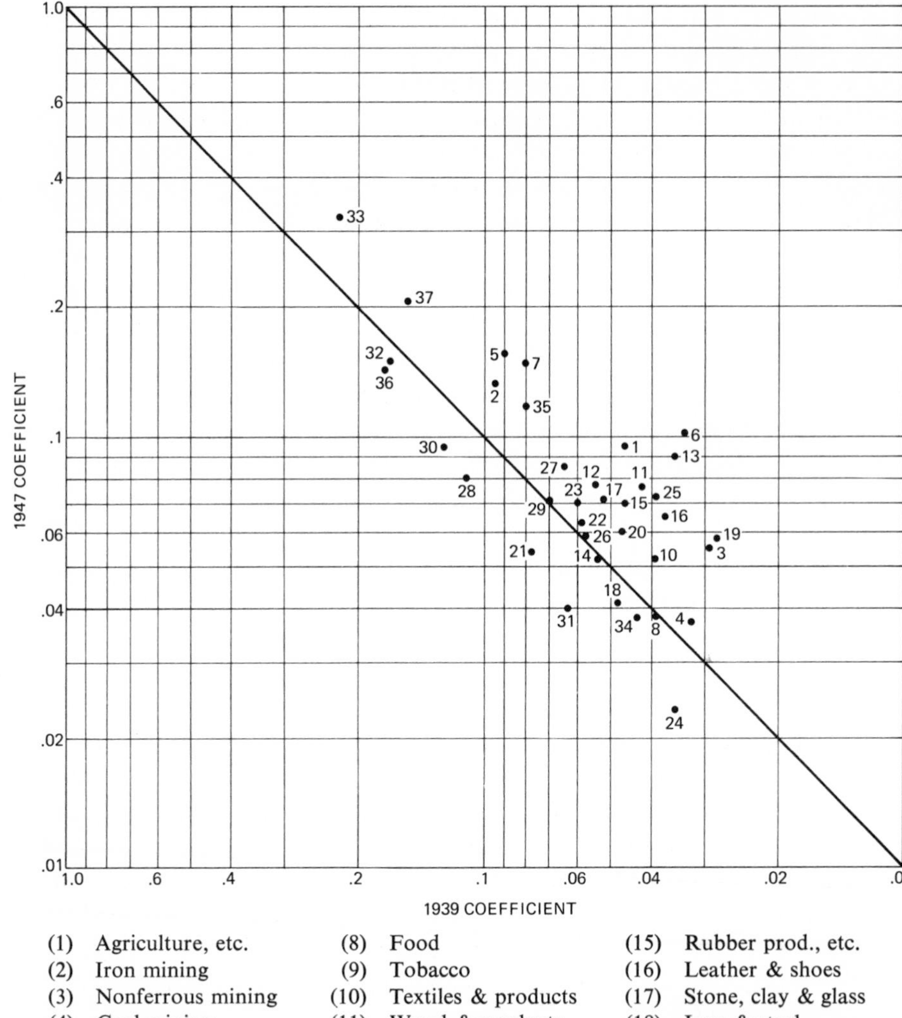

(1)	Agriculture, etc.	(8)	Food	(15)	Rubber prod., etc.
(2)	Iron mining	(9)	Tobacco	(16)	Leather & shoes
(3)	Nonferrous mining	(10)	Textiles & products	(17)	Stone, clay & glass
(4)	Coal mining	(11)	Wood & products	(18)	Iron & steel
(5)	Petroleum mining	(12)	Paper & publishing	(19)	Nonferrous metals
(6)	Nonmetallic mining	(13)	Chemicals	(20)	Metal forming
(7)	Construction	(14)	Petroleum refining	(21)	Nonelectrical eq.

5.7 Sums of Direct Coefficients for All Services for 1939, 1947, and 1958 (dollars per dollar in 1947 prices). Each point indicates the value of the sum of coefficients for a single 38-order consuming sector in each of two years.

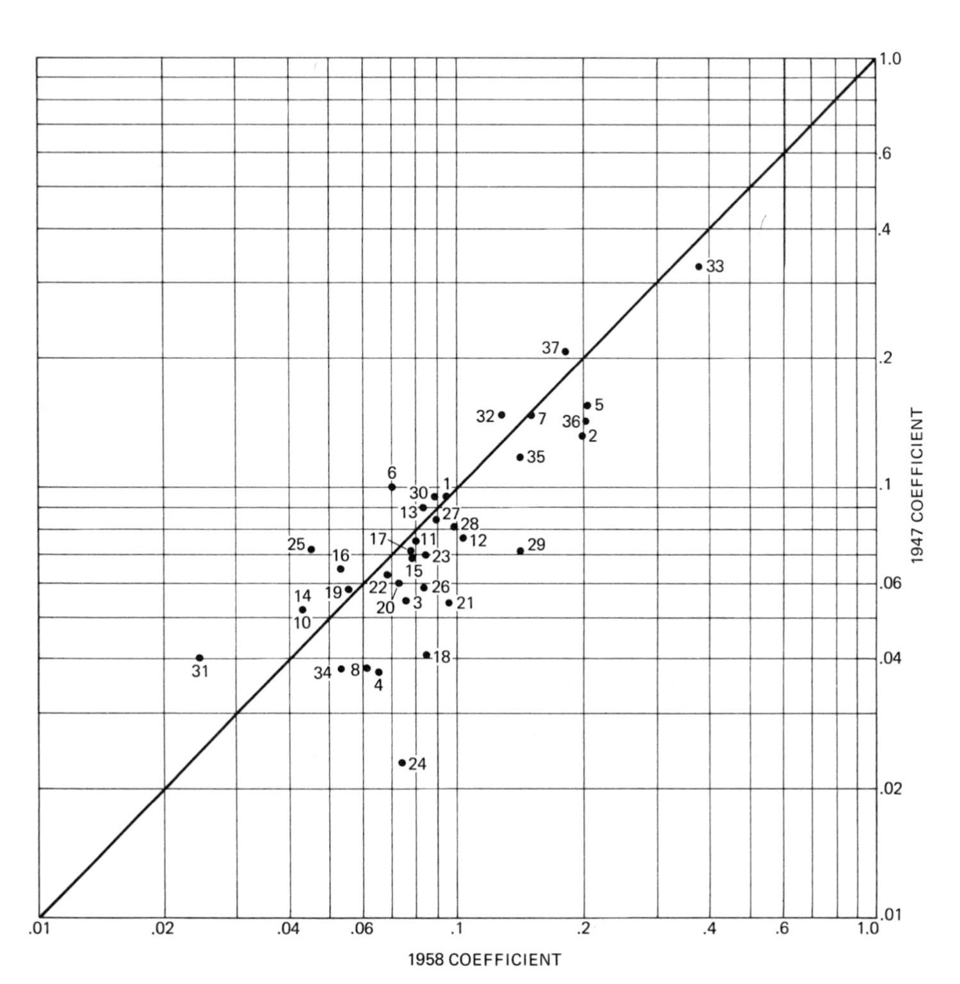

(22) Engines & turbines
(23) Electrical eq., etc.
(24) Motor vehicles & eq.
(25) Aircraft
(26) Trains, ships, etc.
(27) Instruments, etc.
(28) Misc. manufactures
(29) Transportation
(30) Communications
(31) Utilities
(32) Trade
(33) Finance & insurance
(34) Real estate & rental
(35) Business serv., etc.
(36) Auto. repair
(37) Institutions, etc.
(38) Scrap

exploratory and trouble-shooting research on engineering and business problems has been delegated to consulting organizations. Advertising work is more centralized in specialized agencies. Complex equipment, including cars and trucks, is beginning to be rented more frequently rather than being bought outright.

Two major factors bear on this trend in specialization. On the one hand, new technologies of information storage, processing and retrieval, servicing of complex equipment, and even technologies of research itself have been developed. Some of these require specialized knowledge and seem to exhibit considerable economies of scale. A specialized service producer can take advantage of the economies of scale with new techniques of providing services, and thus supply them at competitive advantage. On the other hand, after 1955, the advantage of centralization of service functions may have been accentuated by a growing shortage of technically trained personnel. The wage rate for engineering and professional personnel began to increase much more rapidly than the average in this period (Freeman 1967). Centralization of service functions can make it possible to economize on skilled labor in these fields. Specialized service organizations can often economize skilled and supervisory labor through mechanization and by subdividing complex jobs into simpler ones.

5.4 Changes in Trade Margins

According to the accounting conventions of a producers' value input-output table, trade and transportation inputs into each sector consist of actual trade and transportation charges for transporting and distributing all *inputs* for a given sector, from the establishment in which they are produced to the destination at which they are consumed. For example, transportation inputs into the automobile industry are costs of transporting the steel, glass, rubber, fuels, and so on purchased by automobile manufacturers and also the transportation costs on batteries, motors, tires, and so forth made in some of the industry's establishments and shipped to assemblers. Figure 5.8 shows that trade margins tend to increase for most types of end products over both time intervals. As in the case of the other service-oriented sectors, this represents increased coordination required by a more complex economy. Trade margins have apparently been increasing more for inputs into durable goods than for those into nondurable ones. Economies of direct buying, associated with the development of mass distribution outlets for food and clothing and increased centralization in food processing industries, have reduced trade margins in

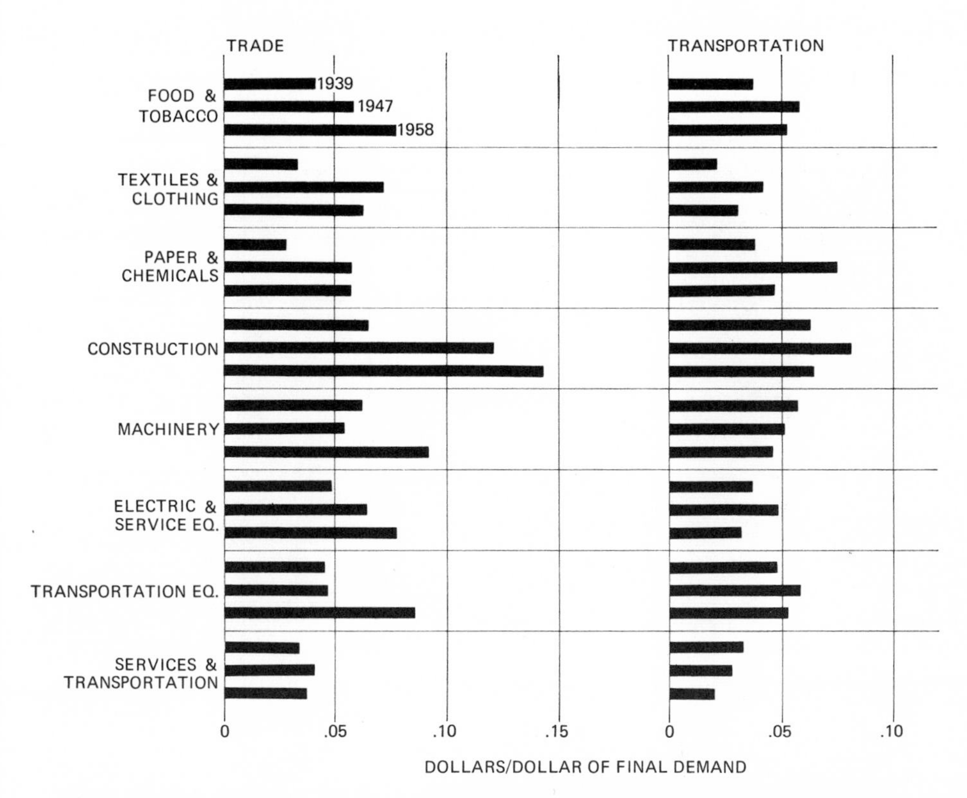

5.8 Trade and Transportation Margins on Inputs Required to Deliver Eight Subvectors of 1961 Final Demand with 1939, 1947, and 1958 Technologies.

these areas. However, development of specialized materials and increased complexity of parts and components has limited economies of mass distribution for inputs into durable goods.

5.5 Changes in Transportation Requirements

For many subvectors of final demand, transportation inputs (figure 5.8) increase between 1939 and 1947 and then return to their 1939 orders of magnitude between 1947 and 1958. The explanation of this pattern requires further study. We know that transportation costs tend to be high when the economy is pushing against capacity limitations. Substantial bottlenecks existed in 1947, and industrial consumers drew on distant materials and components suppliers to fill their needs. Postwar adjustments eased these bottlenecks, and more economic interregional shipment patterns could be

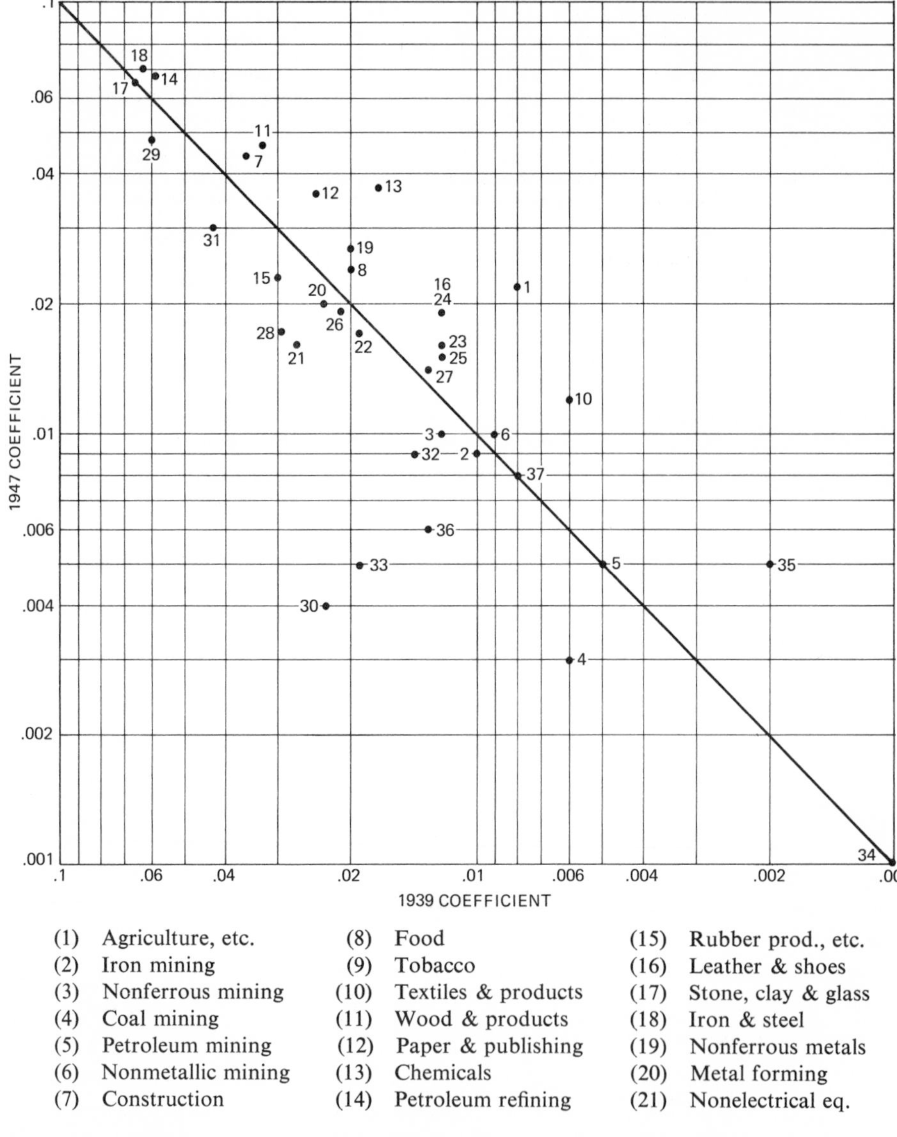

5.9 Producers' Value Transportation Margin Coefficients for 1939, 1947, and 1958 (dollars per dollar in 1947 prices). Each point indicates the value of the coefficient for a single 38-order consuming sector in each of two years.

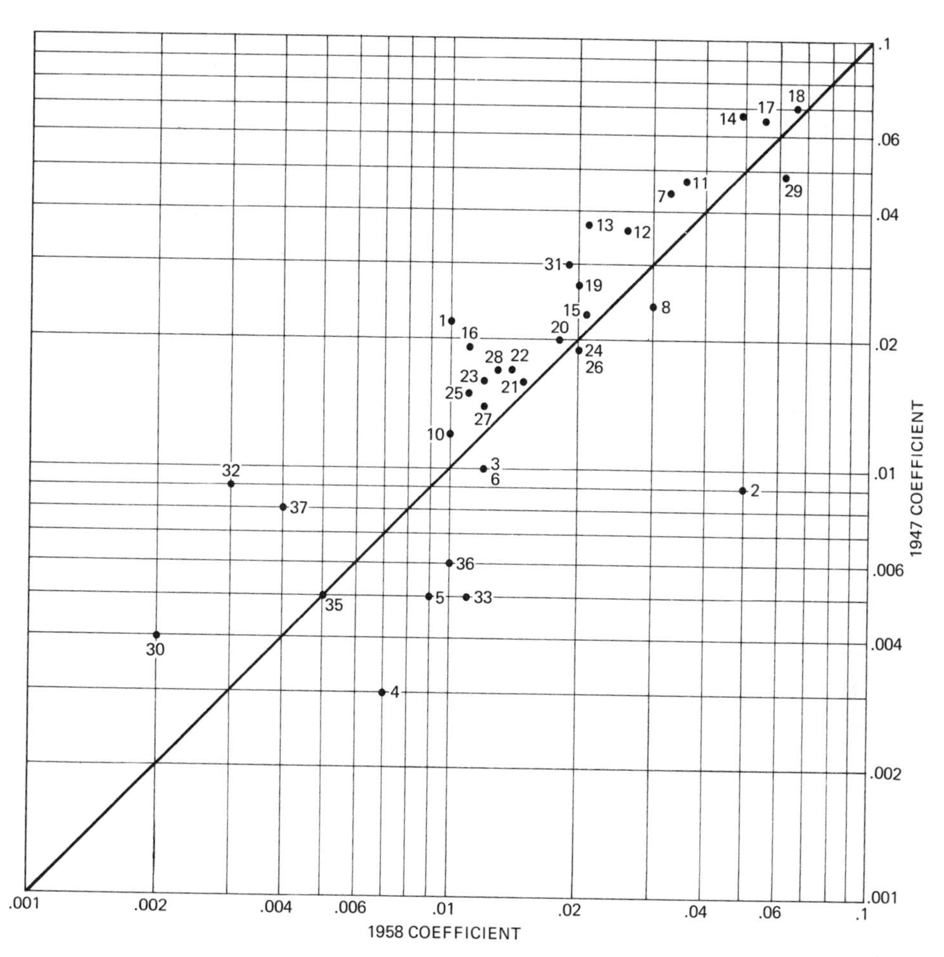

(22) Engines & turbines
(23) Electrical eq., etc.
(24) Motor vehicles & eq.
(25) Aircraft
(26) Trains, ships, etc.
(27) Instruments, etc.
(28) Misc. manufactures
(29) Transportation
(30) Communications
(31) Utilities
(32) Trade
(33) Finance & insurance
(34) Real estate & rental
(35) Business serv., etc.
(36) Auto. repair
(37) Institutions, etc.
(38) Scrap

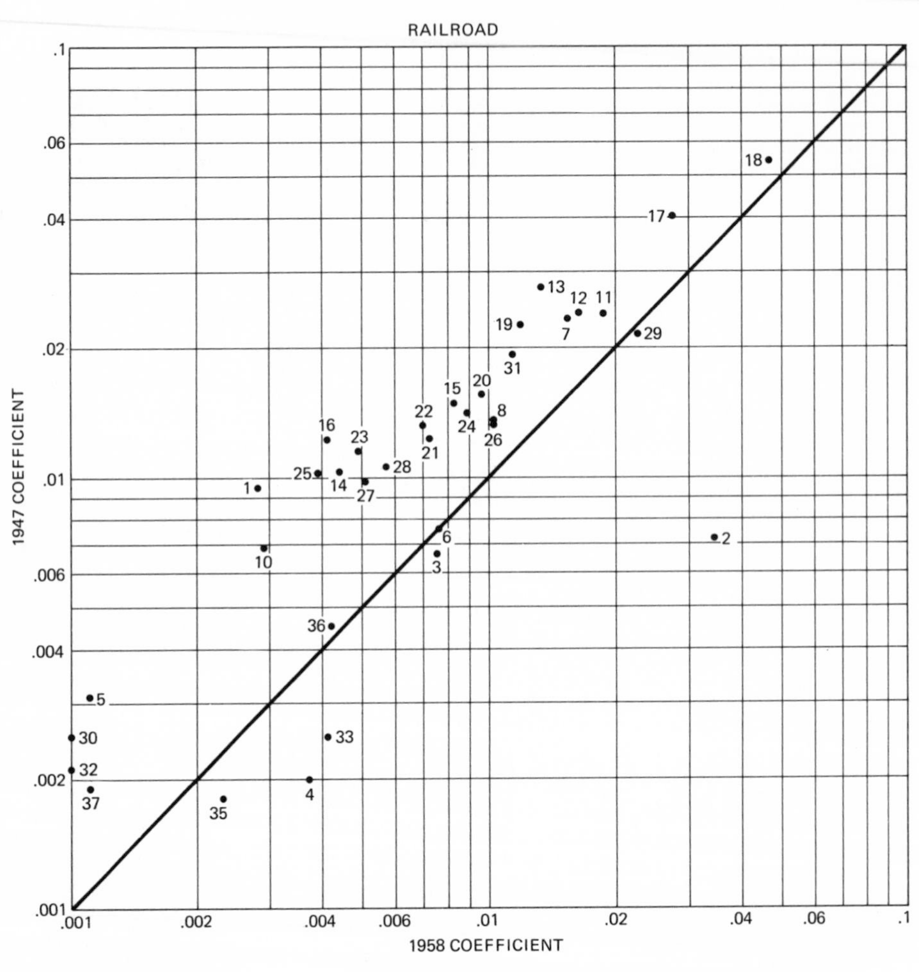

(1)	Agriculture, etc.	(8)	Food	(15)	Rubber prod., etc.
(2)	Iron mining	(9)	Tobacco	(16)	Leather & shoes
(3)	Nonferrous mining	(10)	Textiles & products	(17)	Stone, clay & glass
(4)	Coal mining	(11)	Wood & products	(18)	Iron & steel
(5)	Petroleum mining	(12)	Paper & publishing	(19)	Nonferrous metals
(6)	Nonmetallic mining	(13)	Chemicals	(20)	Metal forming
(7)	Construction	(14)	Petroleum refining	(21)	Nonelectrical eq.

5.10 Producers' Value Railroad and Truck Transportation Margin Coefficients for 1947 and 1958 (dollars per dollar in 1947 prices). Each point indicates the value of the coefficient for a single 38-order consuming sector in each of two years.

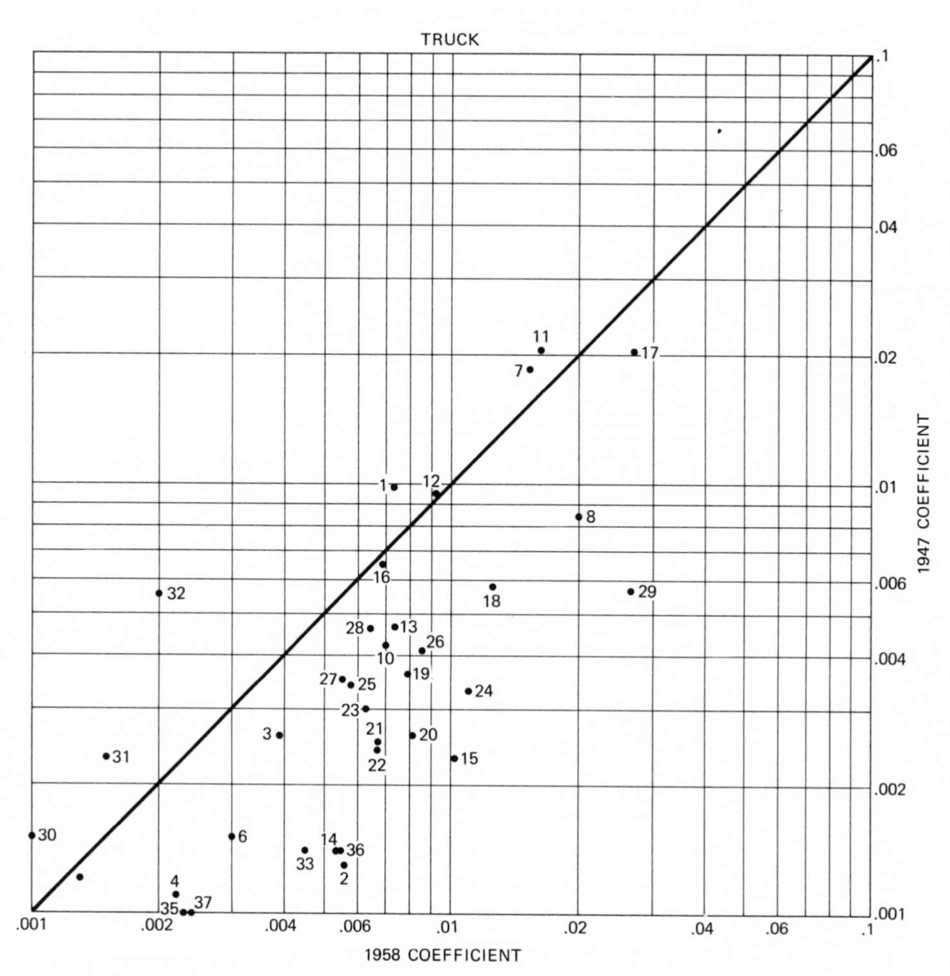

(22) Engines & turbines
(23) Electrical eq., etc.
(24) Motor vehicles & eq.
(25) Aircraft
(26) Trains, ships, etc.
(27) Instruments, etc.
(28) Misc. manufactures
(29) Transportation
(30) Communications
(31) Utilities
(32) Trade
(33) Finance & insurance
(34) Real estate & rental
(35) Business serv., etc.
(36) Auto. repair
(37) Institutions, etc.
(38) Scrap

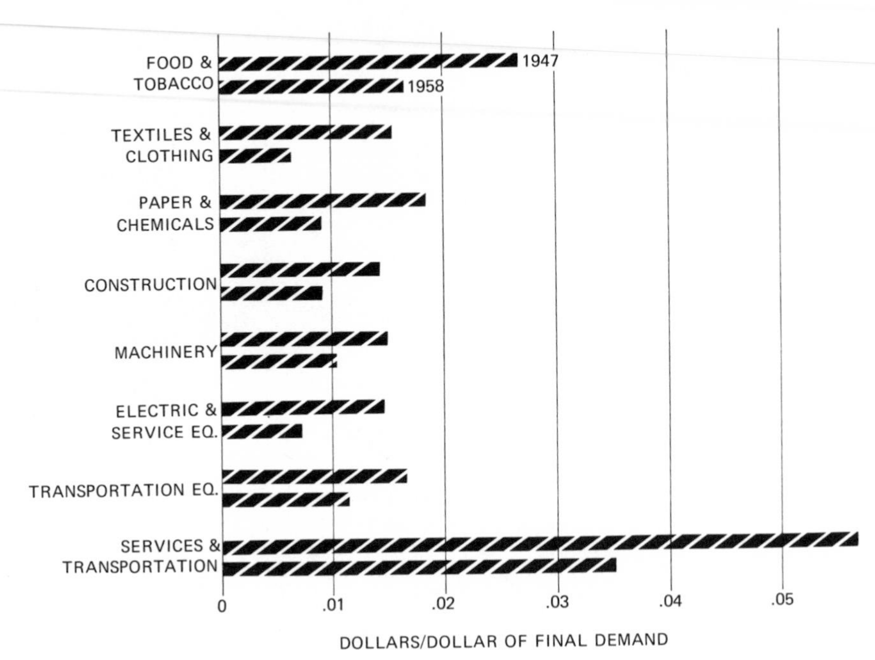

5.11 Intermediate Requirements from Maintenance Construction to Deliver Eight Subvectors of 1961 Final Demand with 1947 and 1958 Technologies.

established. Specific developments in certain industry groups contributed to further economies in transportation. Locally produced stone and clay products displaced steel and wood in construction (see Chapter 6). Smaller, lighter equipment was cheaper to ship (see Chapter 7). Fuel economies meant savings in transportation costs, particularly for fuel-intensive sectors. The growing relative importance of services as compared to tangible goods inputs also decreased transportation requirements. Direct transportation margin coefficients on inputs into individual 38-order sectors (figure 5.9) tell a similar story but in greater detail. The most striking changes affect transportation-intensive sectors. Changing fuel requirements meant smaller margins on inputs into electric and gas utilities (31). Margins on inputs into stone, clay, and glass (17), construction (7), petroleum refining (14), agriculture (1), wood and products (11), chemicals (13) and paper and publishing (12) also decreased noticeably between 1947 and 1958. Reduced transportation costs for producing chemicals result from changes in product mix, with increased emphasis on petrochemicals, which are produced from petroleum and natural-gas–based feedstocks. Petrochemicals are produced relatively near to the sources of raw material supply. Many of the margin coefficients that decreased between 1947 and 1958 increased between 1939 and 1947.

Truck transportation grew in relation to rail between 1947 and 1958 (figure 5.10)* most notably for inputs into food (8), rubber and plastic products (15), and motor vehicles (24); trucking grew in importance as an input to the transportation and storage sector (29) itself. Truck transportation margins remained small, if increasing, in quite a few sectors. The rankings of industries as consumers of rail and truck transportation differ. Motor transport grew primarily through very sizeable switches to trucking in a few heavy industries such as stone, clay, and glass (15), iron and steel (18), nonferrous metals (17), and the metalworking sectors (20) to (26). There were small increases in the use of trucking and decreases in the use of rail transportation by most other sectors.

5.6 Changes in Requirements for Other General Inputs

In the 76-order classification, maintenance construction, printing and publishing, and several packaging sectors are distinguishable as separate industries; at 38 order, they are aggregated with other sectors. Comparison can only be made for the 1947–1958 period for these inputs.

Maintenance construction

Figure 5.11 shows maintenance construction to be one of the few general inputs whose requirements fall for all subvectors of final demand. The explanation is found in the great backlog of delayed maintenance as a wartime heritage in 1947 and in the relative newness of the stock of buildings in use in 1958 (see Jaszi et al. 1962). Maintenance was postponed in 1958 because of the recession. There are, also, long-run factors at work. New methods of industrial construction produce structures with lower maintenance requirements than older types; and in view of the growing bulk of new construction in our industrial building stock, maintenance construction requirements may be expected to level off, if not actually to decrease further. Chandler and Sayles (1959:11–13) report that the proportion of maintenance functions contracted out remained stable in the 1950s. This suggests that maintenance performed within the establishment also declined.

Printing and publishing

Printing and publishing requirements (figure 5.12) increase for some subvectors of final demand and decrease for others, and changes are not very substantial. The most important increases appear in printing and publishing

* The disaggregation of truck and rail transportation was estimated by Jack Faucett Associates, Inc.; see Jack Faucett Associates, Inc. (1968a and 1968b).

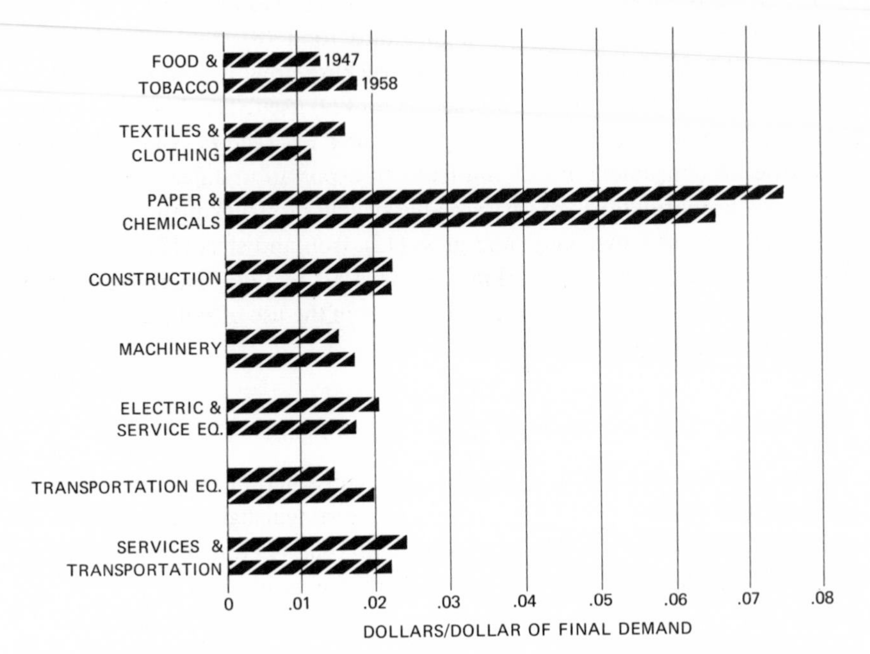

5.12 Intermediate Requirements from Printing and Publishing to Deliver Eight Subvectors of 1961 Final Demand with 1947 and 1958 Technologies.

into food and tobacco and into transportation equipment. In the case of food and tobacco, the increase is easily explained by greater direct outlays for packaging and labeling. The transportation equipment subvector is dominated by automobiles. Their direct purchases from printing and publishing are small (less than 0.001 dollars per dollar of output), but large increases in purchased services imply greater indirect use of printing and publishing.

No scatter diagram of direct coefficient changes is shown here, but a notable decrease appeared in the printing and publishing input coefficient into business services. This may be a result of changing product mix, an increase in relative importance of advertising as a component of the business service sector, and the increased use of television instead of the printed word as an advertising medium.

Packaging

Because of classification and other difficulties, figure 5.13 gives only a crude approximation of changing total packaging inputs at 76 order. Metal containers (39), paper and products (24), wooden containers (21), synthetic materials (28), rubber and plastic products (32), and glass products (35) are classified as packaging sectors for this analysis. Some of these sectors

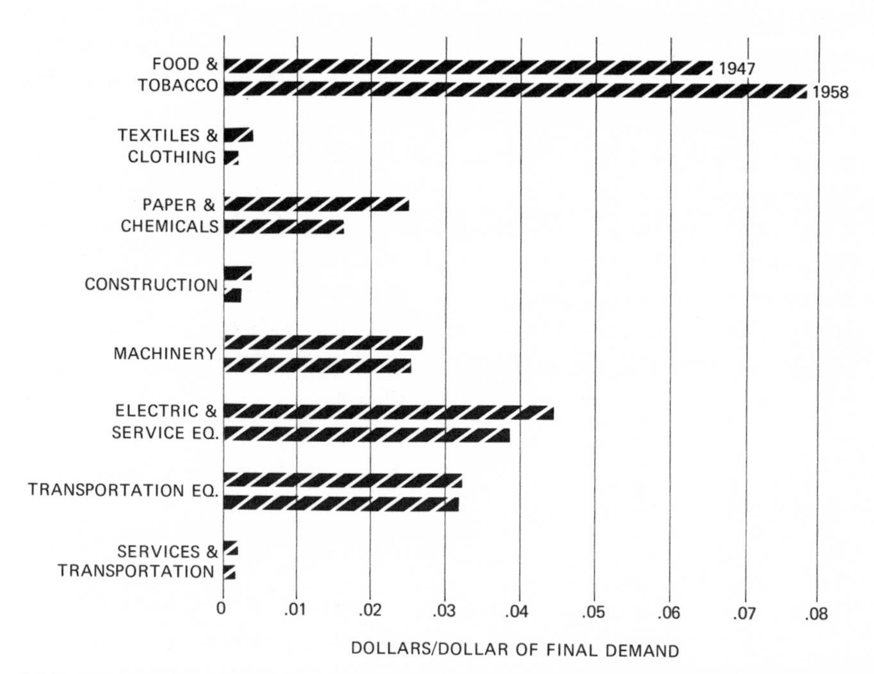

5.13 Intermediate Requirements of Total Packaging Inputs to Deliver Eight Subvectors of 1961 Final Demand with 1947 and 1958 Technologies.

contain much more than pure packaging activities. Glass containers are aggregated with flat glass and other glass products; paper and products includes wood pulp, building paper, printing stock, and stationery, as well as wrapping paper and paperboard boxes. In computing the data for figure 5.13, only those elements were included in each sum that were judged to be used for actual packaging. Thus, the inputs of paper and products into the construction and service subvectors were not considered packaging; and paper could not be distinguished from wood pulp as an indirect input into textiles and clothing (cellulosic fibers) and into paper and chemicals. Hence, all inputs from the paper and products sector into the textiles and clothing and the paper and chemicals subvectors were omitted. Glass containers were included only for the food and tobacco and the paper and chemicals subvectors. Plastic packaging for durables and textiles had to be omitted entirely because it could not be distinguished from plastic components; it is included only for the food and tobacco subvector. Where a particular input was dropped from the packaging category, it was eliminated for both 1947 and 1958. Even though packaging is crudely delimited, the 1947–1958 measures are consistent.

In the particular classification scheme used here, packaging sectors are distinguished by the principal materials they use: *metal* containers, *wooden* containers, and so on. Therefore, the competition among packaging sectors is essentially materials competition of the sort described in Chapter 6. The changing relative importance of three different container sectors is pictured in figure 5.14. Across the board, paper and products increases, while wooden containers decrease, per dollar delivery to each final demand subvector. Requirements for metal containers are growing for food and chemicals, the two largest users. Replacement of metal barrels by paperboard boxes leads to a decline for the other subvectors. The spectacular rise in plastic containers and film, which had already begun during the 1947–1958 era, had competitive impact on all three of the packaging elements explicitly mentioned in figure 5.14, but its growth could not be shown directly.

5.7 Changing Requirements for Chemicals

Chemicals are sometimes specialized, and sometimes general, inputs. Of the four chemicals sectors distinguished at 76 order, three—drugs, soaps, and cosmetics (29), paint (30), and basic chemicals (27)—sell directly to the great majority of industries. Their products perform general functions as aids in cleaning and maintenance. Chemicals also enter many widely different productive processes in analogous ways: they are often auxiliary inputs, representing a very small fraction of total cost, and do not retain their identity as part of the final product. This is true, for example, of fertilizers, catalysts, special textile finishes, and preservatives. The importance of this type of function is increasing everywhere, and, while it assumes many guises, there is certainly some basis for classifying chemicals among general inputs. However, synthetic materials (28) is clearly a materials producer. Basic chemicals, in turn, furnishes the bulk of the latter's materials. Thus, it seemed reasonable to keep the chemicals category separate, rather than to include it in the totals for either materials or general inputs in figure 4.8. The combined, that is, the 38-order chemicals, sector is certainly growing in importance for all end-product groups.

The direct coefficient picture for total chemicals (13) is mixed (figure 5.15). While total chemicals requirements tend to rise for all subvectors of final demand (figure 4.8), some of the direct coefficients decline in each time period. This suggests that rising indirect requirements are an important element in the overall picture. The chemicals complex is also experiencing a rise in interdependence among its own branches.

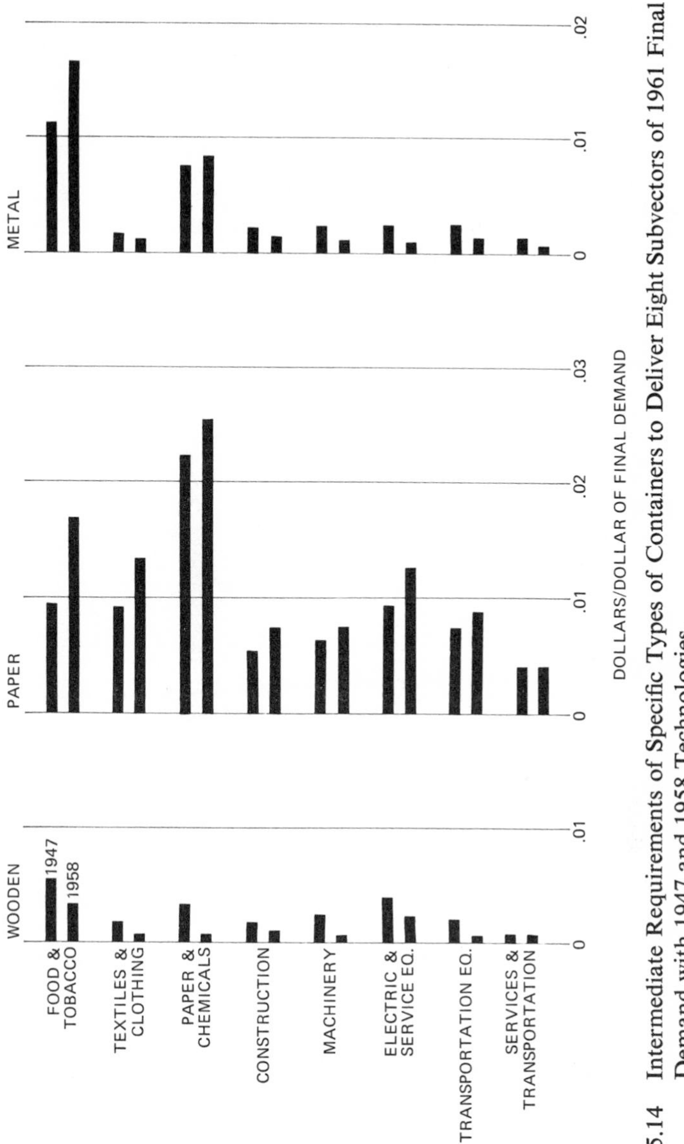

5.14 Intermediate Requirements of Specific Types of Containers to Deliver Eight Subvectors of 1961 Final Demand with 1947 and 1958 Technologies.

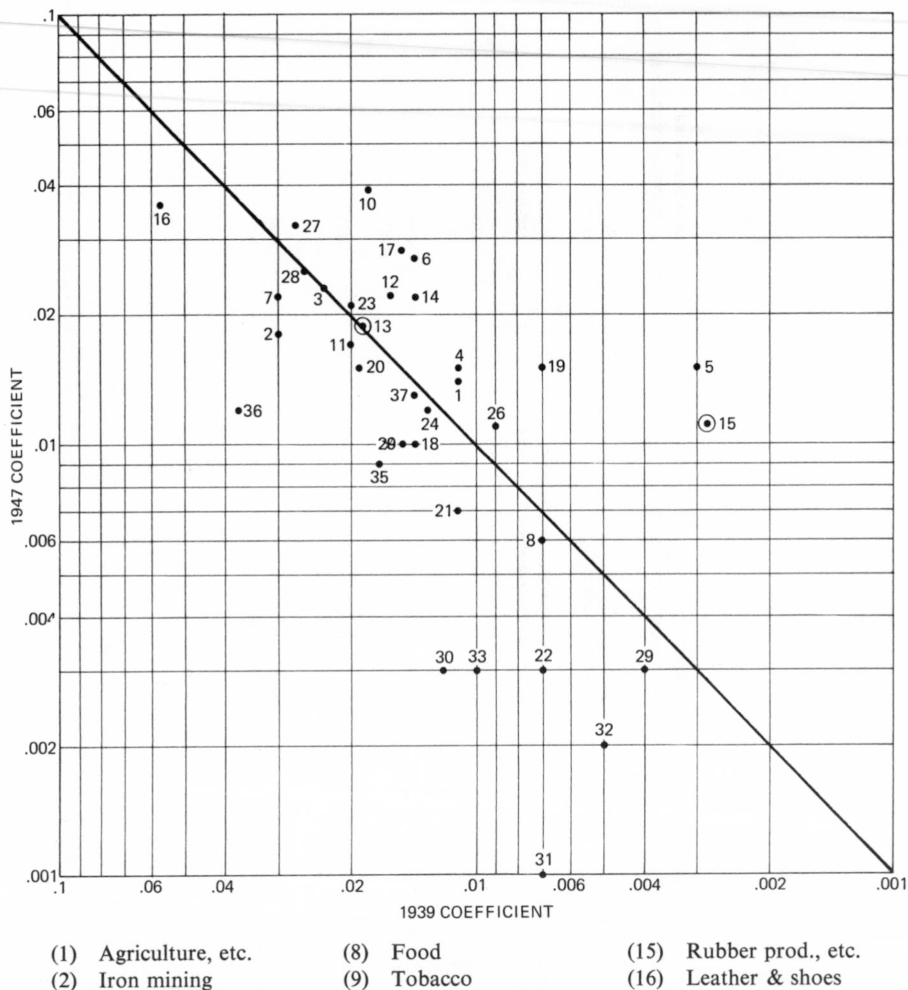

(1)	Agriculture, etc.	(8)	Food	(15)	Rubber prod., etc.
(2)	Iron mining	(9)	Tobacco	(16)	Leather & shoes
(3)	Nonferrous mining	(10)	Textiles & products	(17)	Stone, clay & glass
(4)	Coal mining	(11)	Wood & products	(18)	Iron & steel
(5)	Petroleum mining	(12)	Paper & publishing	(19)	Nonferrous metals
(6)	Nonmetallic mining	(13)	Chemicals	(20)	Metal forming
(7)	Construction	(14)	Petroleum refining	(21)	Nonelectrical eq.

5.15 Direct Chemicals Coefficients for 1939, 1947, and 1958 (dollars per dollar in 1947 prices). Each point indicates the value of the coefficient for a single 38-order consuming sector in each of two years (for circled points, multiply scales by 10).

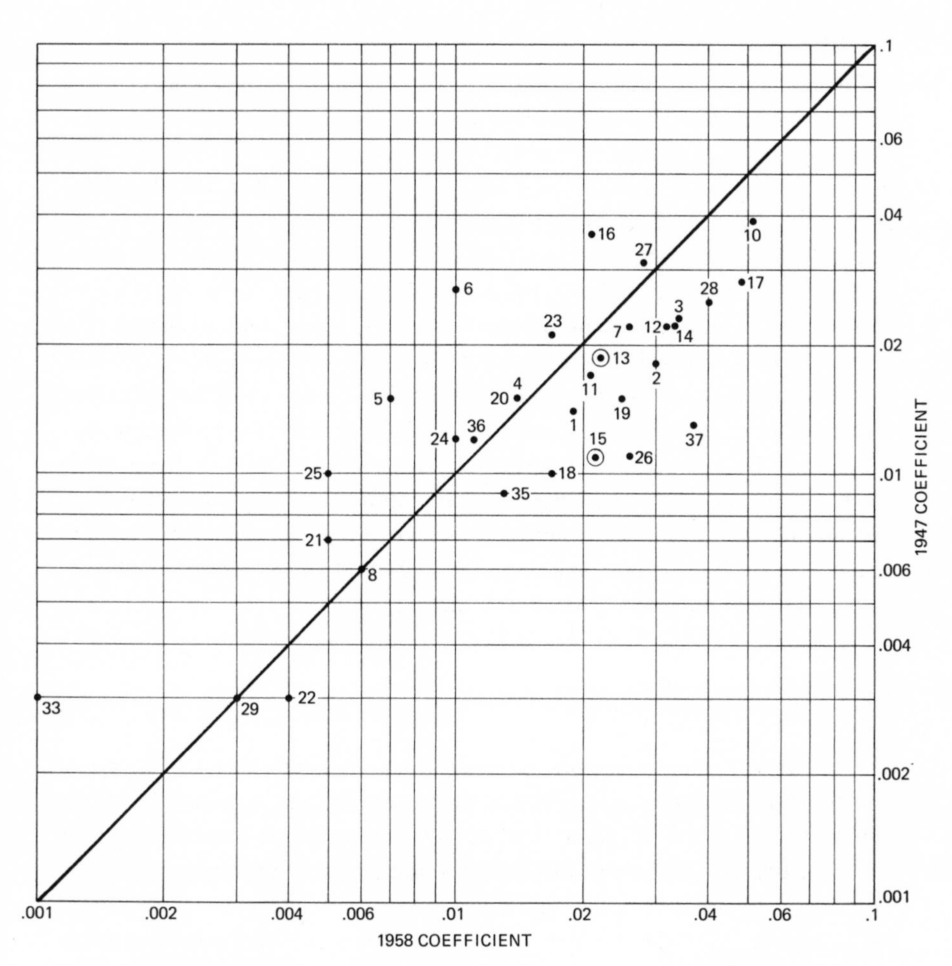

(22) Engines & turbines
(23) Electrical eq., etc.
(24) Motor vehicles & eq.
(25) Aircraft
(26) Trains, ships, etc.
(27) Instruments, etc.
(28) Misc. manufactures
(29) Transportation
(30) Communications
(31) Utilities
(32) Trade
(33) Finance & insurance
(34) Real estate & rental
(35) Business serv., etc.
(36) Auto. repair
(37) Institutions, etc.
(38) Scrap

General chemical inputs

Changes for the three particular 76-order chemicals sectors that produce more general, rather than materials, inputs are shown in figure 5.16. Requirements for basic chemicals (27) seem to increase most dramatically, perhaps because the general portion was not successfully disentangled. Increases pictured in figure 5.16 measure the growth of chemicals as auxiliary inputs in many manufacturing processes—chemical treatment of goods and paper products, for example, and also the materials component of the rising use of plastics and synthetics. Requirements for drugs, soaps, and cosmetics (29) increase moderately for most subvectors, with the exception of electric and service equipment and textiles and clothing. This latter exception occurred because a large volume of natural fibers requiring chemical processing was supplanted by synthetics. Requirements from sector (29) and from basic chemicals for the paper and chemicals subvector are disproportionately high because the input and final demand categories overlap. Paint (30) requirements decline with lower maintenance requirements and the substitution of precolored plastics for painted wood and metal products.

5.8 Specialization and Transfer of Function

The growth of general sectors in relation to the economy as a whole is analogous to the growing importance of managerial and auxiliary functions within the firm. Finer division of labor calls for specialists in administration and coordination, and corps of managers and technical experts grow in disproportion to the traditional "production worker" category—or so one is led to believe. Actually, it is very difficult to substantiate this general impression with statistical evidence (see Horowitz and Herrnstadt 1966).

The rise in general industries parallels the growing specialization of function within the firm: some of the auxiliary and coordinating functions are themselves most efficiently performed in separate, specialized establishments. Witness the growth of technical and public relations consultants and of insurance. Furthermore, when more specialized establishments are set up, they must be linked by networks of trade, transportation, and communication; for these reasons the rise of general sectors may be an essential feature of economic development.

One very important question remains to be answered factually: To what extent does the growth of general sectors represent a transfer of functions from inside the product-oriented firm or establishment itself to specialized service and allied sectors? Gauging the extent of this transfer of function calls

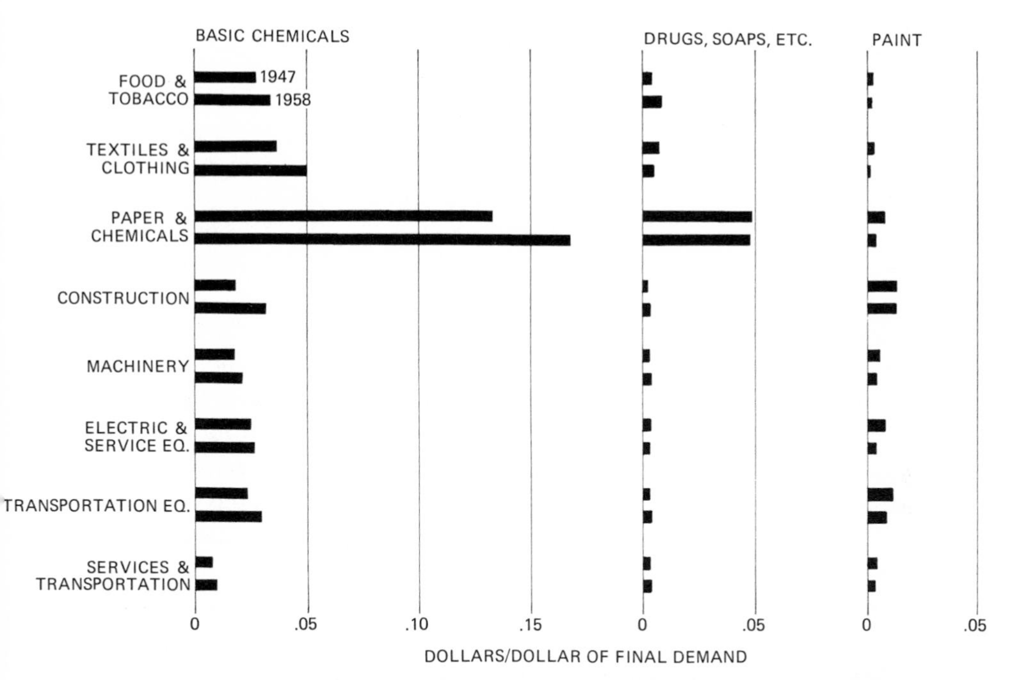

5.16 Intermediate Requirements of Three Specific Types of Chemicals to Deliver Eight Subvectors of 1961 Final Demand with 1947 and 1958 Technologies.

for direct information on changing employment in clerical and service occupations within each industry's establishments and on their packaging, transportation, and communications operations. There may be information at this fine level of detail for some industries, but it is sparse and certainly has not yet been reconciled with the classification schemes on which present input-output tables are based. Thus, no attempt is made here to answer this question systematically for the economy as a whole, or for any major group of sectors. Yet, this question is of prime importance from a policy point of view. Long-range plans for continued high-level employment in a changing economy stress the importance of training workers for jobs in growing industries (see National Commission on Technology, Automation, and Economic Progress 1966:17–31). Service sectors are counted on as a promising source of jobs. It is important to find out whether the projected growth of these sectors represents mere transfer of functions formerly performed elsewhere, and also whether increased specialization and structural change of the general industries themselves may produce net displacement of workers.

Chapter 6 Competition among Basic Materials

6.1 General Background

This chapter turns from general inputs to a group of materials—metals, plastics, rubber, wood, and stone and clay products—that are major intermediate inputs into construction and the manufacture of durable goods. For a long time, it has been possible to substitute one material for another in particular uses. For example, buildings have been constructed of wood, stone, concrete, steel, aluminum, plastic, and many combinations of these and other materials, depending on specific material supply conditions. Copper and cast iron have long competed in the piping market; boats, toys, pots, and pans have been made from a variety of materials throughout the twentieth century and even earlier. However, before World War II, particular materials often captured and dominated specific product markets for long periods. Steel in machinery and transportation equipment, copper in electrical wiring, wood in household furniture are classic examples. It is difficult to say, in any precise quantitative terms, to what extent the "traditional" specialization of materials was dictated by technological necessity and to what extent, by price considerations. Because of its strength and the quality that allows it to be shaped accurately by metalworking techniques, the prewar advantage of steel in the manufacture of many types of machinery was overwhelming within wide ranges of relative material prices. Copper is an excellent conductor of electricity, rubber, a good insulator. Given the orders of magnitude of prices of possible alternative materials, they had no serious competitors.

Since World War II, aluminum, concrete, and, most dramatically, plastics have been successfully challenging the other materials in many specific markets, and specialization of materials seems much less rigid than it used to be. Changes in the relative price picture certainly have had some bearing. Figure 6.1 gives a rough indication of these relative price trends. It shows that prices of copper, steel, and timber have been increasing relative to those of aluminum, plastics (polyethylene and polyvinyl chloride), and concrete. These changes are, in turn, dependent on structural changes in producing sectors (see Chapter 9). At the same time, qualitative improvements in each of the materials—new alloys, plywood and other laminates, anodizing of

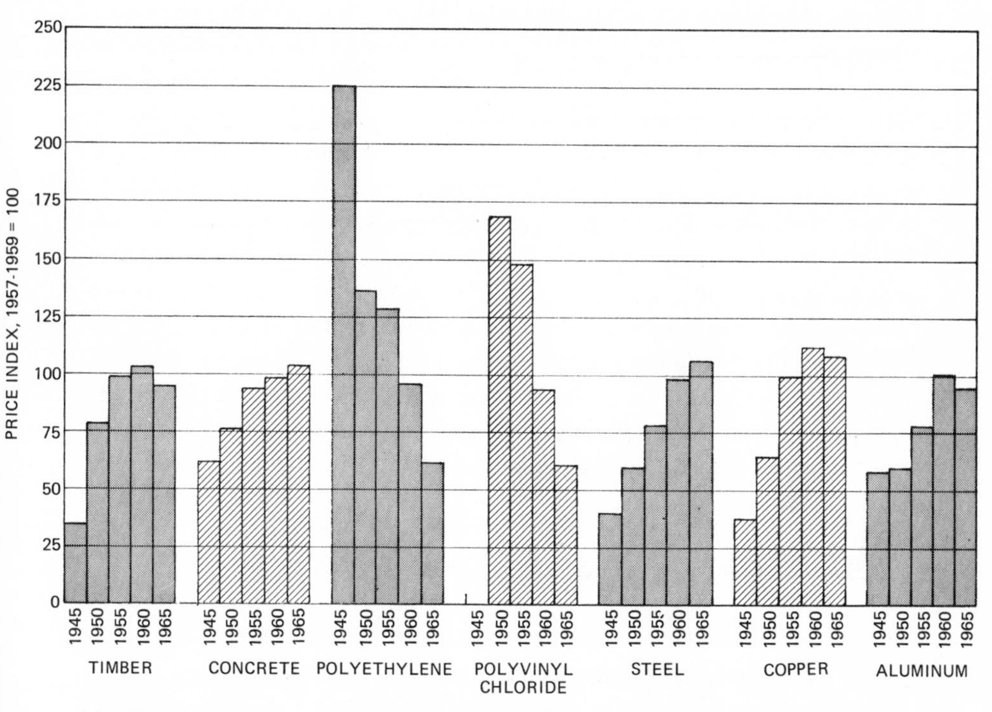

6.1 Time Trends in Specific Materials Prices, 1945–1965 (from Alexander 1967:264).

aluminum, prestressing of concrete, and creation of new plastic materials with special properties—have increased the potential scope of price competition. Corrosion-resistant and stronger aluminum can compete with steel in construction of appliance casings; less brittle, shape-retaining plastic offers competition to both steel and the new aluminum. Thus, there now seems to be a technological basis for greater substitutability as relative price conditions change. We are talking in terms of long-run substitutability. Shifts in materials use require outlays for redesign and retraining; hence, changes are geared to long-run price trends rather than to short-term fluctuations.

In some areas—notably aircraft, aerospace, and instrumentation—technology has even reached the stage where materials are being designed "to order" through research and development. There is also research directed specifically toward the use of low-priced materials, which tends to blur even more the operational distinction between technological development and price substitution.

From figure 4.8, it is already clear that total (deflated) value of materials input requirements has been going down for most types of final demand delivery. This decrease reflects, first of all, the use of a greater proportion of cheaper materials: less than a dollar's worth of plastic replaces every dollar's worth of steel; less than a dollar's worth of aluminum replaces a dollar's worth of copper (even when the values are expressed in 1947 prices). The cost advantages of newer materials were clear even in terms of prewar relative prices, although postwar developments have tended to increase them still further. Two other developments have been significant in effecting reductions in total materials requirements: decreases in the size of equipment items required to perform a given function ("miniaturization") and reduction of scrap generation and spoilage in fabrication. Reductions in scrap come with simplification of design and with improvements in the techniques of fabrication. Neither of these two changes is independent of changes in the qualities of the materials themselves: improved materials made possible the development of lighter and less bulky equipment and of designs involving fewer parts.

6.2 Data for Comparing Materials Requirements, 1947–1958

Within the materials group, aluminum, plastics, and concrete have been growing at the expense of steel, copper, timber, and, to some extent, rubber. Unfortunately, the 38-order classification developed for 1939–1947–1958 comparisons does not permit study of these materials at the necessary level of detail. In particular, at 38 order, plastic and synthetic resins is lumped with basic chemicals, and forestry products, with agriculture; it is virtually impossible to disaggregate nonferrous metals for 1939 and still retain comparability with the 1947–1958 information. We are only able to study changes in individual materials for the 1947–1958 period. There is evidence, however, that postwar changes in the relative importance of individual materials are part of continuing trends, reaching well back into the thirties. Figure 6.2 shows estimated annual production, measured in cubic feet, of six major materials over half a century—from 1910–1961. These changes in production came in response to changes in final demand, as well as to changes in techniques of production and product design. Figure 6.2 shows that production of all materials grows over time; that, since 1930, however, timber production grows negligibly and steel and copper grow rather modestly, while aluminum, plastics and synthetics (polymers), and cement grow much faster.

Computations using the more detailed 1947 and 1958 input-output tables make it possible to separate shifts in materials requirements associated with changes in the structure of production from those attributable to changes in final demand. To further clarify the picture for specific materials, the non-

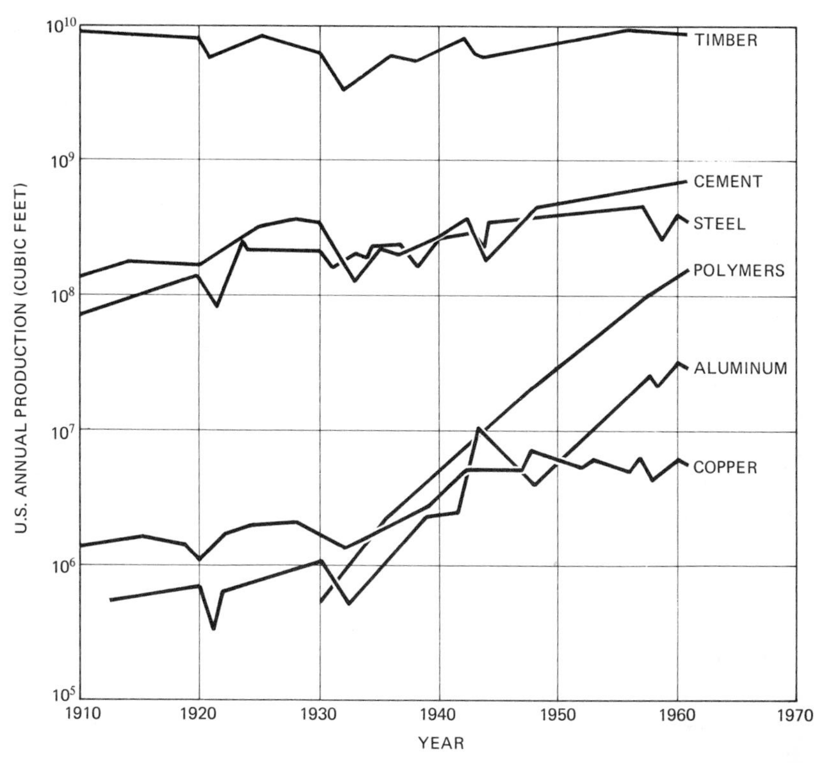

6.2 Annual Production of Six Major Materials, 1910–1961 (from Alexander 1967:256).

ferrous metals manufacturing sector of the 76-order classification was disaggregated into copper, aluminum, and other nonferrous metals; and rubber was distinguished from plastics production. Computations for the study of changing materials requirements between 1947 and 1958 were made with the basic 76-order classification expanded to 79 order to allow for additional detail in materials. The disaggregation of the 1958 nonferrous metals industry is based on information published by the Office of Business Economics (1966). Disaggregation of the rubber and plastics sectors and modification of the published disaggregation of nonferrous metals are based upon the work of A. George Gols and other economists at Arthur D. Little. Disaggregations of nonferrous metals and rubber and plastics sectors for 1947 are based on information in the 450-order 1947 input-output study. To facilitate comparisons with projections for the early 1970s, the data in this chapter are expressed in terms of 1958 prices.

6.3 Scientific American–Arthur D. Little Projections for the Early 1970s

In place of comparisons with 1939–1947 changes, projected changes between 1958 and the early 1970s are introduced in many of the figures that follow. Projected input-output coefficients for materials in the early 1970s were estimated in a large-scale exploratory study by economists and engineers at Arthur D. Little.* The projections of direct coefficients were made for inputs of six specific materials—iron and steel, aluminum, copper, other nonferrous metals, rubber, and plastics—into fifty consuming sectors. All of the major markets for these materials were covered with the exception of the construction industry. Scientific American, Inc. (1968) summarizes the major conclusions. Although it is not possible to discuss the Scientific American–Arthur D. Little study in great detail here, it is important to review enough of its methodology to clarify the meaning of the projections. Most major changes in technology of production or product design can be anticipated by industry specialists five or more years before they are put into actual use. This is not to say that industry experts can be expected to know exactly how much plastic will be used in every type of, say, vacuum cleaner five years hence; but they can anticipate the kinds of plastics that will be available, the methods that will be used to shape them, and any important changes in the design of vacuum cleaners themselves. Given this key product information and general familiarity with price levels and trends in the materials markets, experts are in a position to project materials coefficients, provided that they work at a very fine level of detail. Specific technological projections are often required for business decisions. However, systematic estimates for detailed groups of interrelated inputs and outputs are rare, and there has not been sufficient time and experience to permit appraisal of technological forecasting techniques.

The 1958 input-output coefficients were first disaggregated to a four-digit level of detail on the basis of Office of Business Economics worksheets and supplementary information. Each expert was given a questionnaire, specifying the relevant 1958 detailed coefficients, and asked to estimate the percentage change in physical input-output ratios anticipated between 1958 and the early 1970s. Some of the individuals consulted were specialists in particular products, while others were materials experts. Through comparison and

* The study was sponsored by Scientific American, Inc. under the direction of A. George Gols, at Arthur D. Little, Inc. Clopper Almon, Wassily Leontief, and the author served as consultants in establishing the general outlines of the research and in dealing with methodological problems as they arose. Materials coefficient projections for individual consuming sectors are documented in a series of reports on file at the research office of Scientific American, Inc.

cross-checking of different expert judgments, it was possible to arrive at a consistent set of estimates of requirements for each of the six materials by four-digit SIC consuming sectors. Detailed projections were then aggregated to the 79-order classification used here.

Since the projections were made in 1966, part of the estimation of 1958–early 1970s change consisted in updating the 1958 coefficients in the light of actual 1958–1966 developments, while the remainder required anticipation of future developments. There was some disagreement as to the likely course of future developments, but the most difficult problems faced by the experts were the kinds that also occur in the course of historical comparisons. Among the most important were difficulties in dealing with the enormous volume of product detail subsumed in a sector, even at the four-digit level, and the problem of making projections "in 1958 dollars" (which is analogous to projecting in physical units) when the qualities of inputs and, more particularly, of outputs, were expected to change. If smaller tractors are to be produced, steel coefficients *per tractor* might decrease, while steel coefficients per horsepower of tractors might remain the same, or increase. For equipment items, an attempt was made to use capacity as the bridge unit of output whenever possible. Various approaches to the problem of measuring and pricing capital goods of changing quality are discussed in Searle (1964) and Grose et al. (1966). (Additional questions of methodology will be raised in the course of the discussion of actual projections.)

6.4 Changes in Direct Coefficients for Individual Materials

Three-way scatter diagrams of 1947, 1958, and projected early 1970s materials coefficients (figures 6.3 to 6.7) give some detail on historical and anticipated changes in direct consumption of individual materials in particular uses.* As in the scatter diagrams for many general inputs, points tend to be clustered on one side or the other of the 45-degree line for most materials. Clustering is most striking for aluminum and for iron and steel in both time intervals and clearer for all materials in 1958–early 1970s comparisons. The reasons for greater regularity in the graphs comparing 1958 and the early 1970s are rooted largely in the methodology of the projections. Because the early 1970s coefficients were projected by applying estimated percentage adjustors to the

* No projections for wood or for stone and clay were made in the Scientific American–Arthur D. Little study, and thus no scatter diagrams are shown for these materials. Their major market is in construction. Direct-plus-indirect requirements for these materials for 1947 and 1958 are shown later on in this discussion.

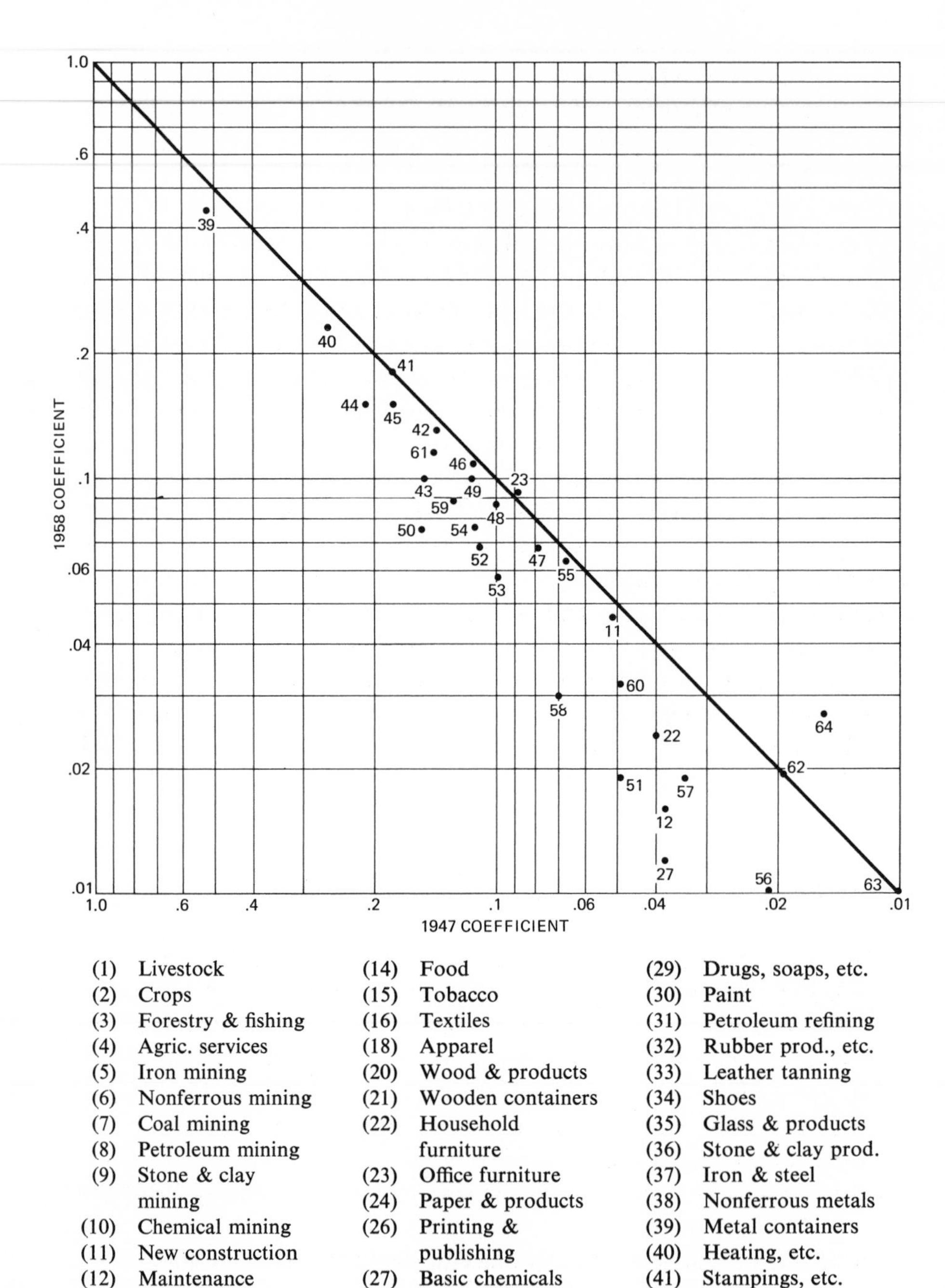

6.3 Direct Iron and Steel Coefficients for 1947, 1958, and Early 1970s (dollars per dollar in 1958 prices). Each point indicates the value of the coefficient for a single 76-order consuming sector in each of two years.

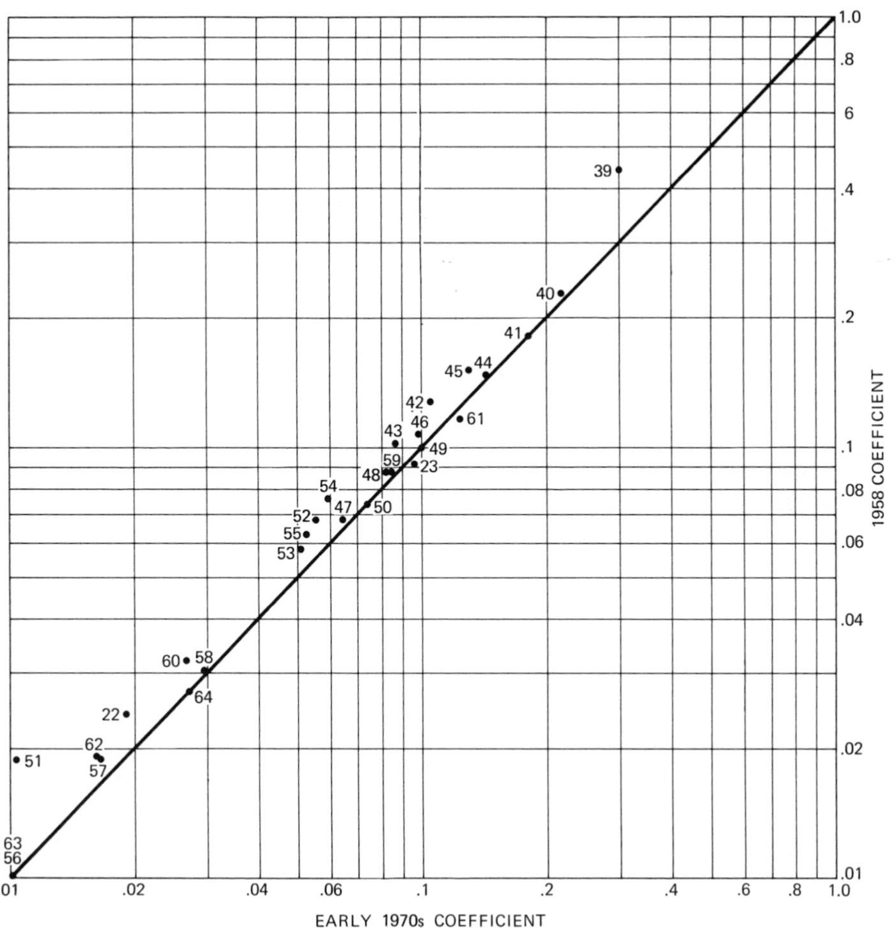

(43) Engines & turbines
(44) Farm equipment
(45) Constr. & mining eq.
(46) Materials hand. eq.
(47) Metalworking eq.
(48) Special ind. eq.
(49) General ind. eq.
(50) Machine shop prod.
(51) Office & comp. mach.
(52) Service ind. mach.
(53) Electrical apparatus
(54) Household appliances
(55) Light. & wiring eq.
(56) Communication eq.
(57) Electronic compon.
(58) Batteries, etc.
(59) Motor vehicles & eq.
(60) Aircraft
(61) Trains, ships, etc.
(62) Instruments, etc.
(63) Photo. apparatus
(64) Misc. manufactures
(65) Transportation
(66) Telephone
(67) Radio & tv broad.
(68) Utilities
(69) Trade
(70) Finance & insurance
(71) Real estate & rental
(72) Hotels & pers. serv.
(73) Business services
(74) Research & dev.
(75) Auto. repair
(76) Amusements, etc.
(77) Institutions
(80) Noncomp. imports
(81) Bus. travel, etc.
(83) Scrap

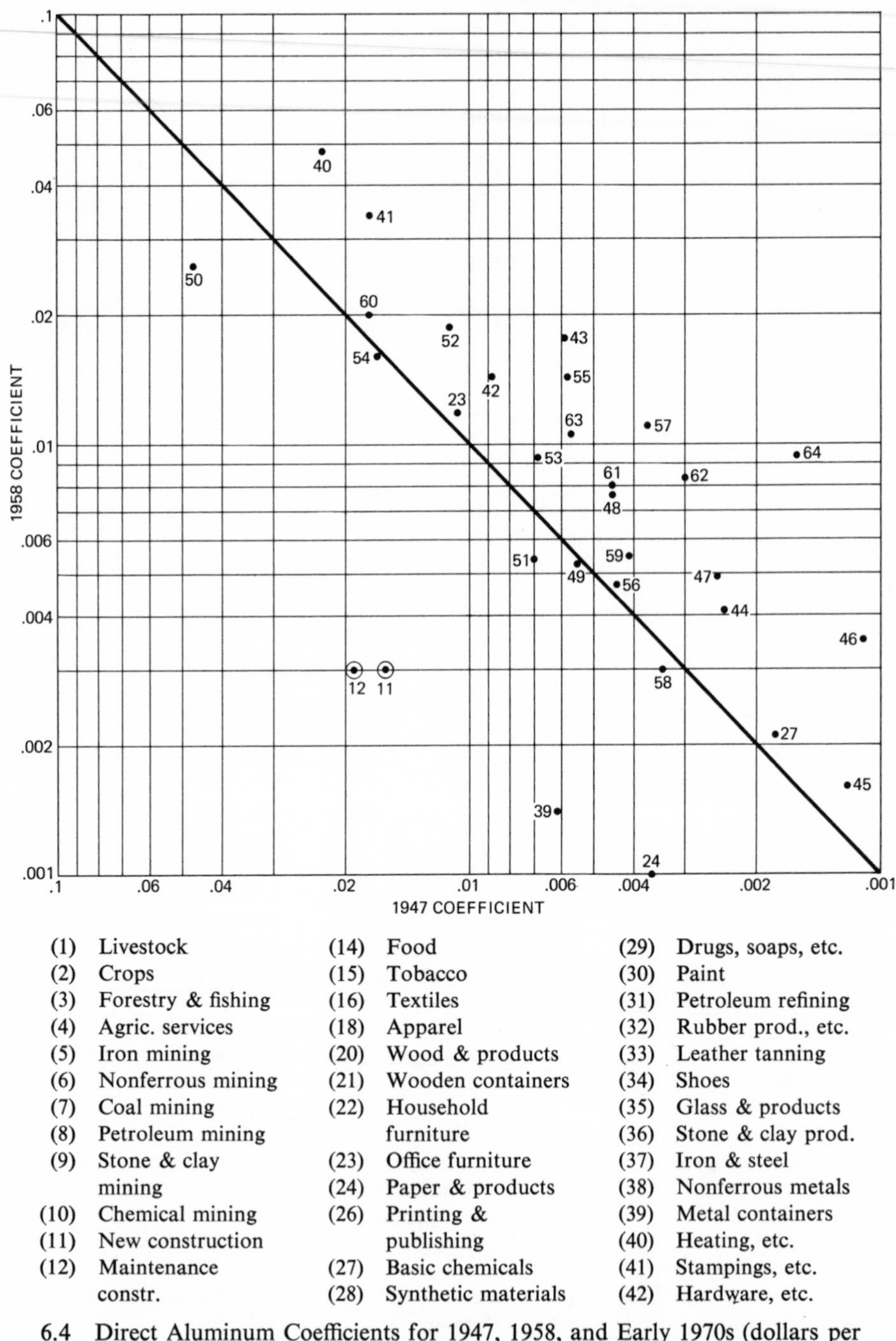

6.4 Direct Aluminum Coefficients for 1947, 1958, and Early 1970s (dollars per dollar in 1958 prices). Each point indicates the value of the coefficient for a single 76-order consuming sector in each of two years (for circled points, divide scales by 10).

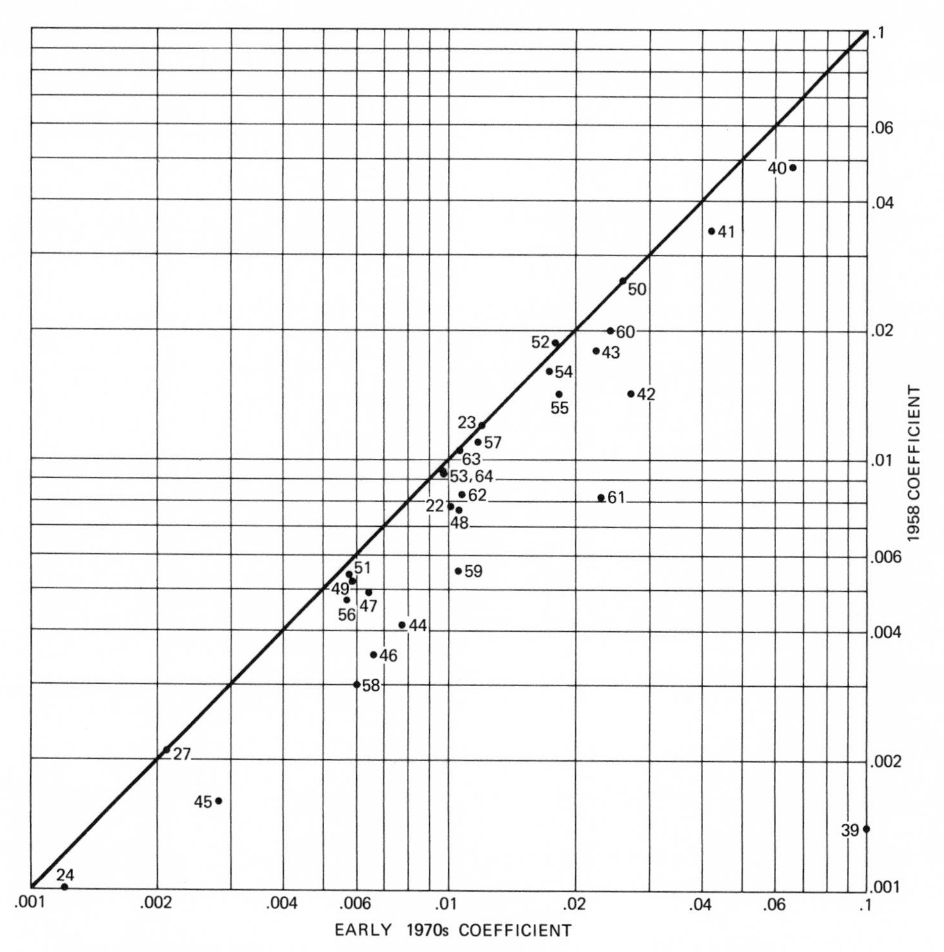

(43) Engines & turbines
(44) Farm equipment
(45) Constr. & mining eq.
(46) Materials hand. eq.
(47) Metalworking eq.
(48) Special ind. eq.
(49) General ind. eq.
(50) Machine shop prod.
(51) Office & comp. mach.
(52) Service ind. mach.
(53) Electrical apparatus
(54) Household appliances
(55) Light. & wiring eq.
(56) Communication eq.
(57) Electronic compon.
(58) Batteries, etc.
(59) Motor vehicles & eq.
(60) Aircraft
(61) Trains, ships, etc.
(62) Instruments, etc.
(63) Photo. apparatus
(64) Misc. manufactures
(65) Transportation
(66) Telephone
(67) Radio & tv broad.
(68) Utilities
(69) Trade
(70) Finance & insurance
(71) Real estate & rental
(72) Hotels & pers. serv.
(73) Business services
(74) Research & dev.
(75) Auto. repair
(76) Amusements, etc.
(77) Institutions
(80) Noncomp. imports
(81) Bus. travel, etc.
(83) Scrap

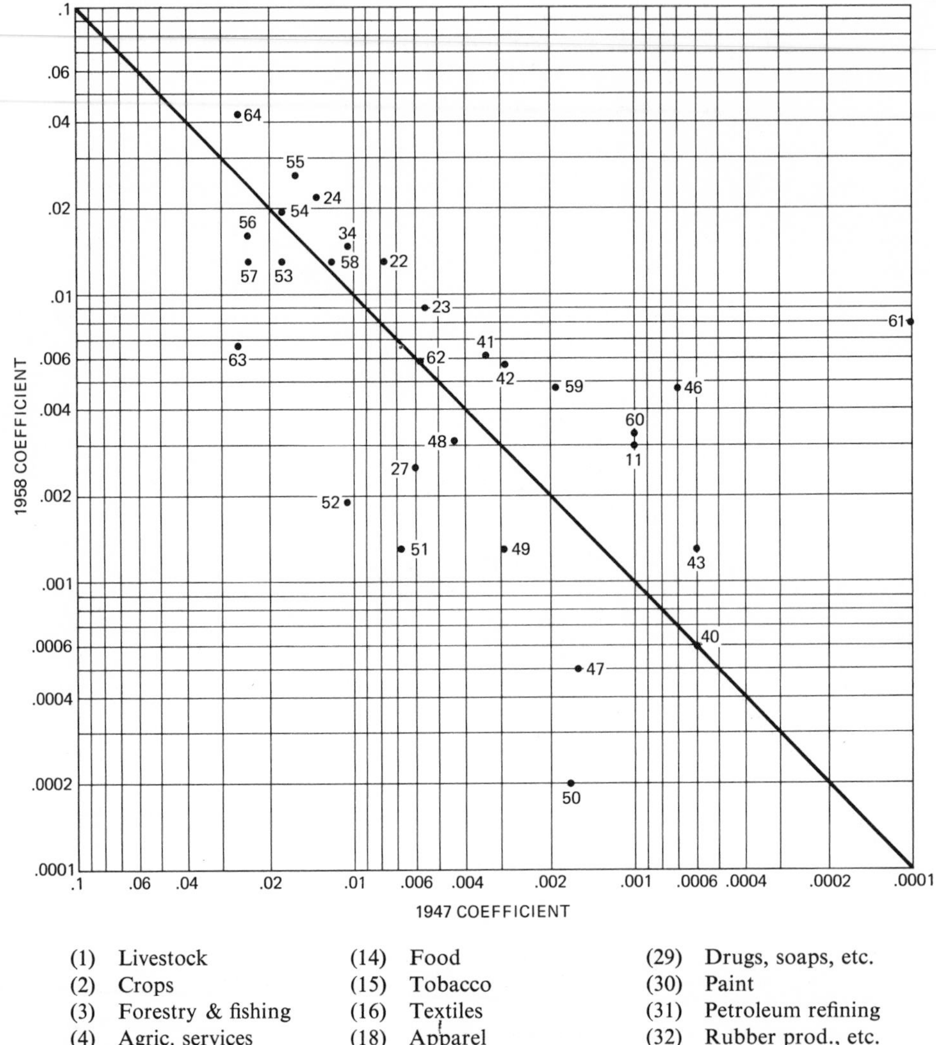

(1) Livestock
(2) Crops
(3) Forestry & fishing
(4) Agric. services
(5) Iron mining
(6) Nonferrous mining
(7) Coal mining
(8) Petroleum mining
(9) Stone & clay mining
(10) Chemical mining
(11) New construction
(12) Maintenance constr.
(14) Food
(15) Tobacco
(16) Textiles
(18) Apparel
(20) Wood & products
(21) Wooden containers
(22) Household furniture
(23) Office furniture
(24) Paper & products
(26) Printing & publishing
(27) Basic chemicals
(28) Synthetic materials
(29) Drugs, soaps, etc.
(30) Paint
(31) Petroleum refining
(32) Rubber prod., etc.
(33) Leather tanning
(34) Shoes
(35) Glass & products
(36) Stone & clay prod.
(37) Iron & steel
(38) Nonferrous metals
(39) Metal containers
(40) Heating, etc.
(41) Stampings, etc.
(42) Hardware, etc.

6.5 Direct Plastics Coefficients for 1947, 1958, and Early 1970s (dollars per dollar in 1958 prices). Each point indicates the value of the coefficient for a single 76-order consuming sector in each of two years.

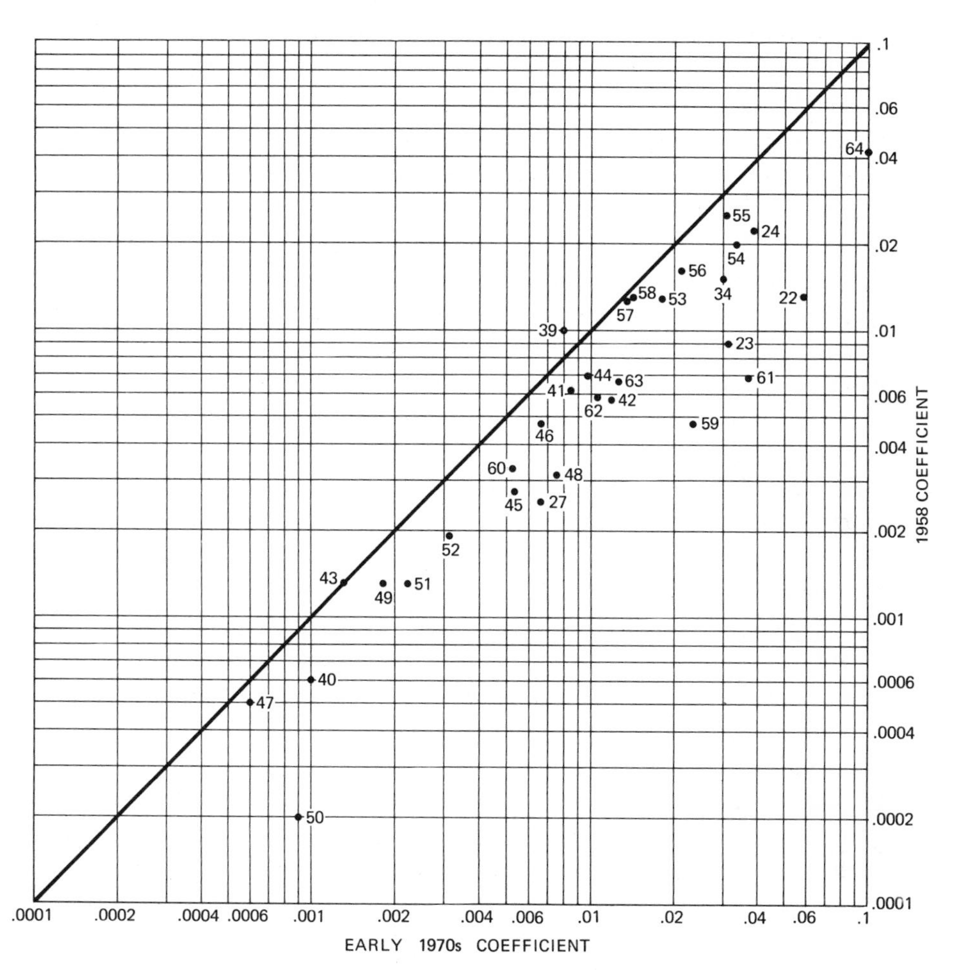

(43) Engines & turbines
(44) Farm equipment
(45) Constr. & mining eq.
(46) Materials hand. eq.
(47) Metalworking eq.
(48) Special ind. eq.
(49) General ind. eq.
(50) Machine shop prod.
(51) Office & comp. mach.
(52) Service ind. mach.
(53) Electrical apparatus
(54) Household appliances
(55) Light. & wiring eq.
(56) Communication eq.
(57) Electronic compon.
(58) Batteries, etc.
(59) Motor vehicles & eq.
(60) Aircraft
(61) Trains, ships, etc.
(62) Instruments, etc.
(63) Photo. apparatus
(64) Misc. manufactures
(65) Transportation
(66) Telephone
(67) Radio & tv broad.
(68) Utilities
(69) Trade
(70) Finance & insurance
(71) Real estate & rental
(72) Hotels & pers. serv.
(73) Business services
(74) Research & dev.
(75) Auto. repair
(76) Amusements, etc.
(77) Institutions
(80) Noncomp. imports
(81) Bus. travel, etc.
(83) Scrap

basic 1958 coefficients, these two sets of coefficients are completely comparable with respect to classification and accounting conventions. However, estimates of the 1947 and 1958 coefficients were made independently of each other, and these coefficients are subject to the vagaries of minor classification changes and statistical aberrations. Direct coefficients for materials often vary with the purchaser's shift from making to buying particular components. In making projections, however, the experts tended to eliminate this sort of consideration, implicitly, and instead to concentrate on the material content of the specified product, rather than on the question of who fabricates particular components. The coefficients are projected on the artificial assumption of minimal shifts in specialization among fabricating industries. This assumption produces unrealistic regularity in projections of direct coefficients, but it should not bias projections based on the inverse matrix.

Direct iron and steel purchases, per unit of output (figure 6.3), decreased in virtually all industries between 1947 and 1958; and experts expect this tendency to continue through the early 1970s. The reverse trend is expected for aluminum in most markets. Figure 6.4 conveys general optimism about the competitive prospects for aluminum, an optimism that is not universally shared. Brubaker (1967:69) speculates that the increase in aluminum's share of the container market will be only marginal, whereas the Arthur D. Little projection suggests spectacular inroads. One explanation of the discrepancy lies in the fact that Brubaker is projecting from a 1966 base, while Arthur D. Little is projecting from a 1958 base. Between 1958 and 1966, aluminum did gain heavily in the containers market. Peck (1961:160–162), as well as Brubaker, points out that the competitive gains of aluminum in the postwar period are a lagged reaction to a relative price advantage that aluminum held over steel and copper. This price advantage was established even before World War II. However, the relative price of aluminum to steel is no longer clearly falling, and future developments depend largely on prospective technical advances in the steel- and aluminum-manufacturing sectors.

Changes in direct plastics coefficients are shown in figure 6.5. Between 1947 and 1958, plastics gained in many key sectors, for instance, automobiles, but lost in others. Whatever the course of future developments in the competition between steel and aluminum, the growing importance of plastics is a matter of common agreement. Plastics are cheap relative to most other materials, and their price advantage appears to be increasing with expanded usage. "In the main, the methods used to produce plastic raw materials allow for large cost reductions whenever the outlets for any particular material reach a certain level. The fast-growth users at present are the automotive, building, packaging, and electronics industries, where the conventional materials used have a tendency to be increasing in cost, rather than decreasing. Therefore, provided the part can be designed to an available plastic material, its introduction can only be a question of time. It must be realized also that the very methods used to fabricate plastics tend to reduce manufacturing costs to the absolute minimum. In this respect, especially in the moulding sphere, the production of a complex part is reduced to a single operation" (Denton 1967:69).

During the 1947–1958 interval, about as many rubber coefficients rise as fall (figure 6.6). In the smaller coefficients, changes may reflect new overall product designs, rather than explicit competition with alternate materials. Since 1958, rubber has been subject to increasing competition from plastics, particularly in electrical insulation.

While copper coefficients have been decreasing for some sectors and increasing for others (figure 6.7), the tendency toward decrease predominates as a result of the growing worldwide copper shortage and the related rise in copper prices. Aluminum is being substituted for copper in electrical transmission systems, as well as in radiators. At the same time, requirements for copper wire have been rising with the increased use of complex automatic devices for regulating and controlling equipment.

6.5 Changing Total Requirements for Specific Materials

The direct coefficient scatter diagrams do not distinguish between large and small sectors, and they do not take into account indirect materials consumption. For example, motor vehicles are represented by only a single point on each scatter diagram, but they account for 17 percent of intermediate consumption of iron and steel used to deliver all final demand in 1958. Similarly, use of rubber for automobiles accounts for 12 percent of all intermediate rubber consumption. Obviously, changes in materials used for motor vehicles will outweigh changes in many smaller sectors in the total materials picture. Furthermore, indirect effects of changes should not be ignored in evaluating overall competition among materials. Particularly in the 1947–1958 comparisons, substitution of one material for another in the general fabricating sectors may either offset or reinforce direct substitutions. Direct coefficients for aluminum into new construction actually fell by 81 percent between 1947 and 1958, but direct-plus-indirect aluminum requirements increased by 52 percent because an increased proportion of aluminum was used in the heating and in the plumbing and structural metals sector in the later year.

In order to use the Arthur D. Little projections to estimate direct-plus-indirect materials requirements, it was necessary to imbed their projections

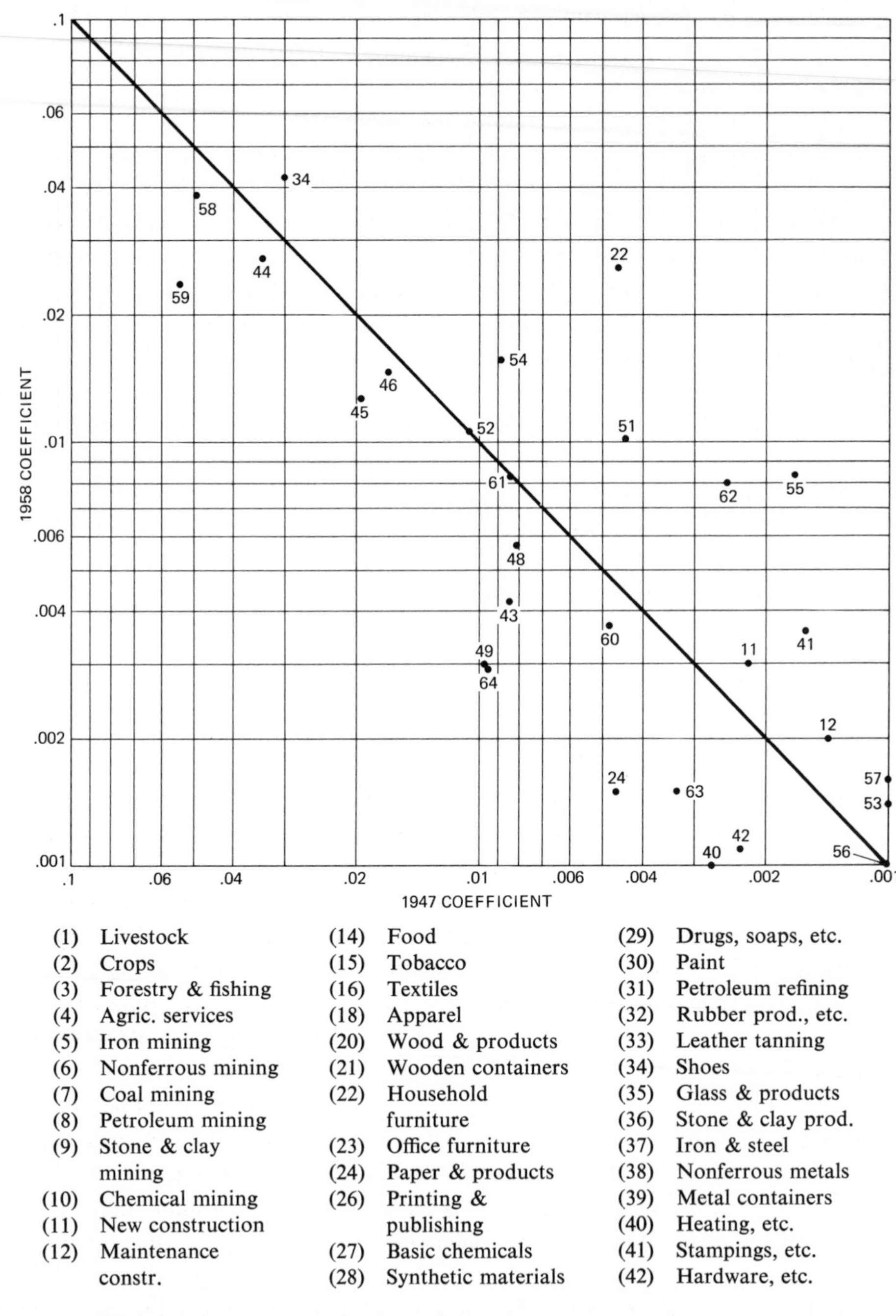

6.6 Direct Rubber Coefficients for 1947, 1958, and Early 1970s (dollars per dollar in 1958 prices). Each point indicates the value of the coefficient for a single 76-order consuming sector in each of two years.

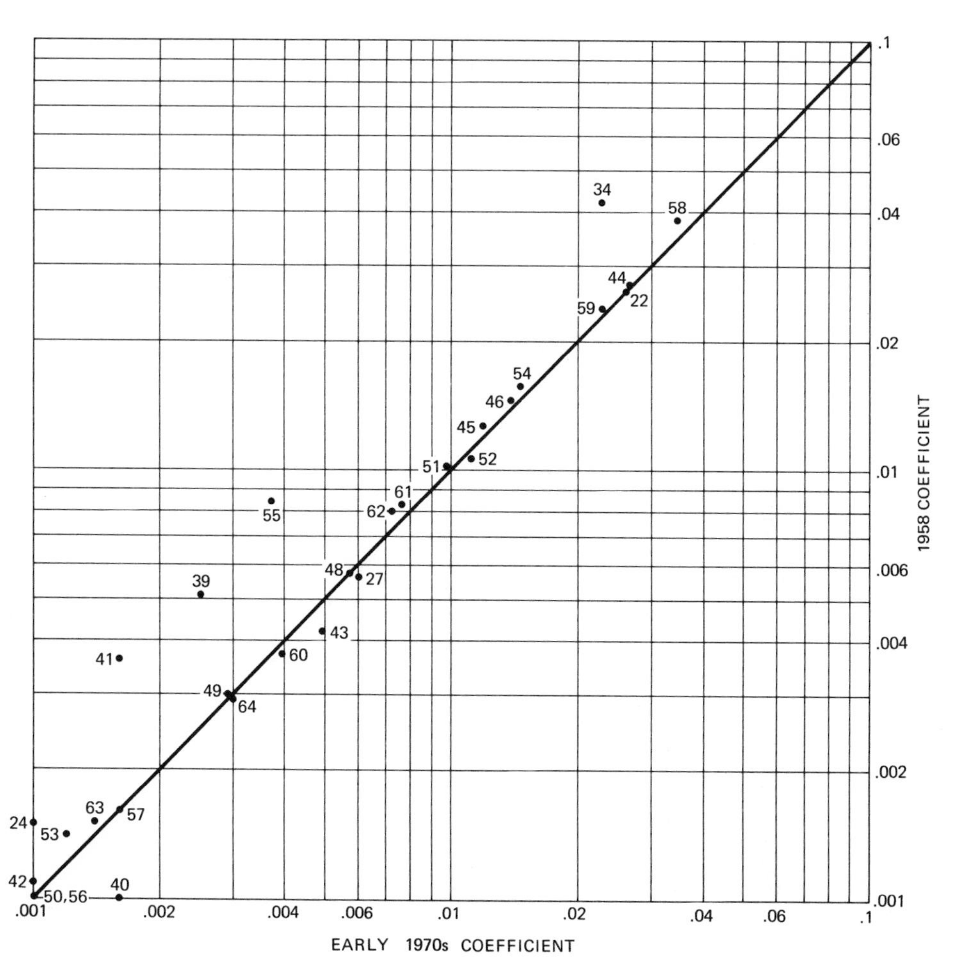

(43) Engines & turbines
(44) Farm equipment
(45) Constr. & mining eq.
(46) Materials hand. eq.
(47) Metalworking eq.
(48) Special ind. eq.
(49) General ind. eq.
(50) Machine shop prod.
(51) Office & comp. mach.
(52) Service ind. mach.
(53) Electrical apparatus
(54) Household appliances
(55) Light. & wiring eq.
(56) Communication eq.
(57) Electronic compon.
(58) Batteries, etc.
(59) Motor vehicles & eq.
(60) Aircraft
(61) Trains, ships, etc.
(62) Instruments, etc.
(63) Photo. apparatus
(64) Misc. manufactures
(65) Transportation
(66) Telephone
(67) Radio & tv broad.
(68) Utilities
(69) Trade
(70) Finance & insurance
(71) Real estate & rental
(72) Hotels & pers. serv.
(73) Business services
(74) Research & dev.
(75) Auto. repair
(76) Amusements, etc.
(77) Institutions
(80) Noncomp. imports
(81) Bus. travel, etc.
(83) Scrap

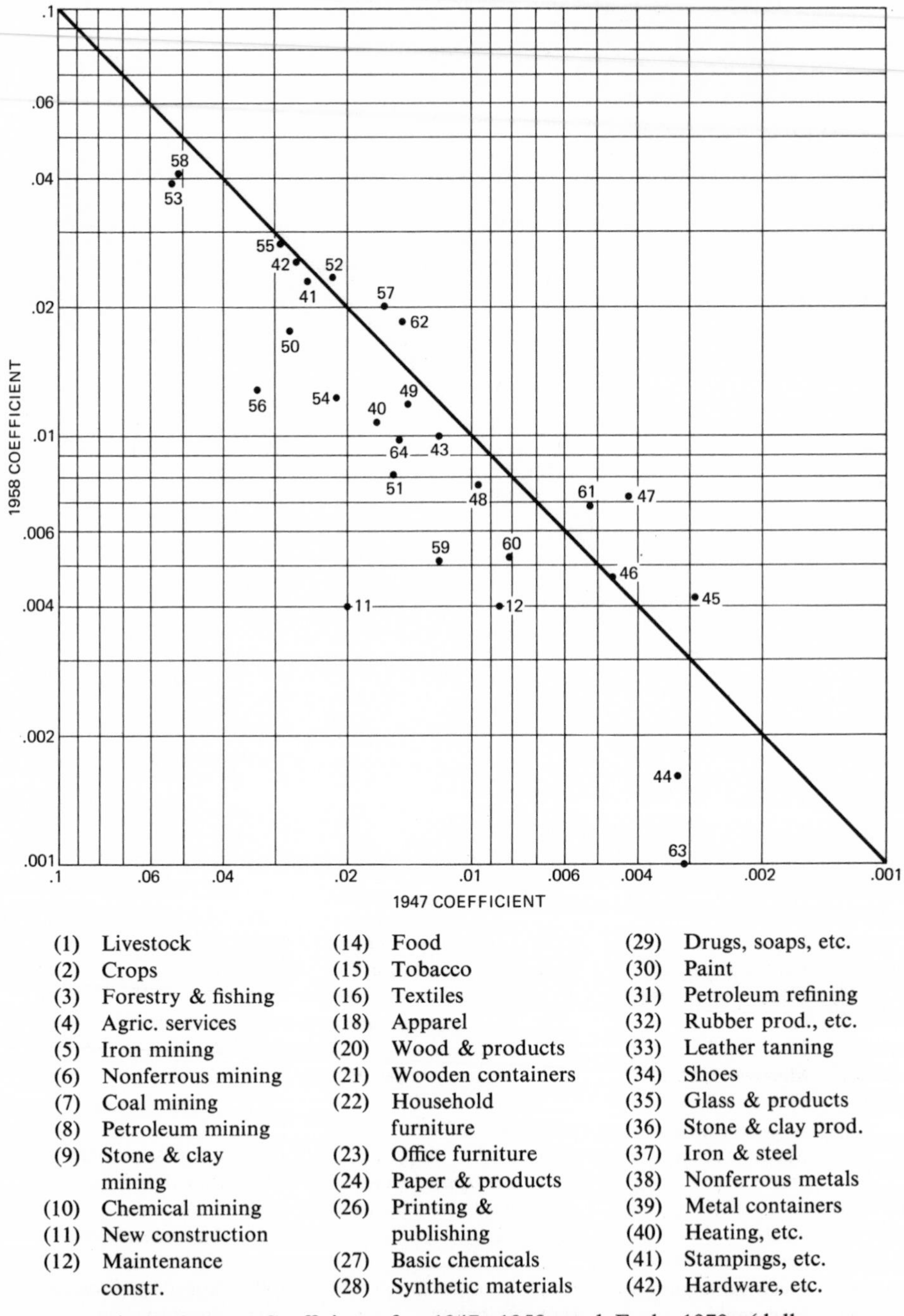

6.7 Direct Copper Coefficients for 1947, 1958, and Early 1970s (dollars per dollar in 1958 prices). Each point indicates the value of the coefficient for a single 76-order consuming sector in each of two years.

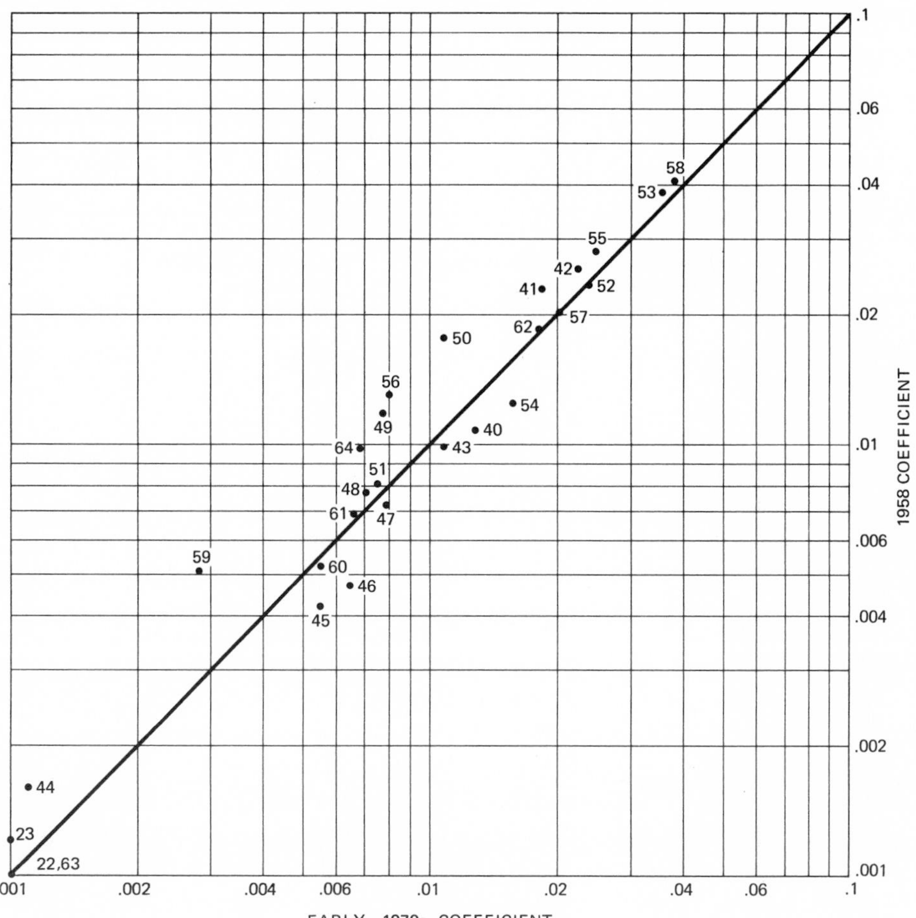

(43) Engines & turbines
(44) Farm equipment
(45) Constr. & mining eq.
(46) Materials hand. eq.
(47) Metalworking eq.
(48) Special ind. eq.
(49) General ind. eq.
(50) Machine shop prod.
(51) Office & comp. mach.
(52) Service ind. mach.
(53) Electrical apparatus
(54) Household appliances
(55) Light. & wiring eq.
(56) Communication eq.
(57) Electronic compon.
(58) Batteries, etc.
(59) Motor vehicles & eq.
(60) Aircraft
(61) Trains, ships, etc.
(62) Instruments, etc.
(63) Photo. apparatus
(64) Misc. manufactures
(65) Transportation
(66) Telephone
(67) Radio & tv broad.
(68) Utilities
(69) Trade
(70) Finance & insurance
(71) Real estate & rental
(72) Hotels & pers. serv.
(73) Business services
(74) Research & dev.
(75) Auto. repair
(76) Amusements, etc.
(77) Institutions
(80) Noncomp. imports
(81) Bus. travel, etc.
(83) Scrap

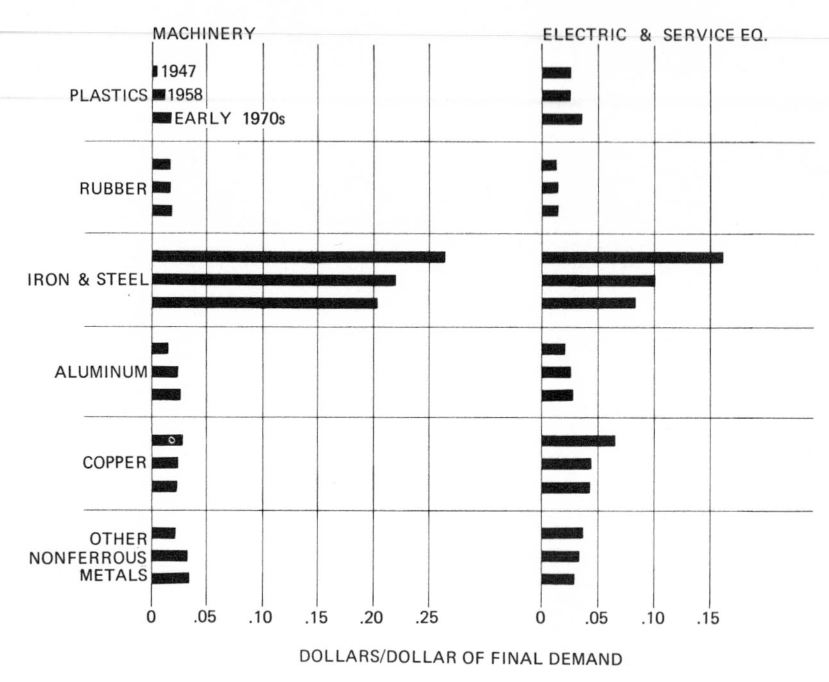

6.8 Intermediate Requirements of Specific Materials to Deliver Five Subvectors of 1961 Final Demand with 1947, 1958, and Early 1970s Technologies.

for the six rows of direct coefficients into a complete matrix; that is, to combine them with appropriate values for all the nonmaterials coefficients as well. Values for nonmaterials coefficients were taken from the 1970 coefficient matrix released by the Bureau of Labor Statistics Interagency Growth Study (Bureau of Labor Statistics 1967). The projected input-output structure for the early 1970s used in the computations described below is a combination of Arthur D. Little projected materials rows and Interagency Growth Project estimates for all other coefficients.* Actually, the choice of the Interagency Growth Project 1970 matrix as a context for materials rows projections was not crucial; for the particular problem at hand, it made very little difference whether the nonmaterials rows were represented by 1947, 1958, or 1970 coefficients. This was ascertained by a side computation: sets of direct early 1970s materials coefficients were imbedded successively in 1947, 1958, and

* The Interagency Growth Project constructed its 1970 matrix by replacing many major 1958 coefficients with specific projections for those cells. Some of these estimates of 1970 coefficients were made by the Interagency Growth Project and some were supplied by other research agencies working under contract to them. Across-the-board projection methods were used to project the remaining cells.

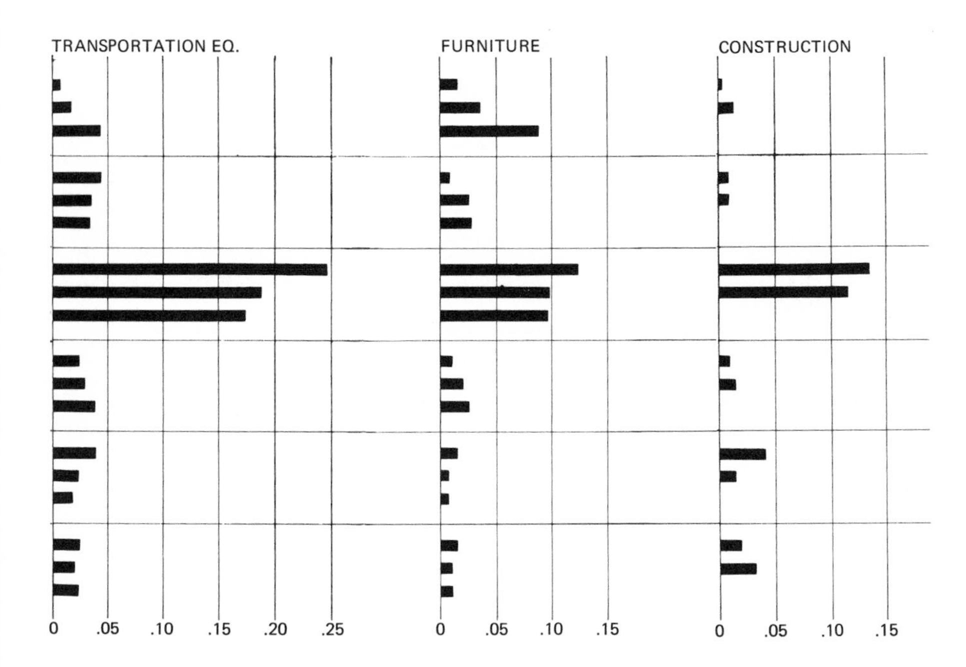

1970 surroundings and total requirements were computed for individual materials with each of the three "hybrid" matrices. Changing the context with a given set of materials coefficients and bill of goods did not change the level of any of the estimated materials requirements by more than 8 percent. It had hardly *any* effect on estimated rates of change over time in requirements for particular materials.

Figure 6.8 shows changing requirements for each of the specific materials to produce particular subvectors of final demand with 1947, 1958, and the early 1970s technologies. Requirements are shown only for major consumers of these materials, that is, for major durable goods subvectors. Direct coefficients for materials into construction were not projected in the Arthur D. Little study, nor were any coefficients for wood nor for stone and clay products. Only the 1947–1958 comparisons are shown in the graph for construction. Changing total requirements for each subvector of final demand bear out the tendencies indicated in the direct coefficient scatters. For each major subvector of final demand, steel and copper requirements decrease while requirements for aluminum and plastics rise. Rubber requirements increase for some subvectors of final demand, and decrease for others. Since other nonferrous metals is a heterogeneous aggregate, the picture varies with the particular metal—nickel, titanium, lead, and so on, or combination of metals—dominant in each subvector's bill of materials.

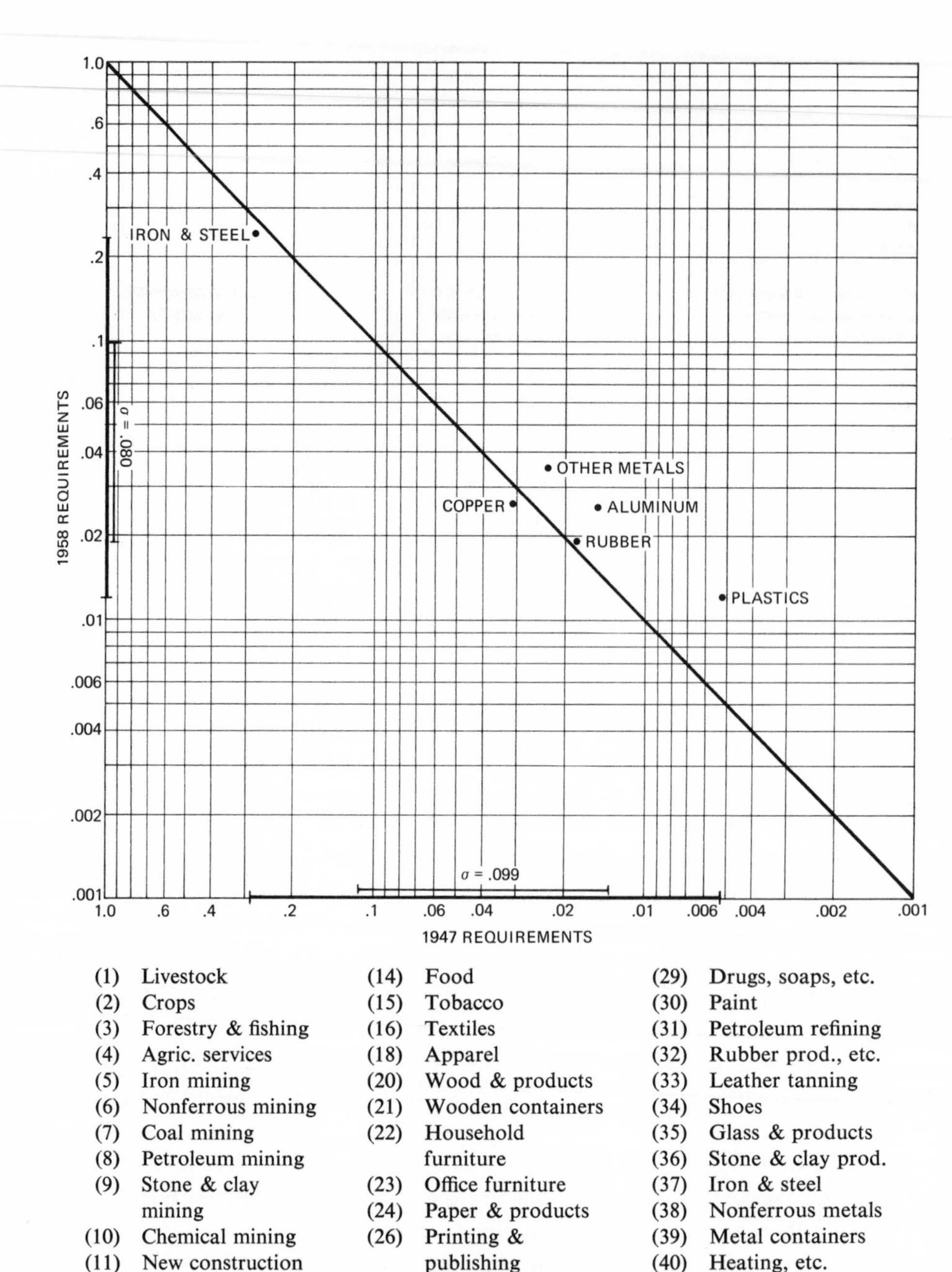

6.9 Intermediate Requirements of Six Specific Materials to Deliver the Machinery Subvector of 1961 Final Demand with 1947, 1958, and Early 1970s Technologies (dollars times 10^{10} in 1958 prices). Each point represents total intermediate requirements for a particular material.

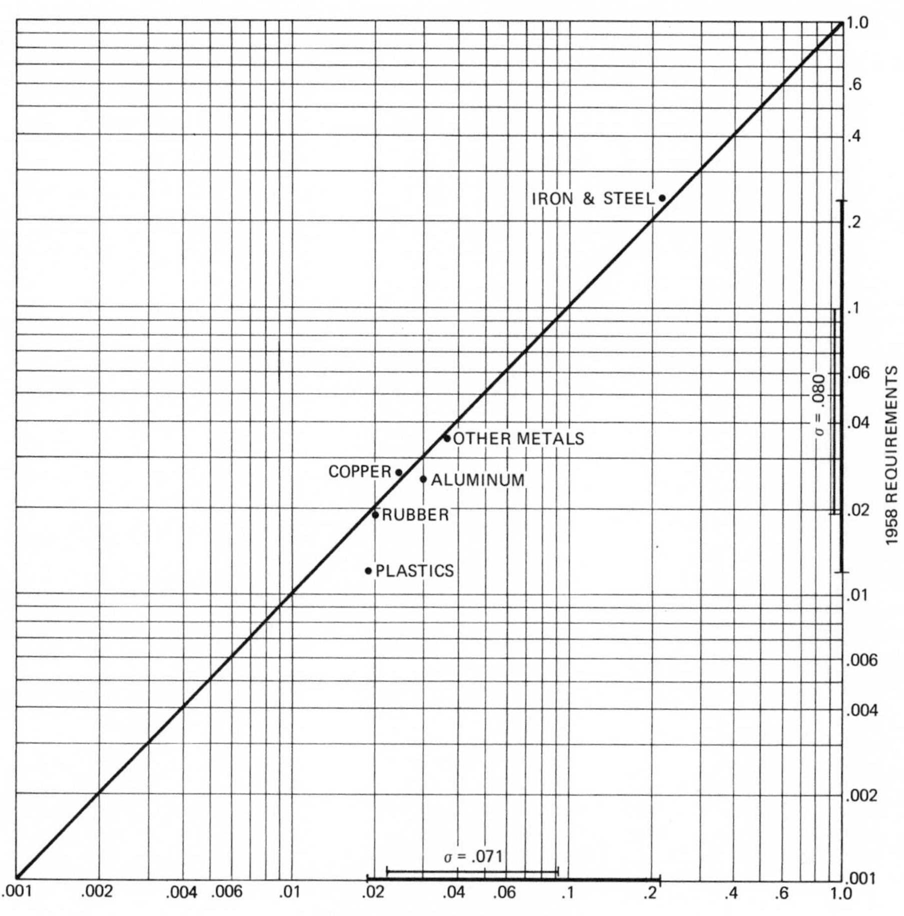

(43) Engines & turbines
(44) Farm equipment
(45) Constr. & mining eq.
(46) Materials hand. eq.
(47) Metalworking eq.
(48) Special ind. eq.
(49) General ind. eq.
(50) Machine shop prod.
(51) Office & comp. mach.
(52) Service ind. mach.
(53) Electrical apparatus
(54) Household appliances
(55) Light. & wiring eq.
(56) Communication eq.
(57) Electronic compon.
(58) Batteries, etc.
(59) Motor vehicles & eq.
(60) Aircraft
(61) Trains, ships, etc.
(62) Instruments, etc.
(63) Photo. apparatus
(64) Misc. manufactures
(65) Transportation
(66) Telephone
(67) Radio & tv broad.
(68) Utilities
(69) Trade
(70) Finance & insurance
(71) Real estate & rental
(72) Hotels & pers. serv.
(73) Business services
(74) Research & dev.
(75) Auto. repair
(76) Amusements, etc.
(77) Institutions
(80) Noncomp. imports
(81) Bus. travel, etc.
(83) Scrap

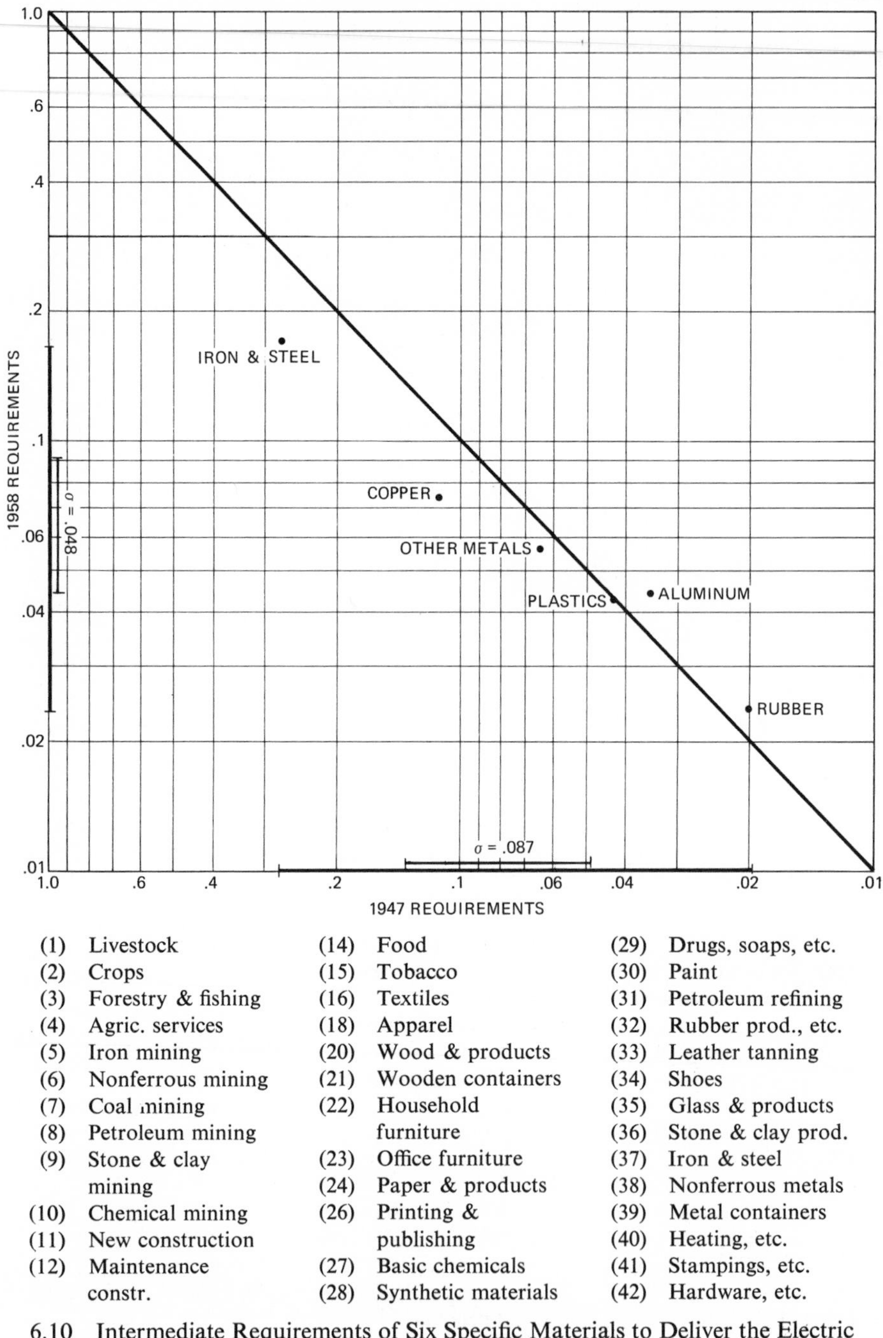

6.10 Intermediate Requirements of Six Specific Materials to Deliver the Electric and Service Equipment Subvector of 1961 Final Demand with 1947, 1958, and Early 1970s Technologies (dollars times 10^{10} in 1958 prices). Each point represents total intermediate requirements for a particular material.

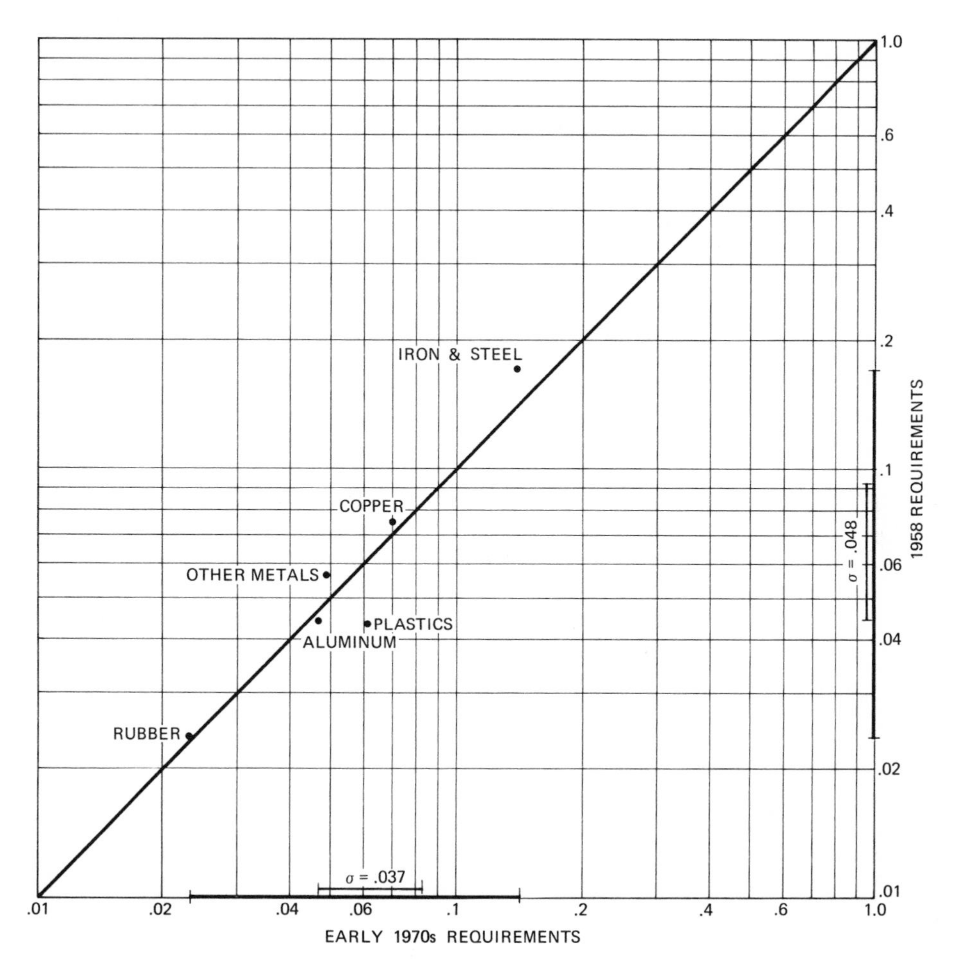

(43) Engines & turbines
(44) Farm equipment
(45) Constr. & mining eq.
(46) Materials hand. eq.
(47) Metalworking eq.
(48) Special ind. eq.
(49) General ind. eq.
(50) Machine shop prod.
(51) Office & comp. mach.
(52) Service ind. mach.
(53) Electrical apparatus
(54) Household appliances
(55) Light. & wiring eq.
(56) Communication eq.
(57) Electronic compon.
(58) Batteries, etc.
(59) Motor vehicles & eq.
(60) Aircraft
(61) Trains, ships, etc.
(62) Instruments, etc.
(63) Photo. apparatus
(64) Misc. manufactures
(65) Transportation
(66) Telephone
(67) Radio & tv broad.
(68) Utilities
(69) Trade
(70) Finance & insurance
(71) Real estate & rental
(72) Hotels & pers. serv.
(73) Business services
(74) Research & dev.
(75) Auto. repair
(76) Amusements, etc.
(77) Institutions
(80) Noncomp. imports
(81) Bus. travel, etc.
(83) Scrap

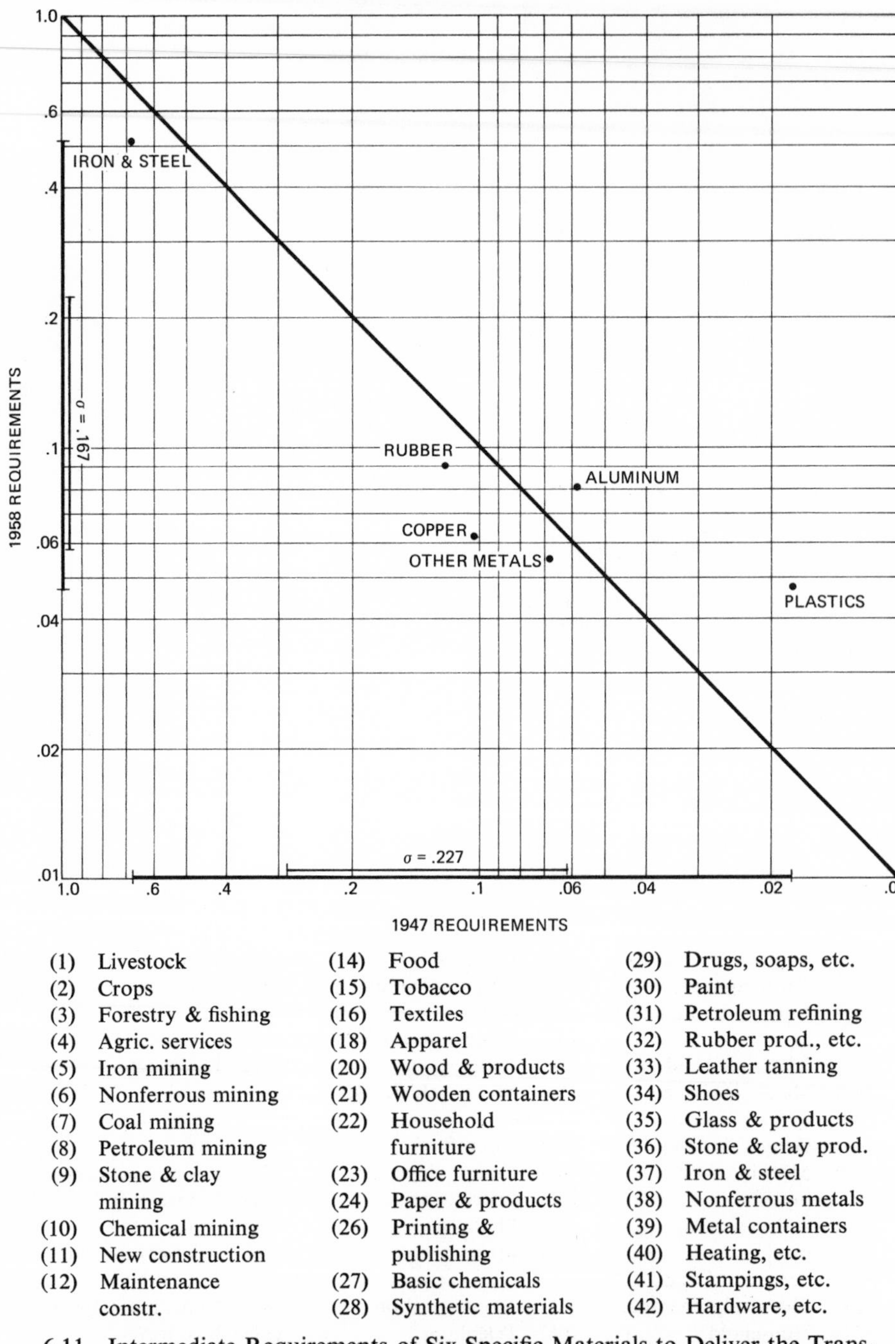

6.11 Intermediate Requirements of Six Specific Materials to Deliver the Transportation Equipment Subvector of 1961 Final Demand with 1947, 1958, and Early 1970s Technologies (dollars times 10^{10} in 1958 prices). Each point represents total intermediate requirements for a particular material.

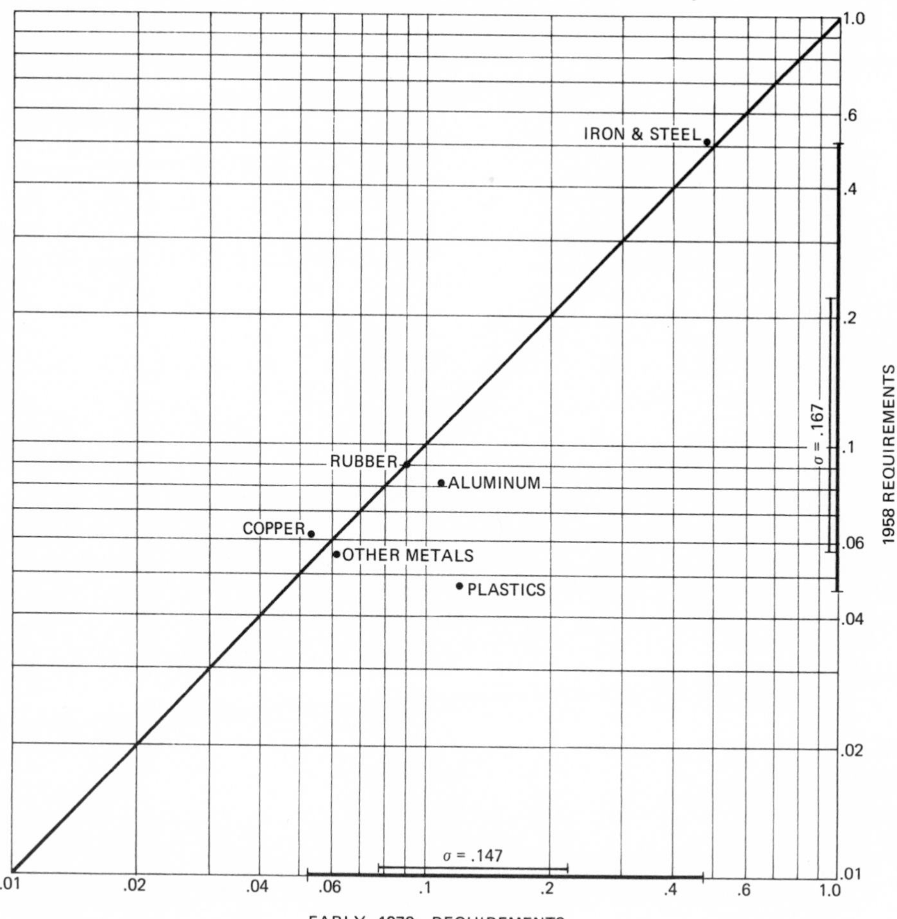

(43) Engines & turbines
(44) Farm equipment
(45) Constr. & mining eq.
(46) Materials hand. eq.
(47) Metalworking eq.
(48) Special ind. eq.
(49) General ind. eq.
(50) Machine shop prod.
(51) Office & comp. mach.
(52) Service ind. mach.
(53) Electrical apparatus
(54) Household appliances
(55) Light. & wiring eq.
(56) Communication eq.
(57) Electronic compon.
(58) Batteries, etc.
(59) Motor vehicles & eq.
(60) Aircraft
(61) Trains, ships, etc.
(62) Instruments, etc.
(63) Photo. apparatus
(64) Misc. manufactures
(65) Transportation
(66) Telephone
(67) Radio & tv broad.
(68) Utilities
(69) Trade
(70) Finance & insurance
(71) Real estate & rental
(72) Hotels & pers. serv.
(73) Business services
(74) Research & dev.
(75) Auto. repair
(76) Amusements, etc.
(77) Institutions
(80) Noncomp. imports
(81) Bus. travel, etc.
(83) Scrap

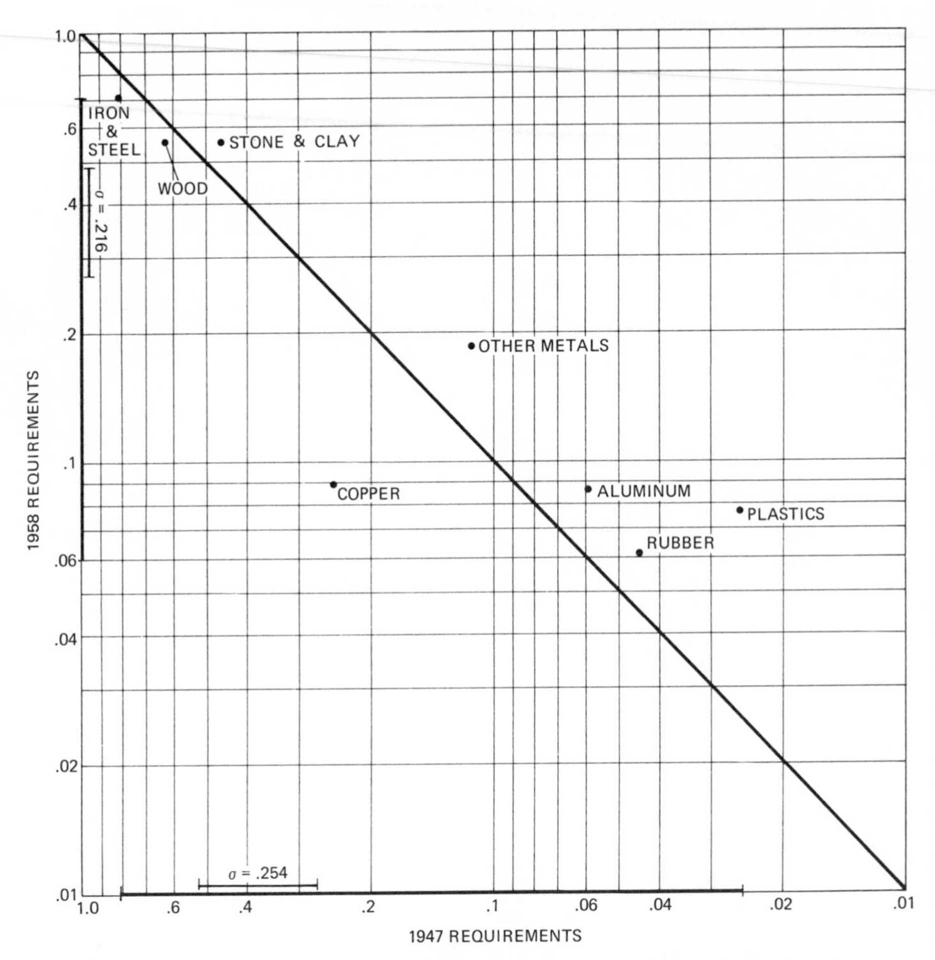

6.12 Intermediate Requirements of Eight Specific Materials to Deliver the Construction Subvector of 1961 Final Demand with 1947 and 1958 Technologies (dollars times 10^{10} in 1958 prices). Each point represents total intermediate requirements for a particular material.

6.6 Diversification of the Materials Base for Broad End-Product Groups

The traditional dominance of a single major material in each end-use category is breaking down. Steel still constitutes the principal material for many durables, but its share has been declining, while that of formerly minor materials has been growing. The result is smaller dispersion in relative contributions of each of the material supplying sectors to the delivery of each subvector of final demand. Figures 6.9 to 6.12 illustrate this tendency.

Each figure shows direct-plus-indirect requirements for six (eight for construction) specific materials used to produce the four principal durable goods subvectors of final demand. Each axis measures direct-plus-indirect materials requirements at a particular time: 1947, 1958, and early 1970s. No projections were made for construction, and only 1947–1958 comparisons are cited for that subvector.

In these four figures, points are more tightly clustered in the direction of the 1958 than in the direction of the 1947 axis and still more tightly clustered in the direction of the early 1970s axis. To crystallize this impression, the range and standard deviation of the points are indicated for each year. The range of the six direct-plus-indirect materials requirements in each year is measured by a darkened segment along each axis; the standard deviation is shown by a line segment, centered about the mean, just above it. Note that a logarithmic scale is used. Each of these measures of dispersion decreases over time for all four durable goods subvectors of final demand.

Parallel diversification of materials also took place in textiles. Synthetics now contribute a larger proportion of the materials dollar than natural fibers, but neither synthetics nor natural fibers can claim the position of overwhelming dominance that natural fibers held prior to 1947. Similarly, aluminum, paper, and plastics now compete for sizeable portions of the containers market, where steel used to dominate. Of course, there are exceptions. In furniture, the share of plastics approaches that of iron and steel, but contributions of other nonferrous metals and rubber continue to decline, increasing the range of the shares of the different materials.

As technological knowledge accumulates, more and more possible alternatives open up, and potential substitutability often increases. Now, improved aluminum, plywood, and prestressed concrete can compete effectively in areas that were formerly the exclusive province of steel. However, such diversification is not a necessary feature of technical progress. A revolutionary improvement in the properties of any one of the materials—for instance, in steel or aluminum—might tip the competitive balance in favor of its exclusive dominance in many markets. In the present situation, however, properties and techniques of utilization for all materials are improving at the same time. Therefore, competition continues to depend heavily on the prices at which the materials can be supplied. The prices, in turn, rest on techniques of producing the basic materials in question. Competition of materials for the durable goods markets is, then, tied to changing technical and resource conditions in earlier stages of production. The question of interdependence of technological changes in different sectors will be treated more fully in Part II.

Chapter 7 Changing Equipment Requirements and the Internal Structure of the Metalworking Block

7.1 Economic Role of Metalworking Sectors

The so-called metalworking sectors produce the bulk of durable goods, with the exception of construction. They deliver a wide variety of intermediate and final products—stampings, screw machine products, and fasteners; heating, plumbing, and structural metal products; many kinds of industrial equipment; household appliances; electrical equipment; scientific and controlling instruments; and transportation equipment. At 76 order, they include 24 sectors, (40) through (63). Metalworking sectors supply capital goods to themselves and to the rest of the economy; they have a special role in regulating technological choice, replacement, and expansion in all industries' capacities. Leontief and Carter (1966) discuss the function of the metalworking sectors in the context of economic development. Changes in the kinds of capital goods produced by metalworkers underlie changes in the input structures—both capital and current account—of all sectors. Rising labor productivity, materials substitution, and product improvement are all closely dependent on the kinds and amounts of equipment supplied by the metalworking sectors.

Importance of metalworking in the stock of capital

Figure 7.1 shows the direct dependence of major groups of sectors on metalworkers on capital account. It reports the amounts of fixed capital stock contributed by metalworking, construction, and all other industries per unit of capacity for major industrial groups in 1947 and 1958. Each bar is really a capital coefficient for a coarsely aggregated sector, with capital, in turn, subdivided into three industry-of-origin categories. Sources of capital information are cited in Chapter 8. For all but one of these aggregated sectors, stocks of machinery (the output of metalworking sectors) constituted more than 40 percent of the total value of the fixed capital stock in each year.

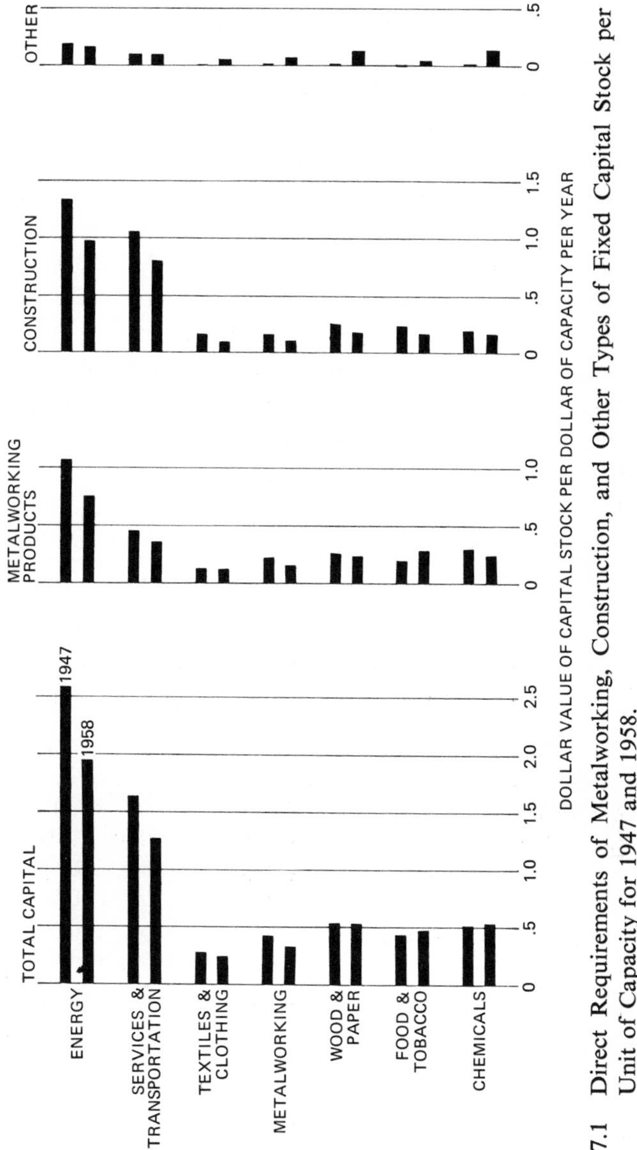

7.1 Direct Requirements of Metalworking, Construction, and Other Types of Fixed Capital Stock per Unit of Capacity for 1947 and 1958.

Capital flows and block triangularity

A number of economists, including Simpson and Tsukui (1965), have noted that current account input-output matrices—for many countries at different stages of economic development—can be decomposed into relatively self-sufficient blocks. This observation is methodologically important because it provides a framework and a rationale for lifting particular industry groups out of the general equilibrium context for separate study. In Section 7.4 of this chapter, the metalworking complex is considered separately in just this way. Identification of self-sufficient blocks is a potentially important substantive generalization about input structure. It is interesting in itself to know whether, in modern economies, certain groups of sectors are virtually independent of other groups of sectors. Carter (1967b) points out that the growing importance of general sectors and the diversification of materials tends to decrease self-sufficiency of blocks on current account.

The absence of flows on capital account from the conventional input-output matrix accounts for some of its apparent block properties. While some groups of sectors exhibit a fair degree of autonomy in their current account

Table 7.1 Effect of Introducing Capital Flows on Current Account: Interdependence of Industrial Blocks for 1958 (percent)

Industrial block	1	2	3	4	5	6	7
1. Agriculture, food & clothing	0.758[a]	0.069	0.018	0.102	0.008	0.002	0.069
	0.728[b]	0.065	0.017	0.096	0.008	0.001	0.059
2. Wood & paper products	0.018	0.539	0.033	0.053	0.124	0.008	0.111
	0.018	0.509	0.033	0.051	0.121	0.006	0.099
3. Metal & metalworking	0.023	0.059	0.629	0.078	0.306	0.043	0.097
	0.043	0.087	0.622	0.108	0.317	0.106	0.158
4. Chemicals	0.040	0.069	0.058	0.446	0.060	0.036	0.022
	0.038	0.065	0.055	0.419	0.058	0.027	0.019
5. Construction	0.010	0.016	0.027	0.038	0.157	0.030	0.102
	0.022	0.033	0.044	0.053	0.156	0.199	0.136
6. Energy	0.018	0.035	0.043	0.085	0.053	0.680	0.095
	0.017	0.033	0.040	0.079	0.051	0.499	0.080
7. Services & trans.	0.134	0.214	0.192	0.197	0.292	0.201	0.503
	0.135	0.208	0.189	0.193	0.289	0.162	0.449

[a] The upper number in each cell is the ratio of current account purchases from the block named at the side to total current inputs purchased by the block named at the top in 1958.

[b] The lower number in each cell is the percentage of direct current account purchases plus capital flows to total purchases on capital and current account, for the same cell in 1958.

transactions, all industries depend on metalworking and construction sectors for their stocks of durable goods. Table 7.1 is designed to illustrate this point. Industries of the economy are grouped into seven blocks and input-output coefficients are aggregated on this basis. The upper figure in each cell is a conventional input-output coefficient; the lower is based on the summation of 1958 capital and current flows. Capital flows are from Bureau of Labor Statistics (1968). They consist essentially of an allocation of the 1958 gross private capital formation vector to individual purchasing sectors. Flows on capital account consist of replacements and additions to capital stock. Since one would expect them to be proportional to the rate of growth and change in the sector, rather than to its absolute level of output, they are normally distinguished from current account transactions.

Large diagonal elements and small off-diagonal elements indicate self-sufficiency of blocks. Some blocks, particularly agriculture, food, and clothing (1), metalworking (3), and energy (6), are more self-sufficient than others. The dependence of all blocks on metalworking (3) and construction (5) is increased substantially with the introduction of capital flows. (The diagonal elements for (3) and (5) are exceptions because parts and components for capital goods predominate in those particular cells.) Explicit introduction of flows on capital account not only increases the apparent interdependence among relatively self-contained blocks but also disturbs any triangular ordering of sectors based on current account transactions alone. Flows on capital account, as do the outputs of general sectors, tend to be distributed to a large number of industrial consumers, although the apparent pattern of specialization of capital goods producers depends on the specific industrial classification scheme.

7.2 Changing Composition of the Capital Stock

The changing internal structure of the metalworking block stems from changing external demands placed on these sectors as they interact with shifting technical alternatives within the metalworking sectors themselves. Figure 7.2 shows the changing relative importance of different types of metal-working products in the nation's capital stock between 1947 and 1958. It shows that electrical apparatus and motors; radio, television, and communication equipment; instruments and clocks; and materials handling equipment increase. Figure 7.2 also indicates that sizeable decreases occur in trains, ships, and cycles; in other transportation equipment—where the decrease principally represents the decreased relative importance of railroad equipment in the nation's capital—in construction and mining equipment; and in

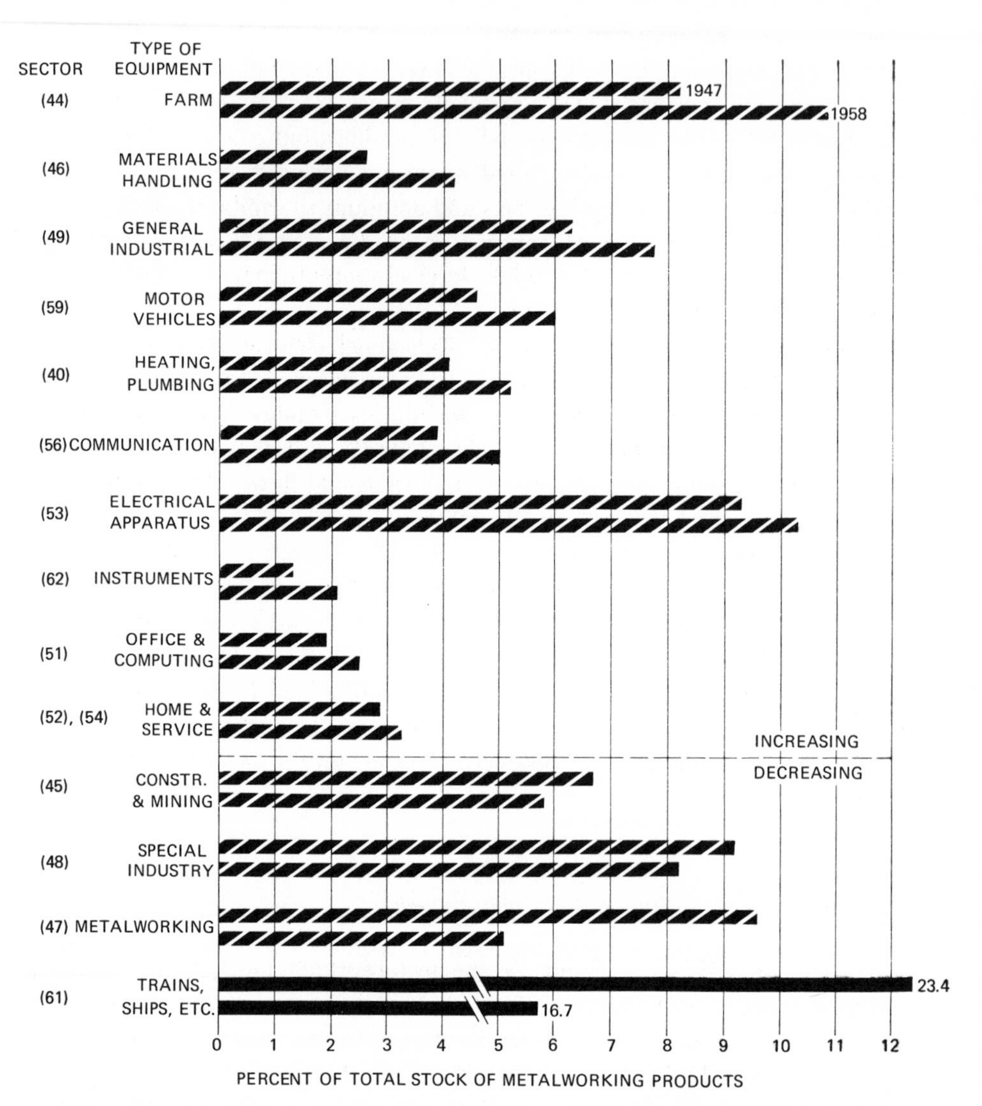

7.2 Changing Relative Importance of Various Metalworking Products in the Nation's Capital Stock from 1947 to 1958 (producers of each type of equipment are identified by the 76-order sector numbers on the left).

Table 7.2 Stocks of Specific Capital Goods Required to Deliver a Dollar's Worth of Each of Nine Subvectors of Final Demand for 1947 and 1958, 76 Order (percent)

	Subvector								
Sector	(1) Food & tobacco	(2) Text. & cloth.	(3) Paper & chem.	(4) Constr.	(5) Mach.	(6) Elec. & serv. eq.	(7) Trans. eq.	(8) Services	(9) Furn.
(46) Materials hand. eq.	0.029[a]	0.008	0.020	0.024	0.030	0.020	0.024	0.006	0.024
	0.044[b]	0.024	0.029	0.026	0.049	0.038	0.049	0.006	0.036
(51) Office & comp. mach.	0.005	0.005	0.009	0.006	0.007	0.007	0.006	0.020	0.006
	0.010	0.009	0.014	0.009	0.011	0.008	0.010	0.018	0.010
(56) Communication eq.	0.007	0.009	0.015	0.012	0.011	0.013	0.012	0.060	0.012
	0.013	0.010	0.019	0.016	0.022	0.014	0.017	0.054	0.016
(59) Motor vehicles & eq.	0.036	0.014	0.021	0.024	0.014	0.013	0.014	0.032	0.025
	0.057	0.020	0.024	0.029	0.028	0.016	0.022	0.034	0.027
(62) Instruments	0.003	0.004	0.018	0.008	0.007	0.006	0.006	0.013	0.005
	0.008	0.008	0.020	0.011	0.012	0.009	0.011	0.011	0.010
(43)–(50) Nonelectrical eq.	0.410	0.308	0.394	0.320	0.562	0.370	0.418	0.166	0.399
	0.501	0.221	0.274	0.283	0.371	0.209	0.291	0.118	0.270
(61) Trains, ships, etc.	0.091	0.071	0.107	0.134	0.086	0.079	0.095	0.135	0.116
	0.072	0.047	0.061	0.095	0.073	0.053	0.075	0.102	0.076

[a] The upper figure in each cell measures direct-plus-indirect requirements for a specific capital good for 1947.
[b] The lower figure measures direct-plus-indirect requirements for a specific capital good for 1958.

metalworking equipment. Direct-plus-indirect requirements of stocks of specific kinds of equipment to deliver various subvectors of final demand were also computed. The stock of the product of industry i per dollar delivery of final demand for subvector g was computed as follows:

$$ {}_g\kappa_i^t = \mathbf{B}_i^t \mathbf{Q}^t {}_g\mathbf{y}^{61} \quad (g = 1, 2, \ldots, 8) \tag{7.1} $$

where

$\mathbf{B}_i^t$ = row of the capital stock coefficient matrix, measuring the value of the stock of the output of industry i required per unit of capacity of each consuming sector in year t (1947 or 1958)

${}_g\kappa_i^t$ = total stock of i required to deliver the gth subvector of 1961 final demand with the technology of year t

${}_g\mathbf{y}^{61}$ = gth subvector of 1961 final demand

To put equation 7.1 on a "per unit of final delivery" basis, both sides were divided by ${}_g\phi^{61} = (1, 1, \ldots, 1) = {}_g\mathbf{y}^{61}$ as in equation 4.6.

Table 7.2 shows these changes in the amounts of specific kinds of capital goods required to deliver different subvectors of final demand with 1947 and 1958 current input and capital structures. Across-the-board increases in requirements for materials handling equipment, office and computing machines, instruments and clocks, and motor vehicles are shown. The rise in requirements for automobiles and trucks is more than counterbalanced by decreases in railroad equipment for all types of end-product deliveries. Nonelectrical equipment requirements decrease for all but the food and tobacco subvector. Increase for this subvector reflects increased mechanization of agriculture and of processing of foods during the period.

7.3 Changing Composition of Additions to Stock: Gross Private Capital Formation

Information about changes in the gross private capital *formation* vector over the period 1947–1958 reinforces the general impression based on capital stock matrices themselves. The gross capital formation vector is one of the major components of final demand in an input-output flow table; it tells the amount of each sector's output that is invested, that is, *added* to the capital stock in a particular year. The vector subsumes the values of newly produced capital goods for replacement or addition to capacity in all sectors—it shows what is happening to the capital stock "at the margin." The percentage contributions of various metalworking sectors to gross capital formation in

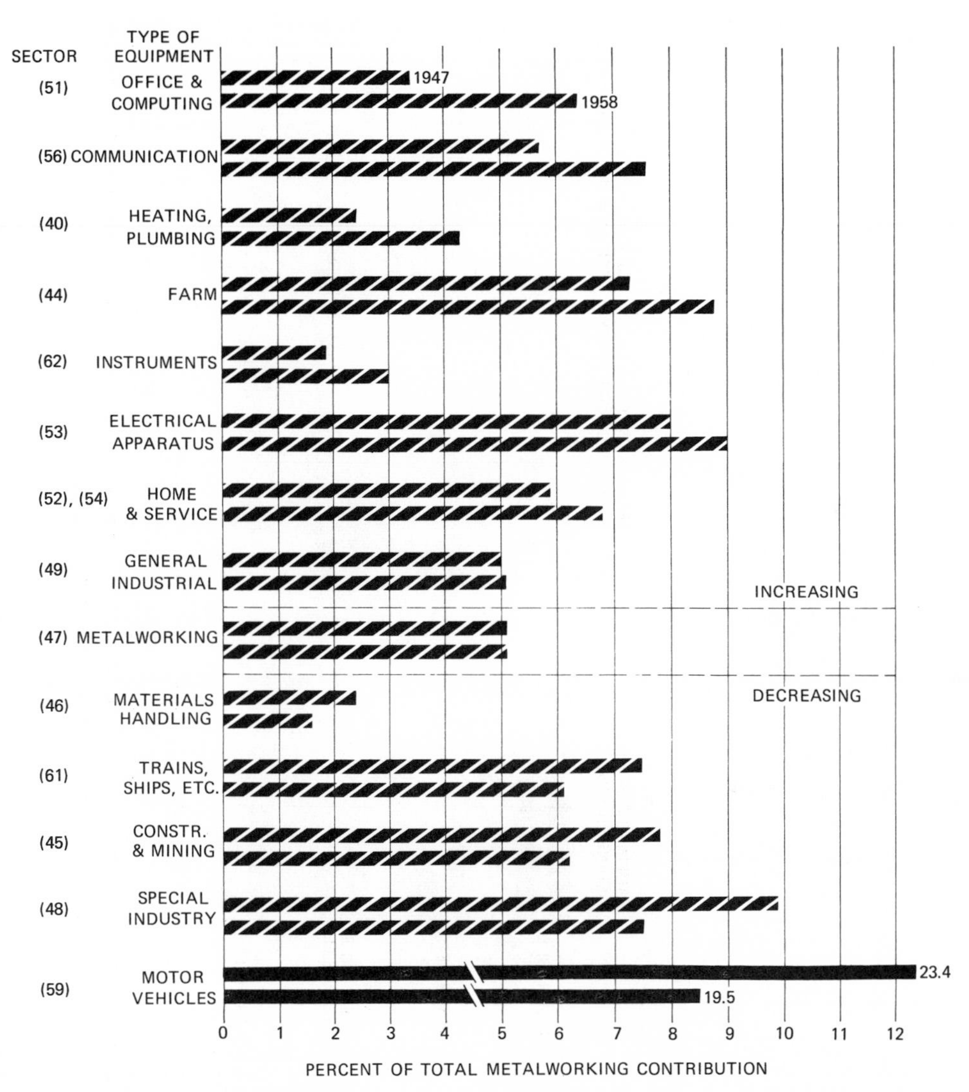

7.3 Percentage Contributions of Various Metalworking Sectors to Gross Private Capital Formation for 1947 and 1958 (producers of each type of equipment are identified by the 76-order sector numbers on the left).

1947 and 1958 are compared in figure 7.3. The figure shows, as one would expect, that relatively larger amounts of office and computing machines; radio, television, and communication equipment; and instruments and clocks were added to the nation's gross capital stock in 1958 than in 1947. However, nonelectrical and transportation equipment still contribute the great bulk of

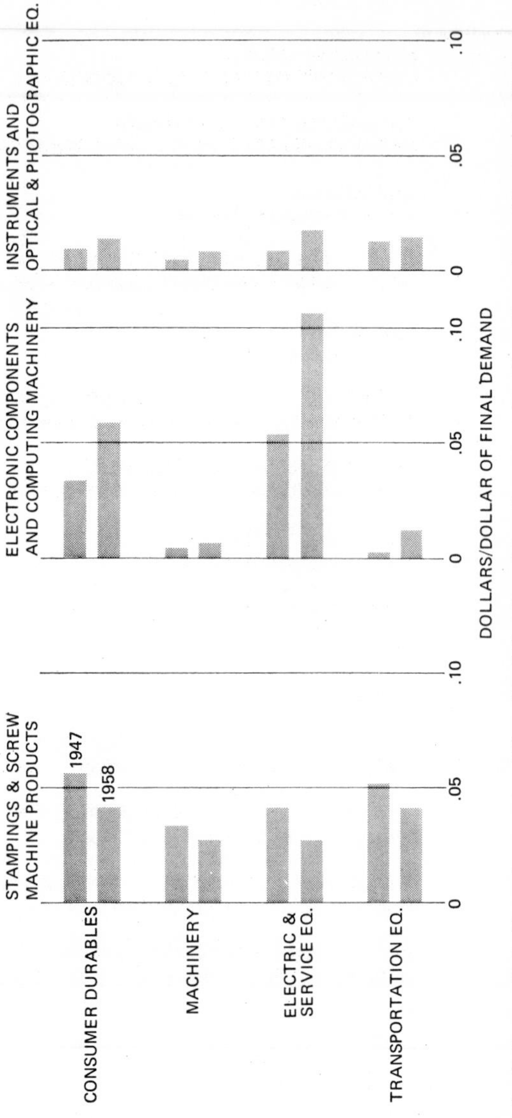

7.4 Intermediate Requirements of Specific Metalworking Inputs to Deliver Four Durable Goods Subvectors of 1961 Final Demand with 1947 and 1958 Technologies.

gross additions to capital stock in 1958. Although automobiles constituted a larger portion of capital stock in 1958 than in 1947, they contributed a smaller percentage of additions, that is, of gross capital formation.

7.4 Qualitative Changes in Capital Goods and Metalworking Specialization

Changes in the relative importance of different final capital goods were accompanied by changes in their design. Increased instrumentation and automation in all kinds of equipment led, in turn, to changes in the current account interdependence of metalworkers. Figure 7.4 shows, for example, that substantially larger amounts of instruments and electronic components and computing machinery were required on current account to deliver a unit of each of four broad types of durable goods to final demand with 1958 structures than with 1947 structures. As these items become more important as capital goods, they also become more important as current account inputs, that is, as parts and components for other types of capital goods.

The phenomenal growth in requirements of electronic components is a change in the division of labor among metalworkers that should be studied in greater depth. The year of 1958 was close to a peak one for this particular area of specialization. By 1961, further advances in the development of printed, and then integrated, circuits began to cut into the advantage of specialized production of components with subsequent assembly. Between 1947 and 1958 the rate of growth of requirements for electronic components to produce the 1961 bill of final demand was roughly 8.5 percent per year. Between 1958 and 1961 this growth rate for producing the same 1961 final demand decreased to 1 percent per year. An account of recent technical developments in this field is given by Lynn et al. (1966).

A more detailed view of the changing internal structure of the metalworking sectors appears in tables 7.3 and 7.4. These describe the current account structure within the metalworking complex—purchases of metalworking sectors from each other—in 1947 and 1958 at 76 order. Transactions are described by direct primary input coefficients. Within the metalworking complex, industries are arranged in roughly triangular order. Industries that specialize in inputs and components for other metalworking industries are placed near the bottom of the table, while those that specialize primarily in final metal products are located near the top. Final metal products are divided into three major groups—transportation equipment, electrical equipment, and nonelectrical equipment—and the sectors included in each group are enumerated in the tables. Industries listed near the top of each group or block, such as office and computing machines and materials handling

Table 7.3 Internal Structure of Metalworking: 1947 Input-Output Coefficients,[a] 76 Order (dollars per thousand dollars)

	Trans. eq.			Electrical eq.							Nonelectrical eq.						Instr.		General metalworking					
Sector	(60)	(61)	(59)	(51)	(52)	(54)	(56)	(58)	(55)	(57)	(46)	(48)	(45)	(44)	(43)	(50)	(63)	(62)	(53)	(47)	(49)	(42)	(41)	(40)
Transportation eq.																								
(60) Aircraft	23																							
(61) Trains, ships, etc.		77																						
(59) Motor vehicles & eq.	2	7	268	1	1	3	2	3	1	2	3	4	6	5	4	2	1	2	2	3	2	1	1	2
Electrical eq.																								
(51) Office & comp. mach.				22																				
(52) Service ind. mach.		3		1	46	25	6					1	2				4					3	6	
(54) Household appliances					8	9																		
(56) Communication eq.	5	2	1	3			100	2		77							1	2	3					
(58) Batteries, etc.	3	6	13			1	1	17	15	1	10		4	5	18		1	1	1	1	1			1
(55) Light. & wiring eq.	1	3	1	2	2	4	9	20	55	14	4	1	1		1	1	5	2	3	1	1		1	1
(57) Electronic comp.	1			1			96	1		1							1	1	3					
Nonelectrical eq.																								
(46) Materials hand. eq.		1		1							27		2	1						1				
(48) Special ind. eq.				1		1						18	1					2		1	1	1	1	
(45) Constr. & mining eq.		2									2		28	13	2									
(44) Farm equipment														54	2									
(43) Engines & turbines		28	1	2	1						13		37	55	39				5		6	1		1
(50) Machine shop prod.	2	19	4	5	8	5		2		6	8	12	11	4	8	8		2	3	5	9			1
Instruments																								
(63) Photo. apparatus							3					1	1		1		66				1			1
(62) Instruments, etc.	15	2	3	2	8	9	1	2	3	2	1	2	1		1	1	1	101	2	1	3	1	1	8
General metalworking																								
(53) Electrical apparatus	2	25		15	76	54	6	11	27	17	48	29	20	4	8	4	10	19	52	25	32	1	1	16
(47) Metalworking eq.	12	7	23	10	11	4	2	4	4	3	11	8	23	12	12	14	4	7	7	47	13	15	13	10
(49) General ind. eq.	6	11	9	3	6	6		11	1	1	24	32	33	32	33	7	3	5	9	21	33	1	1	7
(42) Hardware, etc.	16	19	26	26	24	21	8	6	19	10	18	11	9	11	6	12	10	12	9	8	10	26	13	23
(41) Stampings, etc.	21	9	41	17	59	54	17	10	36	25	16	14	18	30	25	9	25	34	17	11	10	24	18	26
(40) Heating, etc.	2	26	1	1	3	6	2	2	6	2	3	5	6	3	13	10	1	4	5	2	18	11	4	25

[a] Excluding fictitious transfers. Blank cells have entries of less than 50 cents per thousand dollars.

Table 7.4 Internal Structure of Metalworking: 1958 Input-Output Coefficients,[a] 76 Order (dollars per thousand dollars)

Sector	Trans. eq.			Electrical eq.							Nonelectrical eq.						Instr.		General metalworking					
	(60)	(61)	(59)	(51)	(52)	(54)	(56)	(58)	(55)	(57)	(46)	(48)	(45)	(44)	(43)	(50)	(63)	(62)	(53)	(47)	(49)	(42)	(41)	(40)
Transportation eq.																								
(60) Aircraft	193														1			1						
(61) Trains, ships, etc.		65																						
(59) Motor vehicles & eq.	3	8	300					4						6	1									
Electrical eq.																								
(51) Office & comp. mach.				95								1						7						
(52) Service ind. mach.		3	2		48	18	1																	2
(54) Household appliances		7				11								2										
(56) Communication eq.	36	2	6				62			1								6						
(58) Batteries, etc.	4	1	14					35	20		3		2	8	18									
(55) Light. & wiring eq.	1	3	4	3	5	5	7	16	40	3	1		1	1		1	2	3	8		1	1	2	1
(57) Electronic comp.	8		1	39			149	7		78								29	21					
Nonelectrical eq.																								
(46) Materials hand. eq.		3									38				1					1				
(48) Special ind. eq.												51								1				
(45) Constr. & mining eq.		4									1		59	1										
(44) Farm equipment													1	39										
(43) Engines & turbines		25	2		1						8		26	45	90				1		3		1	
(50) Machine shop prod.	11	4	6	2	1	1	1	2	2	1	15	3	5	14	32	69		5	2	2	4	2	3	3
Instruments																								
(63) Photo. apparatus	3																52	1						
(62) Instruments, etc.	14	1	4		10	23	2	1		1		1		1	1	1	1	70	1	1	4	1		7
General metalworking																								
(53) Electrical apparatus	3	36	1	18	75	29	7	7	16	3	52	38	13	6	10	5	8	20	64	29	41	4	2	9
(47) Metalworking eq.	19	4	10	7	2	5	3	11	5	4	11	11	14	14	15	14	3	10	9	58	14	22	7	4
(49) General ind. eq.	11	14	5	5	4	6		16		1	50	43	54	54	21	2		4	8	31	70	2		6
(42) Hardware, etc.	10	20	34	4	18	22	8	3	15	10	11	12	10	4	3	11	8	10	6	4	13	39	18	22
(41) Stampings, etc.	18	3	29	8	27	35	12	19	21	18	12	8	9	25	18	2	4	15	12	7	9	12	27	10
(40) Heating, etc.		49			3	2					10	9	12					1	3	1	7			19

[a] Excluding fictitious transfers. Blank cells have entries of less than 50 cents per thousand dollars.

equipment, sell little or nothing to other metalworking sectors on current account. Below them are listed sectors (such as electronic components and lighting and wiring equipment) that provide current inputs to electrical equipment producers at later stages and a sector (engines and turbines) that produces components for nonelectrical and transportation equipment manufacturers. The bottom rows of the table consist of industries that perform metalworking functions not specialized to a particular final metal product: stampers, makers of ball and roller bearings, and so on. These general metalworkers provide components for all the later stages of metalworking production.

Note the block character of the electrical and nonelectrical machinery sectors. These blocks buy relatively little from each other on current account, although both groups purchase from the general metalworkers detailed at the bottom of the tables. Transportation equipment manufacturers do not form a self-contained block. They purchase from both the electrical and the nonelectrical blocks, as well as from each other.

Similarities in structure between the two years are more impressive than are the differences. The clustering of coefficients within the electrical and nonelectrical subblocks and in the general metalworking group persists between 1947 and 1958, although the transportation equipment industries become more dependent on the other blocks. At the same time, some of the functions performed by general metalworkers in 1947 appear to have been transferred into specialized blocks in 1958. Individual column sums of coefficients do not change appreciably between the two years. However, for each consuming sector, the sum of general metalworking coefficients tends to decrease while intrablock sums increase. This is shown graphically in figure 7.5. The decrease in requirements from one of the principal general metalworkers—stampings, screw machine products and fasteners—was already pointed up in figure 7.4.

The changing organization of metalworking just discussed stems from changing design of durable goods and is also a facet of materials substitution discussed in Chapter 6. Direct metal consumption declined, for most metalworking sectors, between 1947 and 1958 (table 7.5). In addition, early stage semifabricated products, which are the most metal-intensive, decline in importance relative to other metalworking outputs. The cost incentives to substitute aluminum and plastic for steel reduced not only direct steel purchases but also purchases of steel-intensive semifabricated products. Where plastic is substituted for steel, it is shaped either in the plastic products sector or within the final metal product industry itself. Plastic parts are usually fewer, for a given function; they require less hardware, nuts and bolts, and other metal fasteners. Overall improvements in product design and in fasten-

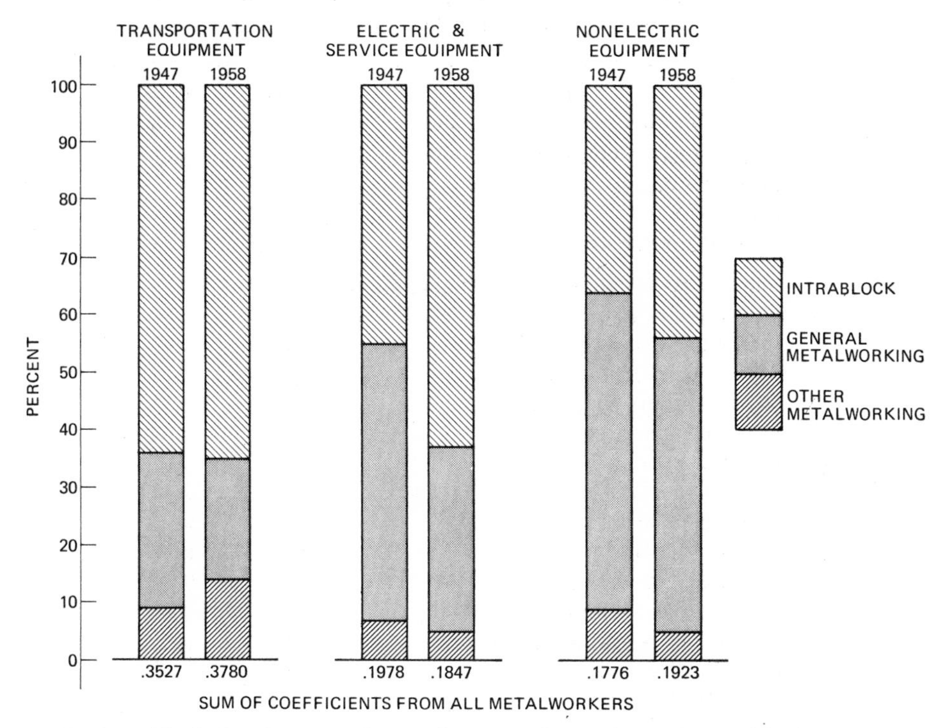

7.5 Intrablock Purchases and Requirements from General Metalworkers per Dollar of Output for Three Groups of Metalworking Sectors with 1947 and 1958 Technologies.

ing technology have decreased the demand for metal fasteners even in the manufacture of all-metal products.

Concern with the impact of the changing internal structure of metalworking has led to more detailed study of the changes by industry itself. U. S. Steel Corporation (1967) reports the first in a series of studies that disaggregates the stampings, screw machine products, and fasteners sector into its three major components and analyzes changes in the demands of metalworkers for these products. Inputs of all three categories are declining across the board. Of the three components, screw machine products constitutes the smallest and suffers the greatest relative decline over the period. This is due primarily to replacement by cold forgings, which use a cheaper grade of steel and involve less waste than screw machine products performing the same function. Fasteners, that is, nuts and bolts, screws, nails, staples, and so on, constitute the next largest component and experience the next most severe decline. It results from the inroads of welding, nylon fasteners, and

Table 7.5 Direct Metal Consumption by 76-Order Metalworking Sectors for 1947 and 1958 (dollars per dollar of output)

Sector	1947	1958
Group I: Basic metal products		
(40) Heating, etc.	0.22	0.22
(41) Stampings, etc.	0.21	0.23
(42) Hardware, etc.	0.19	0.18
Group II: Heavy equipment and appliances		
(43) Engines & turbines	0.17	0.14
(44) Farm equipment	0.19	0.14
(45) Constr. & mining eq.	0.16	0.15
(46) Materials hand. eq.	0.11	0.12
(47) Metalworking eq.	0.09	0.10
(48) Special ind. eq.	0.12	0.13
(49) General ind. eq.	0.14	0.12
(50) Machine shop prod.	0.19	0.15
(52) Service ind. mach.	0.11	0.10
(53) Electrical apparatus	0.16	0.12
(54) Household appliances	0.12	0.09
(55) Light. & wiring eq.	0.10	0.11
(58) Batteries, etc.	0.18	0.15
(59) Motor vehicles & eq.	0.11	0.08
(61) Trains, ships, etc.	0.13	0.11
Group III: Electronic and other complex products		
(51) Office & comp. mach.	0.06	0.03
(56) Communication eq.	0.05	0.03
(57) Electronic compon.	0.08	0.06
(60) Aircraft	0.11	0.06
(62) Instruments, etc.	0.06	0.07
(63) Photo. apparatus	0.03	0.03

design changes that call for fewer parts. Stampings, the largest component, declines least, but still substantially, for virtually all consuming sectors.

Most observed changes in the current account structure of metalworking sectors seem to be explained by materials substitutions and changing product design. Radically new techniques of metal cutting and forming—in particular, the numerically controlled machine tool—appeared during the 1947–1958 period but made only negligible inroads into actual metalworking practices (see Bureau of Labor Statistics 1965:15). Automatic assembly began to make inroads even later than numerical control. Schwartz and Prenting (1966) analyze these new technological developments in the metalworking sectors, citing effects of specific techniques on labor, skill, capital, and material requirements. Numerical control is particularly advantageous where lot sizes are small and flexibility is important, whereas automatic assembly is best suited to mass production situations. Although new metal-fabricating techniques can ultimately save metal through reduction of scrap and spoilage, the primary impact of these techniques will be on direct labor and capital requirements. Eventually, these new techniques will also affect current account input structures through their influence on materials competition and product design. New metal-fabricating techniques work toward restoring the relative cost advantage of metals as materials, and they realign the cost structures associated with alternative product designs.

Part II Structural Change and Economic Efficiency

Chapter 8 Changing Labor and Capital Requirements

8.1 Labor and Capital as Primary Factors

Part I dealt with changing industrial division of labor, that is, with a description of changing requirements for specific intermediate inputs. Many of these changes could be explained by concrete technological and organizational developments in particular areas of the economy. Changing costs or prices of specific inputs—such as plastics, aluminum, steel, coal, and natural gas—were introduced ad hoc to explain shifts in industrial specialization; no systematic analysis of the cost aspect of changing structure was undertaken. In Part II we analyze changes in input-output structures in terms of primary factor requirements. Explicit attention is given to labor and capital as real costs, and we analyze the impact of observed structural changes on the efficiency of the economic system as a whole. This, in turn, makes it easier to relate the particular findings of this study to the mainstream of current economic literature on technological change.

Throughout this analysis, we remain within the self-imposed limits of an open static model: the circuit linking labor supply to its production by households and the one linking stocks of capital goods to capital-goods production are not closed. Instead, consumer expenditures and delivery of new capital goods are considered (part of) the final product of the economy, and labor and capital are treated as exogenous inputs. In such a setting, it makes some sense to evaluate the performance of the economy in terms of the amounts of these primary factors required to deliver a given bill of final demand. An economy capable of producing a given final demand with less labor and capital than another, or of delivering a greater final demand of given composition with a given input of primary factors, may be judged technologically superior to the first. Changing natural resource conditions are occasionally mentioned to explain particular findings; but, principally for lack of information, natural resources are not introduced systematically along with labor and capital as primary factors. Resource problems are still a frontier of economic research. Such studies as Landsberg et al. (1963) offer promising data and insight for eventual incorporation of natural resources into a general equilibrium framework.

In a closed dynamic formulation, of course, the present treatment of labor and capital as distinct primary factors must be modified. Fixed capital goods are themselves stocks of products of the system, and materials and components have an inventory or capital, as well as a current account, aspect. Trained labor may also be viewed as human capital, with education and experience representing special kinds of economic accumulation. Now, however, we do not inquire into the origins of labor and capital but view their supplies as data. A number of arbitrary simplifications are required by the theoretical properties of this static treatment. These will be discussed in the course of the analysis.

In contrast to changing industrial specialization, changing labor and capital requirements have recently been studied, rather intensively, both for individual sectors and for the national economy as a whole. The National Bureau of Economic Research studies of Creamer et al. (1960); Tostlebe (1957); Ulmer (1960); and Kendrick (1961) are outstanding, as is Hickman (1965). The Cobb Douglas and CES production functions also deal with labor and capital. To be useful in the present context, however, labor- and capital-output ratios must be specified for the same fine industrial classification grid used in the basic input-output tables. For each sector, the labor coefficient measures labor input per dollar of output and the capital coefficient, value of capital stock per dollar of annual capacity output. Labor input is measured in man-years and, where the analysis requires it, in wage costs. As in the case of intermediate input coefficients, labor and capital coefficients are based on gross value of output or capacity, rather than value added, denominators.

Like intermediate inputs and output, labor is measured as an (annual) rate of input, and labor coefficients are the quotient of an input rate and an output rate. Capital, however, is measured as the value of a stock of goods. Thus, capital coefficients—the structural parameters describing capital requirements—are stock/flow ratios (value of capital stock per unit of annual capacity output). If output were converted from a per year basis to, say, a monthly basis, consistency would require that capital coefficients be increased to twelve times their original values.

To the extent that some portion of the capital stock is idle, observed capital/output ratios will be higher than technical requirements—hence, the convention of describing capital coefficients as capital requirements per unit of *capacity* output. Capacity is hard to measure (see Klein 1960); and, therefore, it is not possible to derive capital/capacity ratios from observed stock and output data with any exactness. Furthermore, as measured here, the capital stock is an aggregate of heterogeneous durable goods, measured in value units. For want of information, inventories of raw materials, goods in process, and finished goods are neglected; and capital coefficients measure requirements of fixed capital stocks only—essentially equipment and buildings —per unit of capacity.

Working with gross output-based labor and capital coefficients in an input-output framework, we distinguish between two concepts of primary factor productivity. Direct labor or capital coefficients ($\mathbf{l}^t$ or $\mathbf{b}^t$) measure labor inputs or value of capital stock required per unit of output (or capacity) of a sector in year t. Given a complete structural description of the economy, the total (direct and indirect) primary factor content of sectoral deliveries to final demand can also be computed.

$$\mathbf{m}^t = \mathbf{l}^t\mathbf{Q}^t \tag{8.1}$$

and

$$\mathbf{k}^t = \mathbf{b}^t\mathbf{Q}^t \tag{8.2}$$

are vectors of man-years and capital stock required, respectively, to deliver a unit of each sector's product to final demand, with the technology of year t prevailing in all sectors ($\mathbf{Q}^t$ is the inverse matrix for year t). Changes in $\mathbf{m}^t$ and $\mathbf{k}^t$ measure changes in the overall labor requirements of an economy in delivering various kinds of final output. Such changes are the net result of changes in direct labor and capital coefficients of many sectors and of shifting division of labor among sectors. Concurrent analysis of changes in direct and in total primary factor requirements for particular outputs gives some notion of the importance of shifting industrial specialization in the changing productivity picture.

Both labor and capital, but particularly labor, inputs tend to be large relative to any specific intermediate input. In the engines and turbines (22) industry in 1958, for example, the direct labor coefficient (wage rate times man-hours per dollar of output) is about 0.23; the capital coefficient (value of fixed capital stock per unit of capacity), 0.34; with annual capital charges at 15 percent (to cover depreciation plus interest), 0.05. The two largest intermediate input coefficients are only 0.12 and 0.10, for inputs from itself and iron and steel, respectively. The size of labor and capital coefficients relative to other input requirements depends, of course, on the classification scheme. The category "labor" subsumes many different kinds of services performed by workers with differing combinations of skills. Detailed information identifying specific types of labor would permit disaggregation of each labor coefficient into many considerably smaller ones. Similarly, when fixed capital requirements are subdivided into types of capital by industry of origin, they are essentially disaggregated from a single large lump to many smaller coefficients.

8.2 Sources of Labor and Capital Data

While primary factor inputs are of central importance to this analysis, basic information about them is not readily available in terms of the input-output classification. For the most part, we rely on unpublished "working" series, released with many qualifications by government agencies. Official information for 1939 was lacking, and we fell back on crude estimates assembled at the Harvard Economic Research Project. A set of 1958 labor coefficients, in man-years per dollar of 1958 output, was based on employment statistics published by the Interagency Growth Project (Bureau of Labor Statistics 1966:152–153). Coefficients for 1947 and 1961 were derived from these 1958 coefficients, using a set of indices of man-hour requirements developed, but not published, by the bureau. Labor coefficients are given in man-year units, but changes are proportional to man-hour requirements; that is, the unit is adjusted man-years. The time series of labor coefficients were carried back to 1939 on the basis of information developed by Alfred H. Conrad (1956). Wage coefficients—wage costs per dollar of output—were also available for 1939, 1947, and 1958, but the task of reconciling the series from diverse sources was formidable. Wage coefficients are introduced into the discussion when it is necessary to give some value weighting to labor as compared with capital and other costs (as in figure 4.6); for testing the sensitivity of technological choice to the wage structure in Section 10.4; and for computing total factor content, that is, for combining labor and capital charges. In general, comparisons based on wage coefficients obtained from different sources are avoided, and constant (1947) sectoral wage rates are used to transform man-year to wage coefficients when analysis requires them.

Information about capital requirements is even scarcer and more tenuous than data about labor. Most economists agree on the crucial role of information in this area; understandably, they hesitate to publish their, necessarily, rough estimates of detailed capital coefficients. Capital coefficients were taken from published sources when they were available, but many informational gaps had to be filled in with ad hoc estimates. Waddell et al. (1966) provides 1958 capital coefficients for manufacturing sectors; 1958 coefficients for nonmanufacturing sectors were developed from Bureau of Internal Revenue materials.

Grosse (1953) was the basic source of capital coefficient information for 1947. This source, in turn, draws heavily on the work of Henderson and others at the Harvard Economic Research Project and on the work of economists at the U. S. Bureau of Mines and the U. S. Department of Agriculture. More aggregative estimates in Creamer et al. (1960); Ulmer (1960); Hickman (1965); and Goldsmith (1962) served both to fill in missing pieces of information and to verify the general orders of magnitude of the estimates. Grose et al. (1966) describes the series used to convert capital stock information for 1947 and 1958 from book values in original prices to constant dollars. Constant-dollar estimates were deflated from 1958 to 1947 prices, using the price deflators for the capital goods sectors of the input-output tables.

Aggregate capital stock estimates for the economy, computed with the capital coefficients for 1947 and 1958, agree closely with those in Goldsmith (1962:112–114). The fact that they are higher than those of the Office of Business Economics (1967) can be attributed to implicit differences in the useful lives of capital goods assumed in the various basic data sources. Capital coefficients are based on observed stock-to-capacity ratios: They represent "average," rather than "incremental," technology (see Carter 1957). Qualitative change, discussed in Sections 1.2 and 7.4, poses serious problems of measurement; and, clearly, the last word is yet to be uttered in this field of research. The decision to work with such rough information, and to publish it, rests on the conviction that this is the most effective way to generate correction and improvement in an area where reliable data are extremely scarce.

8.3 Changing Labor Requirements

A characteristic pattern of interrelationships among labor, capital, and intermediate input requirements began to emerge in Chapter 4. Total labor requirements decline relative to capital and intermediate inputs required to deliver a given final demand (figure 4.6). Changes in all three are closely interwoven. New structures that economize increasingly expensive labor often mean a revamping of the entire capital and current account input structure as well. The clearest feature of this network of structural changes is the marked decrease in direct labor coefficients. Figure 8.1 presents direct labor coefficients in man-years per 1947 dollar of output for 1939, 1947, and 1958 for the 38-order classification. In general, decreases in direct labor coefficients are striking in both intervals. No productivity increase is apparent between 1939 and 1947 for aircraft (25), chemicals (13), or wood and products (11). However, (25) and (13) are sectors in which changes in product mix make it particularly difficult to compare productivities over time.

Note that there is considerable variation among sectors in the rates of decline in the labor coefficients. Over both periods, outstanding gains are shown for labor productivity in electric and gas utilities (31), coal mining (4), and agriculture, forestry, and fishing (1). Despite the generally prevalent impression of lagging productivity in service sectors, some of these show above average improvement. Special difficulties in measuring or interpreting

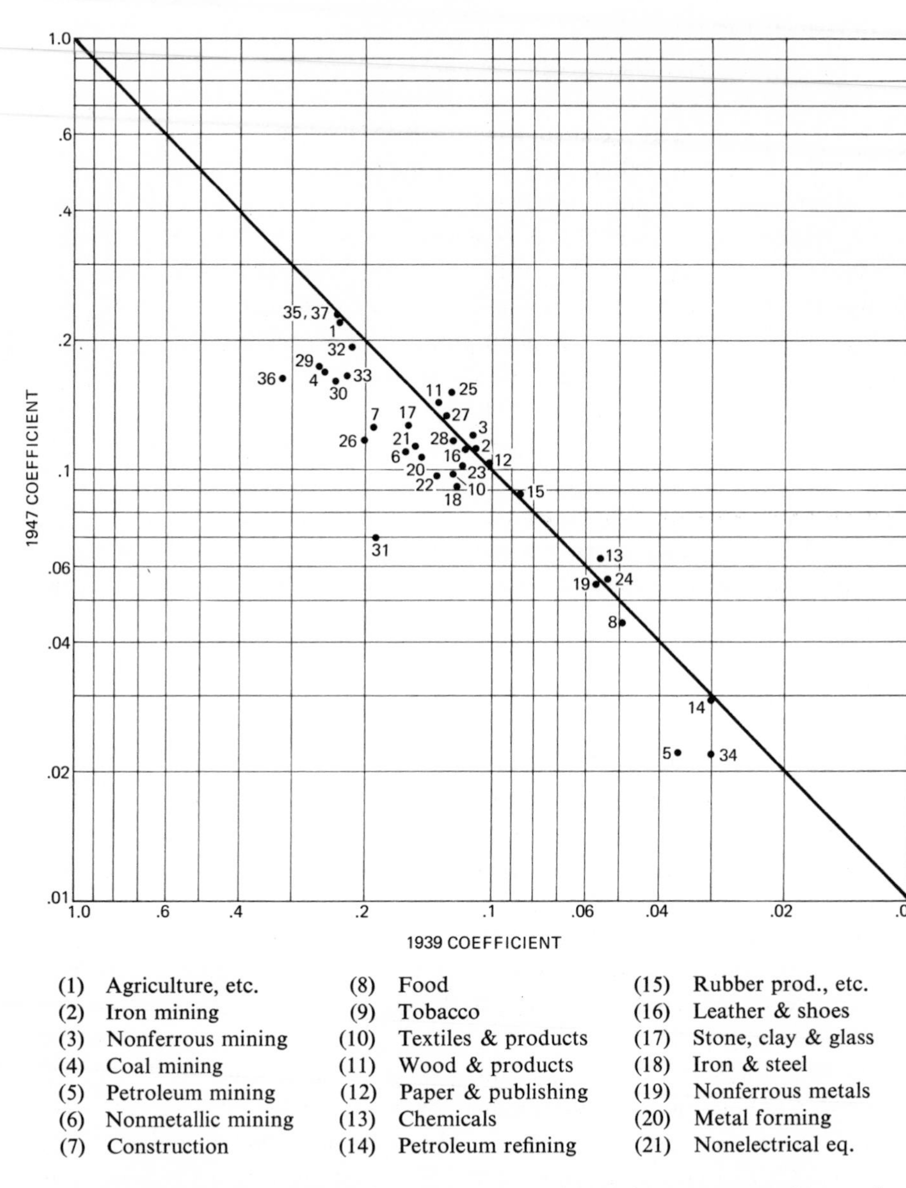

8.1 Direct Labor Coefficients for 1939, 1947, and 1958 (man-years per thousand dollars per year in 1947 prices). Each point indicates the value of the coefficient for a single 38-order consuming sector in each of two years.

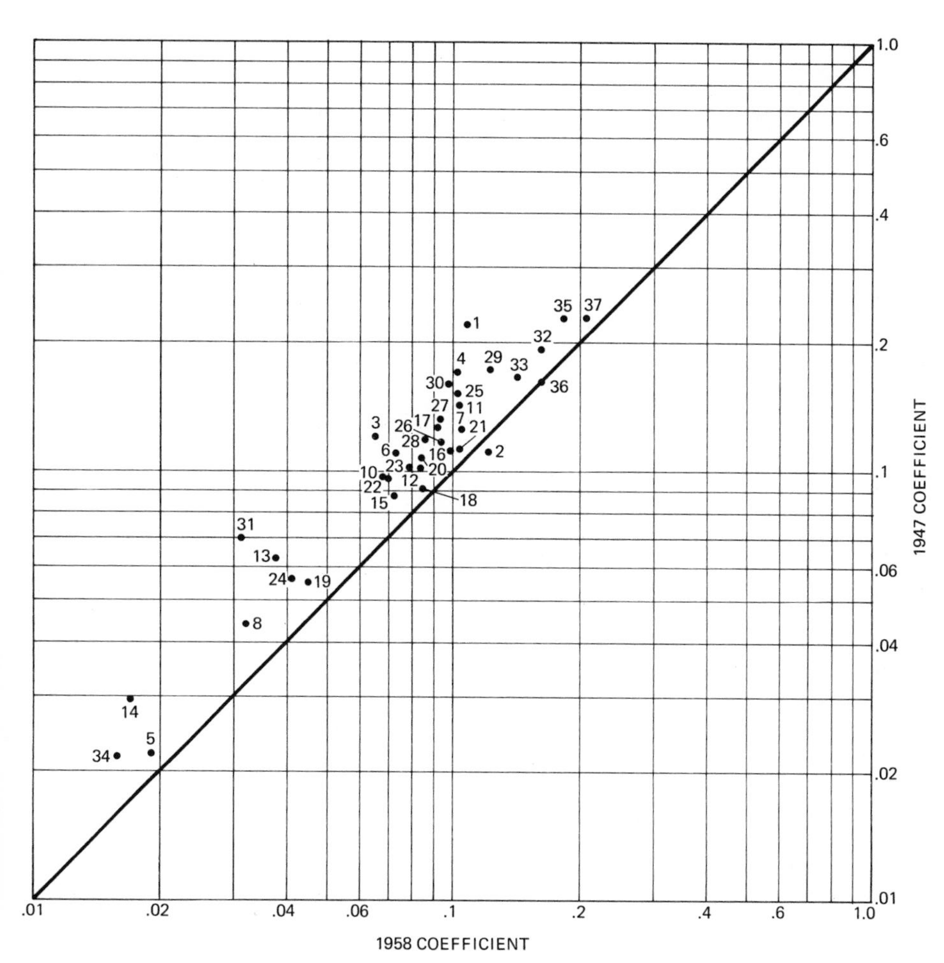

(22)	Engines & turbines	(28)	Misc. manufactures	(33)	Finance & insurance
(23)	Electrical eq., etc.	(29)	Transportation	(34)	Real estate & rental
(24)	Motor vehicles & eq.	(30)	Communications	(35)	Business serv., etc.
(25)	Aircraft	(31)	Utilities	(36)	Auto. repair
(26)	Trains, ships, etc.	(32)	Trade	(37)	Institutions, etc.
(27)	Instruments, etc.			(38)	Scrap

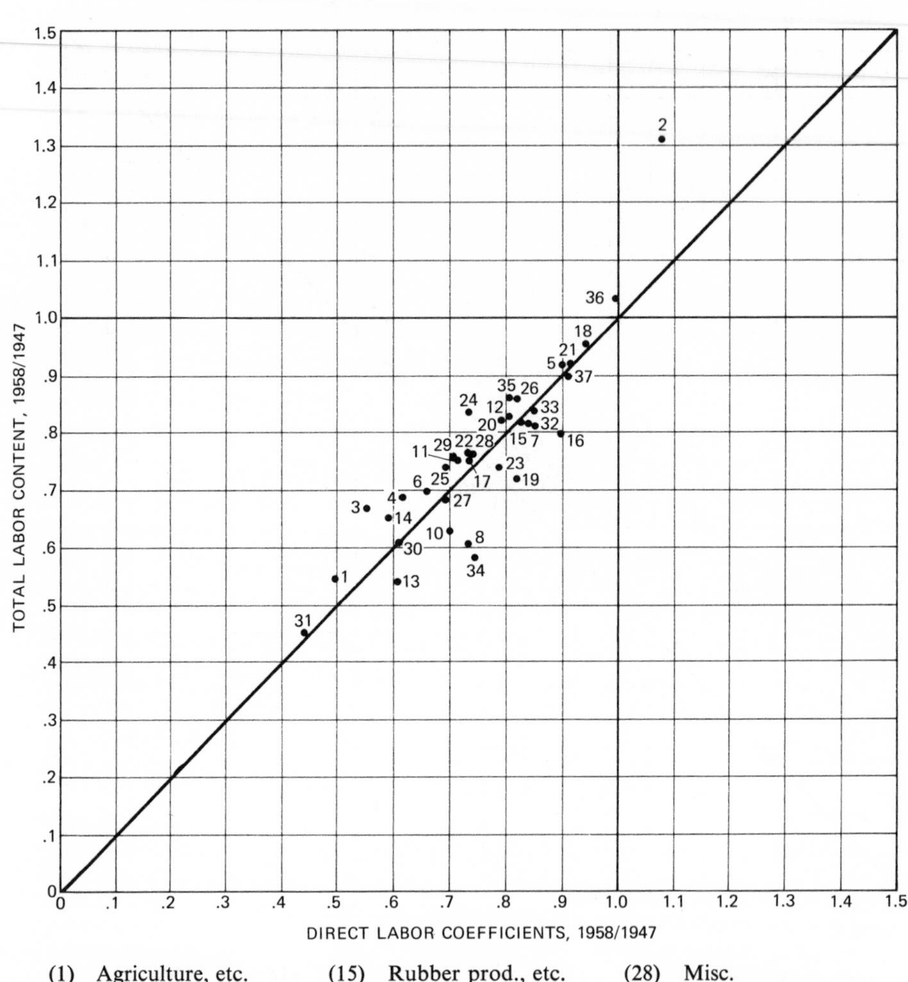

(1) Agriculture, etc.
(2) Iron mining
(3) Nonferrous mining
(4) Coal mining
(5) Petroleum mining
(6) Nonmetallic mining
(7) Construction
(8) Food
(9) Tobacco
(10) Textiles & products
(11) Wood & products
(12) Paper & publishing
(13) Chemicals
(14) Petroleum refining
(15) Rubber prod., etc.
(16) Leather & shoes
(17) Stone, clay & glass
(18) Iron & steel
(19) Nonferrous metals
(20) Metal forming
(21) Nonelectrical eq.
(22) Engines & turbines
(23) Electrical eq., etc.
(24) Motor vehicles & eq.
(25) Aircraft
(26) Trains, ships, etc.
(27) Instruments, etc.
(28) Misc. manufactures
(29) Transportation
(30) Communications
(31) Utilities
(32) Trade
(33) Finance & insurance
(34) Real estate & rental
(35) Business serv., etc.
(36) Auto. repair
(37) Institutions, etc.
(38) Scrap

8.2 Changes in Direct ($\mathbf{l}^t$) and in Total Content ($\mathbf{m}^t$) Labor Coefficients, 1958/1947; 38 Order.

productivity in service sectors are discussed in Fuchs and Wilburn (1967:32–37). Smaller rates of decrease in direct labor requirements are shown for most metalworking sectors (20 through 27). Nonferrous metals (19) shows small changes in labor coefficients over both periods, while iron and steel (18) shows a moderately large decrease for 1939–1947 and a very small decrease for 1947–1958.

Total (direct-plus-indirect) labor requirements to deliver a unit of any given sector's output to final demand depend on that sector's direct labor coefficient and on the labor required to supply all the requisite intermediate inputs as well. Figure 8.2 compares changes in direct labor coefficients, 1958-to-1947, with changes in direct-plus-indirect labor requirements to deliver a unit of the sector's product to final demand, 1958-to-1947. The horizontal axis measures l_j^{58}/l_j^{47}, the change in direct labor coefficients for each sector. The vertical axis measures m_j^{58}/m_j^{47}, changes in total labor requirements. Changes in the latter ratio depend not only on changes in sector j's direct labor coefficient but also on changes in its requirements for intermediate inputs. For most sectors, the ratio of 1958-to-1947 labor requirements is appreciably less than one on both the direct (horizontal axis) and the total (vertical axis) basis. Twenty-three of the 38 sectors are represented by points above the 45-degree line. Total labor requirements per unit of final delivery did not decrease as much as direct labor coefficients for these sectors. Total labor requirements actually increased for iron mining (2) and for automobile repair (36). If labor were considered the sole primary factor, one would conclude that the economy became less efficient in delivering outputs from these two sectors. Direct labor coefficients hardly decreased in either of these sectors, and there is some doubt as to the precise comparability of 1947 and 1958 statistics on automobile repair. Iron mining is heavily resource oriented. Here the apparent deterioration of technology reflects exhaustion of the richest, most accessible layers of our resource endowment (see Chapter 10).

8.4 Changing Capital Requirements

Figure 8.3 shows changing capital coefficients (value of capital stock per dollar of capacity) in 1947 dollars at 38 order for 1939, 1947, and 1958. In general, it appears that capital stock per unit of capacity tended to decrease between 1939 and 1947 and to decrease further between 1947 and 1958. Between 1939 and 1947, capital intensity increased in some sectors, decreased in others. Between 1947 and 1958, it decreased in all but a few. Direct-plus-indirect capital requirements to deliver a unit of any sector's output to final demand were also computed. The 1958-to-1947 ratios of direct capital requirements, b_j^{58}/b_j^{47}, and of total (direct-plus-indirect) capital requirements, k_j^{58}/k_j^{47}, are

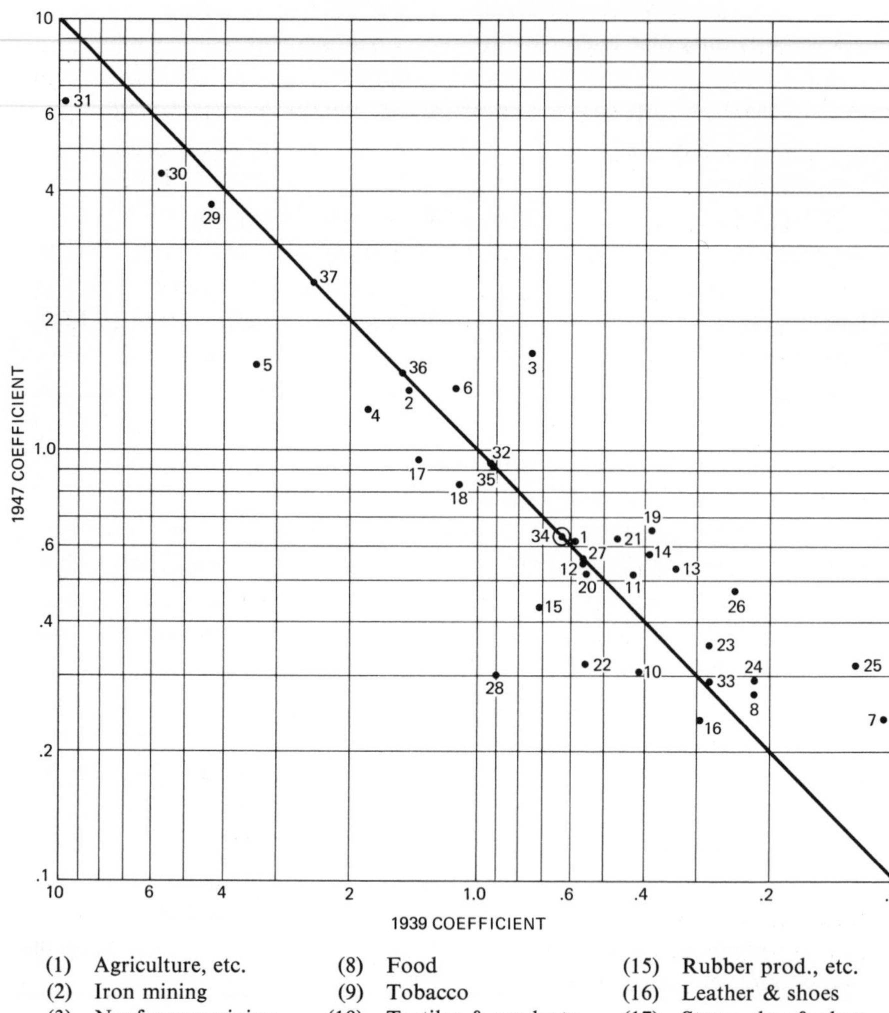

(1)	Agriculture, etc.	(8)	Food	(15)	Rubber prod., etc.
(2)	Iron mining	(9)	Tobacco	(16)	Leather & shoes
(3)	Nonferrous mining	(10)	Textiles & products	(17)	Stone, clay & glass
(4)	Coal mining	(11)	Wood & products	(18)	Iron & steel
(5)	Petroleum mining	(12)	Paper & publishing	(19)	Nonferrous metals
(6)	Nonmetallic mining	(13)	Chemicals	(20)	Metal forming
(7)	Construction	(14)	Petroleum refining	(21)	Nonelectrical eq.

8.3 Direct Fixed Capital Coefficients for 1939, 1947, and 1958 (dollar value of capital stock per dollar of capacity per year in 1947 prices). Each point indicates the value of the coefficient for a single 38-order consuming sector in each of two years (for circled points, divide scales by 10).

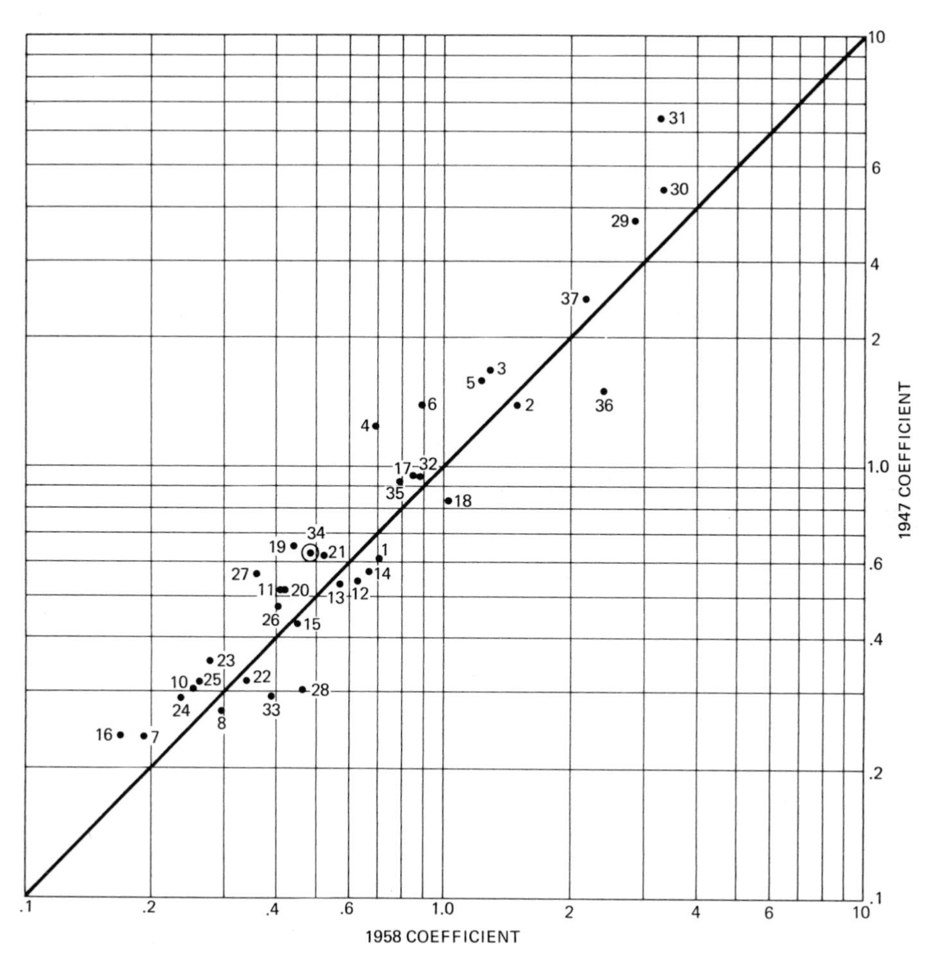

(22) Engines & turbines
(23) Electrical eq., etc.
(24) Motor vehicles & eq.
(25) Aircraft
(26) Trains, ships, etc.
(27) Instruments, etc.
(28) Misc. manufactures
(29) Transportation
(30) Communications
(31) Utilities
(32) Trade
(33) Finance & insurance
(34) Real estate & rental
(35) Business serv., etc.
(36) Auto. repair
(37) Institutions, etc.
(38) Scrap

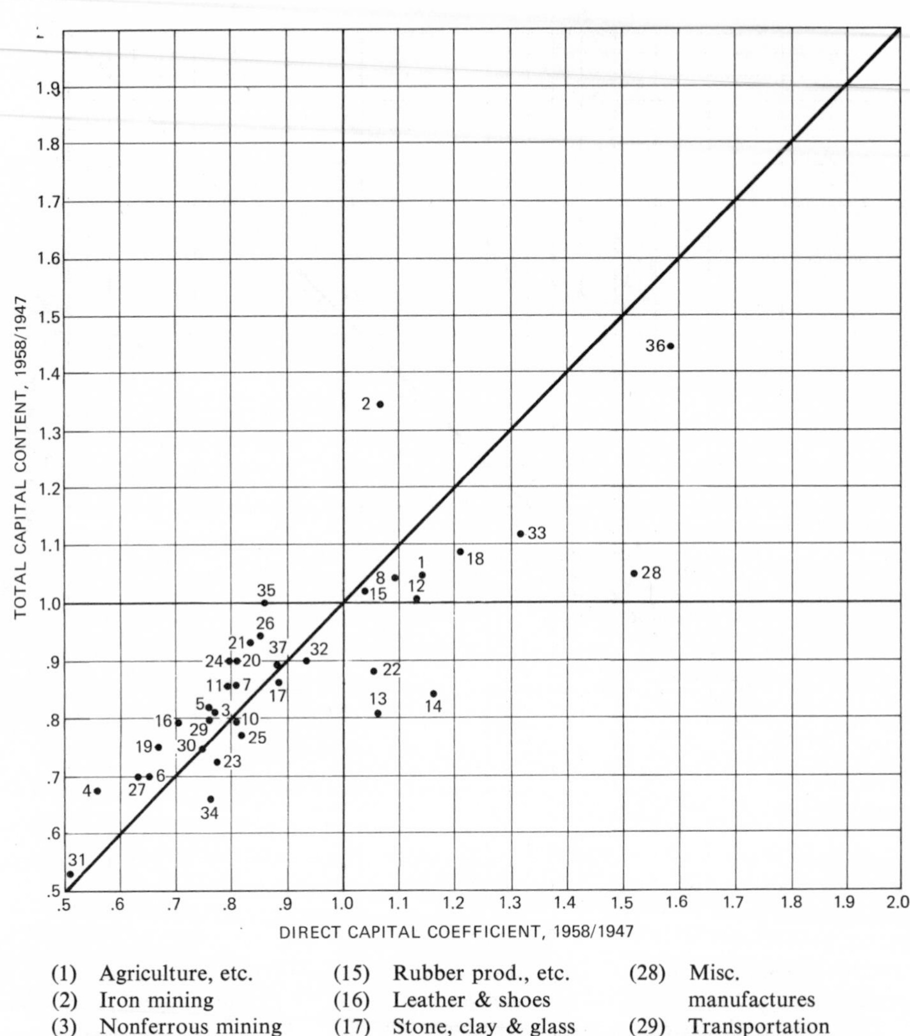

(1) Agriculture, etc.
(2) Iron mining
(3) Nonferrous mining
(4) Coal mining
(5) Petroleum mining
(6) Nonmetallic mining
(7) Construction
(8) Food
(9) Tobacco
(10) Textiles & products
(11) Wood & products
(12) Paper & publishing
(13) Chemicals
(14) Petroleum refining
(15) Rubber prod., etc.
(16) Leather & shoes
(17) Stone, clay & glass
(18) Iron & steel
(19) Nonferrous metals
(20) Metal forming
(21) Nonelectrical eq.
(22) Engines & turbines
(23) Electrical eq., etc.
(24) Motor vehicles & eq.
(25) Aircraft
(26) Trains, ships, etc.
(27) Instruments, etc.
(28) Misc. manufactures
(29) Transportation
(30) Communications
(31) Utilities
(32) Trade
(33) Finance & insurance
(34) Real estate & rental
(35) Business serv., etc.
(36) Auto. repair
(37) Institutions, etc.
(38) Scrap

8.4 Changes in Direct ($\mathbf{b}^t$) and in Total Content ($\mathbf{k}^t$) Capital Coefficients, 1958/1947; 38 Order.

compared in figure 8.4 at 38 order. Note that the range of the ratios is smaller for the direct-plus-indirect (vertical axis) than for the direct (horizontal axis) measure. Changes in total capital requirements are more uniform among sectors than are changes in direct coefficients, because changes in direct capital requirements tend to be offset by changes in the capital required indirectly to supply intermediate inputs.

8.5 Interdependence of Changing Labor and Capital Requirements

Most labor coefficients fell more than the corresponding capital coefficients, and thus the capital-to-labor ratio increased in most sectors. This is true in terms of both the direct and the total measures of labor and capital intensity (see figure 8.7). Figures 8.5 and 8.6 show to what extent the larger decreases in labor requirements are associated with smaller decreases or even increases in capital requirements, and vice versa. Two comparisons are given. Figure 8.5 shows changes in 38-order direct labor coefficients with changes in direct capital coefficients over the 1947–1958 interval. It tells to what extent relative decreases in direct labor coefficients are correlated with relative changes in direct fixed capital coefficients. The 45-degree line is the locus of points with equal 1958-to-1947 ratios for capital and labor. The 1958-to-1947 ratios are greater for capital than for labor in most sectors, and therefore points cluster above the line. The same kind of distinction is made for total requirements in figure 8.6. It shows 1947–1958 changes in total (direct-plus-indirect) labor and capital requirements for delivering products of various industries. It differs from figure 8.5 in that it takes into account for each product labor and capital requirements necessary for furnishing its intermediate input requirements, in addition to its direct requirements. Figure 8.5 includes changes in labor or capital intensity resulting from changes in methods of producing intermediate inputs and from the substitution of more or less labor- or capital-intensive inputs for one another.

Neither figure 8.5 nor 8.6 shows any clear negative correlation between ratios of change in labor and in capital requirements. If anything, there seems to be a positive association, that is, improvements in labor and capital productivities go hand in hand. In most sectors, labor and capital intensities both decline. Capital intensity increases in roughly one third of all sectors. Most of these exhibit only modest improvement in labor productivity, and no clear dependence of increases in labor productivity on the size or direction of change in capital intensity is discernible. Nor, for that matter, should it be expected to exist. Figures 8.5 and 8.6 should not be interpreted as cross-sectional production functions. Each of the changes, represented by a single

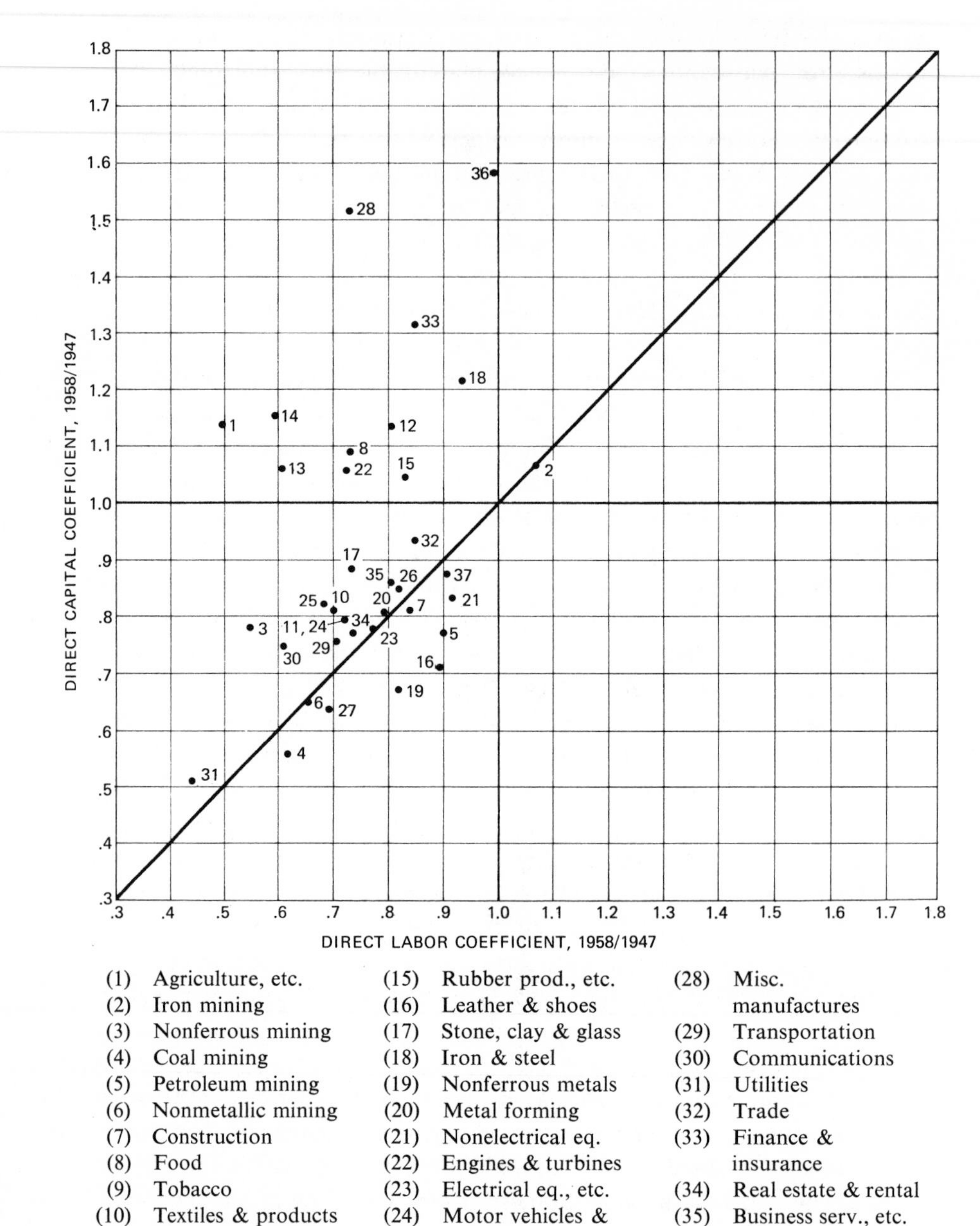

8.5 Changes in Direct Labor and Capital Coefficients, 1958/1947; 38 Order.

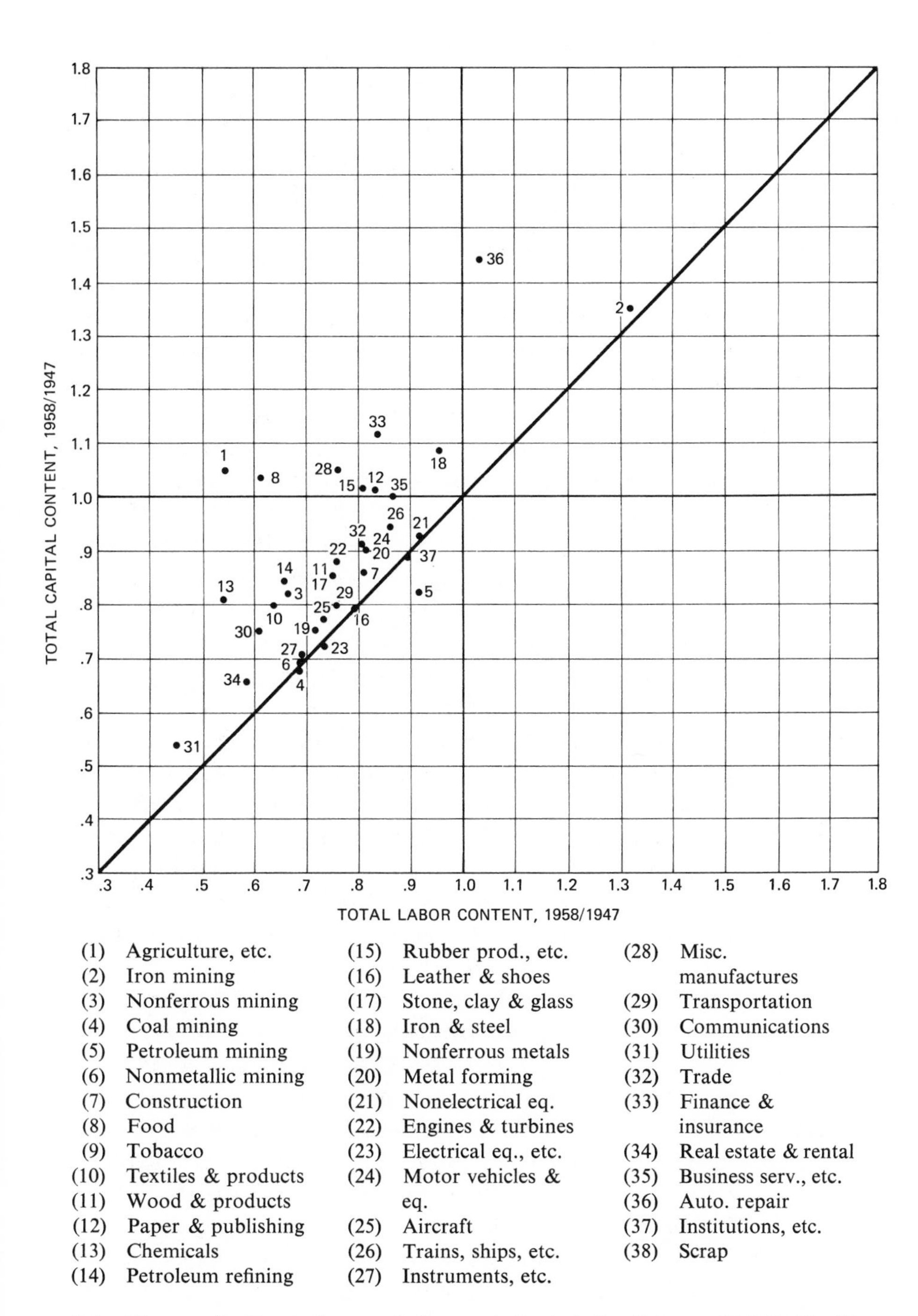

8.6 Changes in Total Content Labor and Capital Coefficients, 1958/1947; 38 Order.

point, reflects the changing technical and economic conditions facing a particular sector or, in the case of figure 8.6, group of sectors. In some industries, or for some products, changing technology has permitted the saving of both labor and capital; for others, it has not. The charts can only give a summary view of where and to what extent changes in capital intensity have accompanied labor saving.

Substantial increases in labor productivity have been achieved along with decreases, or, at least, without substantial increases in capital requirements in most sectors. The conclusion would seem to be that automation has not been synonymous with increased capital intensity. Instead, automation has involved major qualitative changes in the capital goods that comprise the fixed capital stock—increased labor productivity depends on improvements in the kinds of equipment available as well as in methods and organization of production. However, we cannot quantify this improvement and call it "more" or "less" capital. This is a central difficulty in analyzing the impact of changing technology on labor and capital requirements. All the inputs, outputs, and the composition of the capital stock change qualitatively over time. Paradoxically, the problem is easiest to recognize when one works with a very fine classification grid; in very detailed analysis it is difficult to overlook the influx of new products, possible shifts in units, and such. Problems of qualitative change may actually be obscured, but not solved, by aggregation. Primary factor inputs—both labor and capital—are each, in fact, heterogeneous aggregates. Their compositions have changed over time and far from trivially. It is this fact that makes it so difficult to measure changes, particularly in capital requirements. Estimates of capital coefficients are based on statistics of gross capital assets, adjusted to a "constant price" basis. However, little is known about the detailed composition of these assets, either in terms of kinds of capital goods or in terms of their ages, and the adjustments could only be impressionistic. Even had it been possible to make the price adjustments in great detail, taking explicit account of innumerable changes in the composition of the stock, the aggregate figures would still be difficult to interpret. Is a desktop computer "more" or "less" capital than an old-fashioned set of card-sorting machines? Is a communications satellite "more" or "less" capital than a transoceanic cable? This really depends entirely on the valuation conventions adopted for purposes of comparison. Whether the capital stock is "greater" or "smaller" is not nearly so interesting as the fact that it is radically different at different points of time. This conclusion gives some justification to the treatment of structural changes as "capital embodied," but that approach will be postponed until Chapter 12.

The problem of qualitative change also affects the comparison of labor inputs over time. Skill requirements and job environments change radically with new methods of production, and a man-year of work in designing systems is not the same as a man-year of work at a lathe; nor, for that matter, is a man-hour of day-shift and night-shift work equivalent, from the point of view of the individual worker. Although model builders quickly assume that skill mix is constantly being upgraded, attempts to specify such generalizations empirically for individual sectors have, thus far, been inconclusive. Systematic knowledge about the changing job structure of the American economy is still far from adequate. Horowitz and Herrnstadt (1966) point out some of the difficulties of quantifying changes in labor quality. For labor, standard physical units—man-hours or man-years—offer a convenient expedient. Man-hours are indeed heterogeneous. However, the amount of time that is being devoted to production has some meaning, and one can add necessary interpretive qualifications. For capital, the standard unit is a value unit, and it is measured by complex indirect procedures. This kind of measure does not permit straightforward answers to the questions usually posed.

In a dynamic context, such as that described in Section 1.2, it is not necessary to assume any fixed units for heterogeneous capital goods originating in different years. Expansion of capacity in each year can be made to depend on investment levels of preceding years, chainwise, without reference to an aggregate capital stock summed over time; and the need to establish some equivalence among qualitatively different products is averted. Such an approach does not furnish answers to the question of whether more or less capital is being used at different points of time or how much capital goods are substituted for labor over time. Perhaps, however, those questions are not really meaningful and cannot be answered after all.

8.6 Capital-to-Labor Ratios Based on Direct and Total Factor Content and Value Added Coefficients

Changing labor and capital requirements have been discussed a lot in recent economic literature, but the great bulk of modern production function analysis relates labor and capital to output without taking explicit account of intermediate inputs (see references cited in Section 1.1). Time series or international cross-section data on labor, capital, and output are used to estimate statistical relationships between inputs of these primary factors and output, for both the gross national product of an entire economy and for outputs of particular and more narrowly defined sectors. Major emphasis is given to elasticity of substitution between the two primary factors.

Measures of total labor and capital content, $\mathbf{m}^t$ and $\mathbf{k}^t$, take systematic account of shifting industrial specialization in relating primary factor requirements to final output. In this sense, they provide a sounder basis than do direct inputs alone for estimating statistical production functions. The advantage should be particularly important in work with international cross sections, since industrial integration varies from country to country; it may not be negligible in intertemporal work either. However, the requisite information on intermediate input structures has been scarce. The most common expedient for handling changing specialization has been to "net out" intermediate inputs by relating labor and capital inputs to value added, rather than to gross output, in each sector. Some difficult problems of interpreting value added as a measure of output, under changing price conditions, are discussed in David (1962). Where the entire economy is aggregated into a single sector, total value added and gross national product are identical; and total labor and capital inputs are their direct and indirect factor content. For individual sectors or end items, the measures are not identical. The use of value added as an output measure does serve to "insulate" the labor-capital-output relation from changes in industrial specialization. Shifts in an industry's "make or buy" practice—a change from fabricating to purchasing a particular component, for example—will simultaneously change both numerators (labor and capital inputs) and the denominator (value added). However, changes in the ratio of labor (or capital) to value added will not necessarily measure changes in the total labor (or capital) content of an item. Compare the two alternative measures of labor requirements:

$$\mathbf{m} = \mathbf{l}\mathbf{Q} \quad \text{and} \quad \boldsymbol{l}_v = \mathbf{l}\hat{\mathbf{v}}^{-1} \tag{8.3}$$

where $\mathbf{v} = (1, 1, 1, \ldots)(\mathbf{I} - \mathbf{A})$. An element m_j of $\mathbf{m}$ measures total labor content of a given sector. It will reflect a change in any element of $\mathbf{l}$, the direct labor coefficient vector, or in $\mathbf{A}$, the intermediate input coefficient matrix. The term $\boldsymbol{l}_v$ is a vector of value added based labor coefficients. For any particular sector j, $l_{v_j} = l_j(1 - \sum_i a_{ij})^{-1}$. Thus, the value added based, direct labor coefficient l_{v_j} is affected only by changes in the input structure of sector j itself. Only where the whole economy is treated as a single sector are equations 8.1 and 8.3 identical.

Analysis of statistical production functions is beyond the scope of this work. We shall digress only enough to show that changes in capital-to-labor ratios measured in terms of direct inputs do differ appreciably from those measured in terms of total content. Apparent instabilities and contradictions in CES production function studies of the sort cited by Nerlove (1967) might be explained by the sensitivity of direct capital-labor ratios to shifting industrial specialization.

The comparison of changes in direct and total content capital-to-labor ratios is presented in figure 8.7. The horizontal axis measures the quotient of direct capital-to-labor ratios for 1958-to-1947:

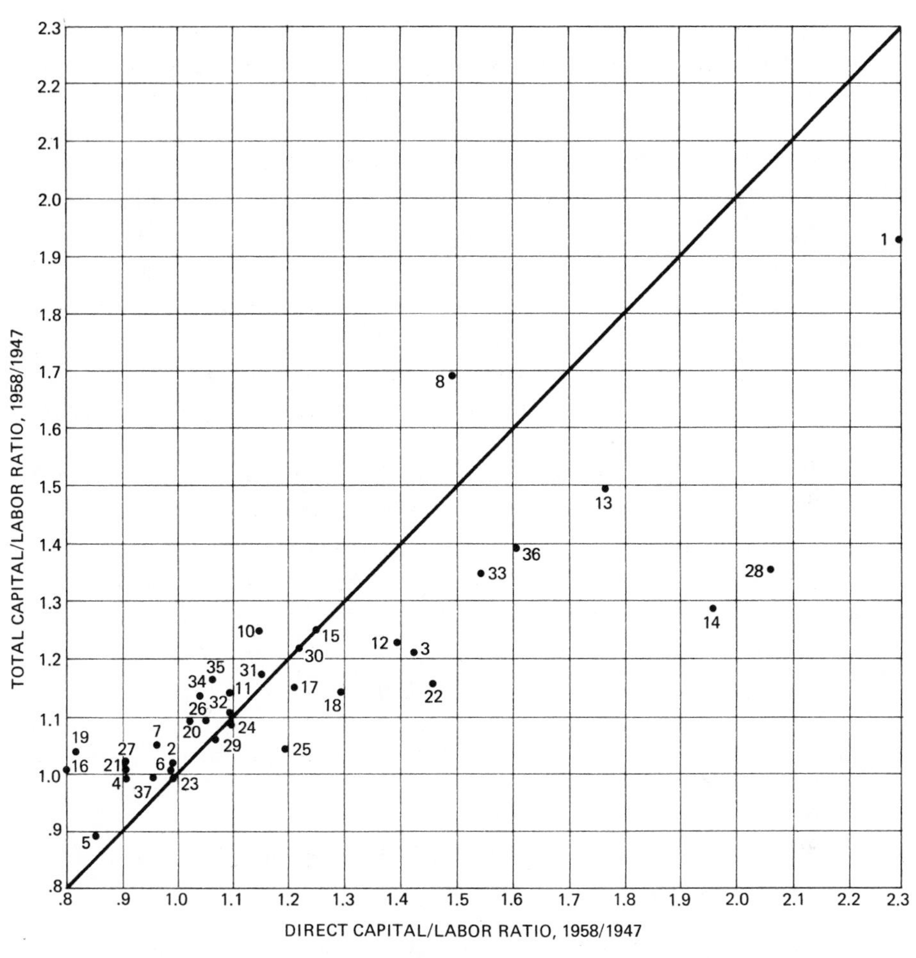

(1) Agriculture, etc.
(2) Iron mining
(3) Nonferrous mining
(4) Coal mining
(5) Petroleum mining
(6) Nonmetallic mining
(7) Construction
(8) Food
(9) Tobacco
(10) Textiles & products
(11) Wood & products
(12) Paper & publishing
(13) Chemicals
(14) Petroleum refining
(15) Rubber prod., etc.
(16) Leather & shoes
(17) Stone, clay & glass
(18) Iron & steel
(19) Nonferrous metals
(20) Metal forming
(21) Nonelectrical eq.
(22) Engines & turbines
(23) Electrical eq., etc.
(24) Motor vehicles & eq.
(25) Aircraft
(26) Trains, ships, etc.
(27) Instruments, etc.
(28) Misc. manufactures
(29) Transportation
(30) Communications
(31) Utilities
(32) Trade
(33) Finance & insurance
(34) Real estate & rental
(35) Business serv., etc.
(36) Auto. repair
(37) Institutions, etc.
(38) Scrap

8.7 Changes in Capital/Labor Ratios Measured by Direct and by Total Content Coefficients, 1958/1947; 38 Order.

$$d_j = (b_j^{58}/l_j^{58})(l_j^{47}/b_j^{47})$$

The vertical axis measures changes based on total labor and capital content:

$$t_j = (k_j^{58}/m_j^{58})(m_j^{47}/k_j^{47})$$

Coefficients b_j and l_j both have the same output denominators, and the direct capital-to-labor ratios would be the same if computed from value added based coefficients or from factor employment statistics for sector j. Similarly, the total content ratios are independent of how sector j's deliveries to final demand might be measured. The only relevant distinction is between the direct and total content measures of primary factor input. The direct measure depends on relative amounts of capital and labor employed in a sector. The total content measure includes labor and capital required to make all intermediate requirements as well.

Figure 8.7 shows how both types of capital-to-labor ratios changed between 1947 and 1958. A value of d_j or t_j greater than 1 signifies an increase in the ratio of capital to labor over the period. The average value of both d_j and t_j is roughly 1.1, but the correspondence between the two measures for individual sectors is far from perfect—changes in direct capital-to-labor ratios are not good proxies for changes in total content ratios (and vice versa). Note that the dispersion of d_j is greater than that of t_j. Capital-to-labor ratios for individual sectors are more variable in terms of the direct than the total measure. This is largely due to the effects of changing industrial specialization.

8.7 Influence of Changing Intermediate Structures and Final Demand on Aggregate Labor and Capital Productivity

The present data also permit us to examine the effects of recent changes in industrial mix on aggregate labor and capital productivity measures. Studies of aggregate productivity, and estimates of aggregate production functions for the economy as a whole, relate labor and capital inputs to the level of gross national product. Tables 8.1 and 8.2 show how labor- and capital-to-output ratios for the economy as a whole are affected by changing industrial mix. Table 8.1 shows labor and capital requirements per unit of final demand computed for different combinations of final demand proportions and technologies of 1939, 1947, and 1958. Entries along the diagonal (1939 final demand with 1939 input structures, 1947 final demand with 1947 input structures, and so on) represent labor- and capital-to-output ratios that would emerge from actual observation of the national aggregates. The off-diagonal elements represent hypothetical combinations of input structure and final demand composition of different years. Note that labor-to-output ratios vary

Table 8.1 Variation of Unit Labor and Capital Requirements with Input Structure and Final Demand Composition

Input-output, labor, and capital coefficients	Final demand proportions		
	1939	1947	1958
1939	0.294[a]	0.294	0.290
	1.668[b]	1.830	1.881
1947	0.253	0.249	0.246
	1.614	1.723	1.756
1958	0.184	0.185	0.182
	1.419	1.502	1.493

[a] The upper figure in each cell measures labor requirements, that is, man-hours per thousand dollars of gross national product, in 1947 dollars.
[b] The lower figure in each cell measures capital requirements, that is, dollars of capital stock per 1947 dollar of gross national product per year.

only slightly with final demand composition, and the aggregate national productivity measures are not changed appreciably by changing final demand weights. However, changing final demand proportions do tend to reduce changes over time in the aggregate capital-to-output ratio. This tendency is consistent with Hickman's findings for the period since World War II (see Hickman 1965:155).

Similarly, one can ask how changing intermediate input structure affects indices of national labor and capital productivity by altering their weighting.

Table 8.2 Variation of Unit Labor and Capital Requirements with Intermediate Input Structures and Primary Factor Coefficients, 1958 Final Demand Proportions

Labor and capital coefficients	Intermediate input structure		
	1939	1947	1958
1939	0.290[a]	0.296	0.292
	1.881[b]	1.980	2.051
1947	0.243	0.246	0.244
	1.672	1.756	1.798
1958	0.178	0.183	0.182
	1.418	1.485	1.493

[a] The upper figure in each cell measures labor requirements, that is, man-hours per thousand dollars of gross national product, in 1947 prices.
[b] The lower figure in each cell measures capital requirements, that is, dollars of capital stock per 1947 dollar of gross national product per year.

Table 8.2 shows how the aggregate capital- and labor-to-output ratios vary with changing intermediate input structure for a fixed (1958) final demand composition. The results are very similar to those found in table 8.1, but differences are somewhat more pronounced. In both cases, changing industrial composition has had little effect on average changes in labor productivity, but it has tended to slow down apparent improvements in average capital productivity. The explanation for the latter lies in the growing relative importance of three of the most capital-intensive sectors—electric and gas utilities (31), communications (30), and transportation and storage (29)—both in the final demand composition and in intermediate structure. Despite the fact that capital-to-output ratios are declining in these three sectors, their unit capital requirements remain very great relative to those in other sectors.

Table 8.2, then, shows that information on changing direct labor coefficients, combined with a fixed input structure, actually gives a good approximation of total labor economies in the system as a whole. This kind of conclusion is not justified for total capital economies.

8.8 Structural Change in Households and Declining Labor Coefficients

Of all the structural changes reviewed thus far, the declines in direct labor coefficients are most pronounced. This impression will be reinforced by the evidence in Chapters 10 and 11. Such pervasive reductions demand systematic explanation. The economy behaves as if labor saving were the goal of technical progress, and most changes in intermediate and capital structure can be justified by reduced direct and, to a lesser extent, indirect labor requirements. In many areas, for inputs with rising relative prices, the economic system tends to substitute specific inputs with falling relative prices. Cheaper materials and fuels displace more expensive ones (see Chapters 5, 6, and 9). In a closed model, direct labor saving might also be explained as a substitution of cheaper inputs for more expensive inputs. The unit cost of labor has been rising relative to that of most other specific inputs, and structural changes that reduce labor requirements have clear economic appeal.

Economic theory provides a generally accepted rationale for substitution of capital for labor within the framework of *given* technological alternatives. However, sophisticated theorists—Fellner (1962), Kennedy (1964), von Weizsäcker (1966), Samuelson (1965)—are still seeking to clarify the relation between *invention* of laborsaving techniques and the rising relative price of labor. To explain the course of invention in terms of prices of inputs, it is necessary to postulate a production function for invention itself—relating specific amounts of "inventive activity" to savings in particular inputs. It does seem, however, unrealistic to view invention as such a structured process that creative effort can be finely tuned toward saving more of one factor or another with changes in their relative prices.

Over the long run, rising trends in wages relative to other costs appear to orient research toward saving labor, and the same is true for other factors with rising relative prices. It is, however, difficult to express this idea rigorously in quantitative terms. A new process that uses even more labor than the old may be economic if savings of other factors are sufficiently large. How can we anticipate the concrete form that a new invention will take? In practice, of course, it is hard to draw a line between invention and the selection among "available" alternatives for development and application. The latter—a more traditional form of substitution—is certainly an important element in observed technical advance.

The rising relative price of labor can itself be viewed as a direct consequence of structural change. Technologies of producing most commodities have been improving; the "structure of households" has not—nor, of course, should we wish it otherwise. If a fundamental objective of the economy is to furnish a larger and larger product to consumers, then a "deterioration" of the household column, that is, an increased standard of living or more consumption per unit of services contributed by households, constitutes progress. The more successful the system is in accomplishing this objective, the greater will be the rise in the price of labor relative to other products, and the greater the incentive to substitute other inputs for labor. This is just another way of restating the familiar problem of maintaining full employment in a society of increasing affluence.

Certainly the classical economists faced this problem and that of the role of capital in relation to it (see Ricardo 1911:263–271 and Salter 1960:35–38). Capital goods are essentially stocks of the outputs of endogenous sectors, and their costs of production are affected by the same structural improvements that affect the production of all intermediate inputs. Therefore, it stands to reason that the prices of capital goods will not rise proportionally to the price of labor in the course of technical progress. Nor is there any reason to expect trends in interest rates to shift the advantage in favor of labor-intensive processes. Thus, the dominance of laborsaving changes in structure may not be mere historical accident but also a systematic consequence of the basic orientation of the economic system.

Chapter 9 Structural Change and Prices

9.1 Adaptive Changes in Intermediate Input Structures

Chapter 8 completed the broad survey of structural changes affecting intermediate, labor, and capital requirements. Given this information, we are in a position to undertake a more systematic analysis of how savings in primary factors have come about. Over the periods 1939–1947 and 1947–1958, direct labor coefficients were declining for most sectors. Capital coefficients were also declining but not as rapidly as labor coefficients. From the point of view of the economy as a whole, direct primary factor savings can be supplemented (or offset) by indirect economies (or diseconomies) achieved through changes in intermediate input structure. Primary factor requirements are reduced indirectly when individual sectors decrease their purchases of intermediate inputs. On the average, total intermediate purchases have tended to remain constant or even to increase (see Chapter 4), but there is variation among sectors. Primary factors are also economized when individual sectors increase purchases of those inputs whose primary factor content is decreasing more, at the expense of those whose primary factor content is decreasing less, over time. This last type of primary factor saving involves adaptation of structural change in any given sector to changes elsewhere in the system, and therefore it may be called adaptive structural change.

This and the two following chapters are concerned with gauging the economic importance of adaptive change in relation to direct primary factor economies. To what extent do the new structures of individual sectors tend to favor those inputs that have been undergoing most improvement in their own input structures? Is the choice of new techniques in one sector significantly affected by differential rates of progress in others? These are really very broad questions, and only partial answers can be found with the evidence at hand. They tie in, at several points, to other current lines of thought about the economics of technical change. Thus, at one level, we are asking how important are "trigger effects," that is, changes in structure induced by new developments elsewhere in the system (Simon 1951). To achieve systematic empirical identification of all trigger effects would require a comprehensive catalogue of the many structural alternatives available at a given time, whereas, in fact, there is just barely enough information to describe a single average structure corresponding to the mix of techniques actually in use in each sector at each point. Still, it is possible to find some answers to this question through the analysis presented in Chapter 11.

The scope of adaptive change can extend well beyond that of choice among given known alternatives, into the theoretically murky area of invention economics. To the extent that economic influences guide the rate and direction of inventive activity, technological change in one sector may encourage invention in others. Improvements in techniques of producing plastics induce the makers of household appliances or motor vehicles to search for new ways to use them; pressures to develop the technology of nuclear power generation are not unrelated to supply conditions for petroleum and natural gas. Schmookler (1966) and Mansfield (1968) give systematic evidence that expected gain exerts a strong influence on the allocation of research efforts and on the volume of inventive activity. Nelson et al. (1967:33) support this point of view, "... efforts to advance technology will tend to be drawn toward reducing cost and increasing product performance in industries and classes of products where demand is rising, and toward saving on factors whose relative cost is rising." If invention as well as innovation can be geared to developments in other sectors, the distinction between substitution and the development of new technologies is blurred (see Section 8.8). Where a large portion of technical modifications is generated by purposive searching, the Schumpeterian distinction between technological change and substitution is not easily drawn. In evaluating the importance of adaptive change, one is inevitably asking about the sensitivity of technological development as well as of traditional economic choice to progress in other sectors.

It is one thing to acknowledge that economic factors influence the allocation of research effort; it is quite another matter to attempt to predict the specific character of new technologies. "Inventions which have not been made, although there is obvious need for them, are much more numerous and just as striking as those which have. It seems quite as hazardous to try to anticipate the broad sweep of innovation as to spot future specific inventions" (Jewkes et al. 1958:229). Schon (1967) points out that the course of technical development has many twists and turns. If invention establishes new combinations and relationships, then it is difficult to conceive of predicting what they will be, short of preinventing them. Available information does not permit distinction between substitution and invention here, but the difference remains meaningful and important in other contexts.

In fact, data are just barely sufficient to support systematic analysis of adaptive change in general, and partial impressions must be pieced together.

We begin by examining the gross superficial evidence on interrelations among structural changes and prices of individual inputs. Under broadly competitive assumptions, changes in direct-plus-indirect factor requirements for any product will change its price. If the price system conveys information on the changing technical structure of the system with reasonable accuracy, then the importance of adaptive change can be judged by responses to changes in the relative prices of specific inputs.

Because input-output data are most reliable and detailed for 1947 and 1958, the analysis is confined to those two years in this and the next two chapters. With minor exceptions, the findings are presented and discussed at 76 order.

9.2 Structural Change and the Price System

The first order of business, then, is to appraise, for given structural change, the correspondence between changes in prices computed with competitive assumptions and price changes actually observed. The interdependence among prices, input structures, and the costs of exogenous inputs (value added) is summarized in the familiar input-output accounting identity:

$$\mathbf{p} - \mathbf{pA} = \mathbf{v} \quad \text{or} \quad \mathbf{p} = \mathbf{v}(\mathbf{I} - \mathbf{A})^{-1} = \mathbf{vQ} \tag{9.1}$$

where

$\mathbf{p}$ = (row) vector of prices of the outputs of each sector
$\mathbf{A}$ = matrix of input-output coefficients
$\mathbf{Q}$ = Leontief inverse
$\mathbf{v}$ = (row) vector of value added per unit of output in each sector

By definition of value added, the cost of all the inputs per unit of output of a sector must add up to its price. In a more detailed formulation, value added might be expressed as a matrix $\mathbf{V}$ of different kinds of labor inputs, capital charges, and profits:

$$\mathbf{V} = \begin{bmatrix} l_1^1 w_1^1 & l_2^1 w_2^1 & \cdots & l_n^1 w_n^1 \\ l_1^2 w_1^2 & & & \\ \vdots & & & \\ l_1^k w_1^k & l_2^k w_2^k & & l_n^k w_n^k \\ b_1^1 c_1^1 & b_2^1 c_2^1 & & b_n^1 c_n^1 \\ \vdots & \vdots & & \vdots \\ b_1^n c_1^n & b_2^n c_2^n & & b_n^n c_n^n \\ r_1 & r_2 & & r_n \end{bmatrix} \tag{9.2}$$

where

l_j^h = labor input coefficient for labor of skill h in industry j $(h = 1, \ldots, k)$
w_j^h = wage of labor of skill h in industry j
b_j^m = the capital coefficient for capital of type m in industry j (when capital requirements are specified by industry of origin, n types are distinguished, that is, $m = 1, \ldots, n$; and the different types of capital inputs may, alternatively, be thought of as elements of the capital coefficient matrix **B**)
c_j^m = rate of return on capital of type m in industry j (it consists of interest plus depreciation)
r_j = profit and other nonwage income paid out by industry j, per unit of its output (operationally, it is often difficult to distinguish profits r_j from return on capital; but this problem need not concern us directly)

Equation 9.1 now becomes

$$\mathbf{p} = (1, 1, 1 \ldots)\mathbf{VQ} \tag{9.3}$$

Changes in product prices result from changes in any of the variables—in the elements of **A**, the endogenous input structures, or in **V**, the value added rows. Equation 9.2 specifies that the value added rows have many components: labor and capital of many kinds and their respective prices and profits. Changes in product prices can result not only from changes in total labor requirements but also from changes in the skill mix or in the wage differentials among sectors. Similarly, changes in the kinds and prices of capital goods, as well as total capital coefficients, and changes in profit levels in any given sector will affect the system of prices as a whole.

Within this framework, one can ask to what extent the price system has reflected the structural changes already described. How well has information about changes in the technology of any given sector been transmitted through the price system to others? From equation 9.2 it should be clear, however, that price changes stem not only from structural change but also from changes in the detailed composition of value added. Some of these changes in value added composition are known, but not all the requisite detail is available. In particular, for 1947 and 1958, only total labor requirements are known; they are not disaggregated by level of skill. Wage statistics have not been made entirely comparable, and changes in industry wage differentials cannot be considered known. Profits and miscellaneous income by individual sectors are not given. Known or not, we treat pricing of value added components as exogenous. What part of observed price change is explained by changes in the "real" elements of capital, labor, and intermediate structure that we have already surveyed? These changes are already expressed in physical (1947 dollars' worth) units. We can compute the hypothetical price effects of these changes with fixed primary factor prices. Before these theoretical prices can be computed, however, it is necessary to specify interest and depreciation rates to convert capital requirements from stock to annual cost terms.

Specification of capital costs: replacement and interest charges

Complete capital coefficient matrices have been estimated for both 1947 and 1958, but the problem of specifying appropriate rates of return, that is, of fixing realistic values for c_j^m, remains to be settled. For present purposes, the unit cost of each item of capital is divided into two components: a replacement element and an interest charge. Since replacement is "postponable" and difficult to distinguish from other components of gross capital formation, it is counted in the gross capital formation vectors rather than with current inputs in the published input-output matrices. From the point of view of individual establishments and sectors, however, replacement is a recurring, nonnegligible cost item. For the analysis in this and the next two chapters, where cost-price comparisons of alternative structures are stressed, replacement requirements are incorporated into the set of current input coefficients. In Chapter 12, where replacement will be viewed explicitly as a vehicle of embodied technological change, it is removed, once again, from the current account matrix. Replacement is discussed in more detail in that chapter.

Replacement rates are based on estimates of average useful lives for different equipment items. Depreciation rates were applied to various types of equipment in each sector in the capital stock matrices in order to yield crude approximations of replacement costs at constant (1947) prices. Annual capital replacement coefficients were then added into the conventional flow coefficient matrix. Depreciation rates allowable by the Internal Revenue Service tended to increase over the period 1947–1958, and there is some evidence that this reflects a real tendency in the direction of shorter equipment lives. However, this shortening of equipment lives is, in large part, a matter of accelerated obsolescence rather than of decreased durability. Since the replacement estimates here are intended to measure the real costs of maintaining a given capital stock, rather than those of improving it, the estimated percentage rates of replacement for given types of capital were held constant over the period 1947–1958 for all but a few items. The replacement matrices are given in Appendix D.

The choice of an appropriate interest rate component of capital charges also involved some judgment. While all economists acknowledge the importance of interest as a capital price, it is difficult to obtain a consensus as to what the relevant interest rate actually is at any given time. Rates of 3 percent for 1947 and 4 percent for 1958 are cited in Hickman (1965:253). The present price computation assumes 1947 prices for value added components, and therefore a single interest rate of 0.03 was applied for both years. Sensitivity tests described in Chapter 10 show that the particular interest rate assumed can be varied within a fairly wide range without changing any major conclusions to be drawn from the computations.

Combining labor and capital into total factor input

The necessity of specifying value weights for capital and labor follows from the open static framework adopted for the present work. In a closed dynamic formulation where both labor and capital goods are themselves treated as products of the system, prices will depend on the current account and capital coefficient matrices; a unique equilibrium rate of interest or return to capital will be implicit in the specified capital and current account structural parameters (see Brody 1966; Solow 1959; and Morishima 1958). In such a context, relative returns of labor and capital are determined endogenously. The static approach used here truncates the more comprehensive analysis by treating labor and capital as exogenous inputs. When labor and capital are exogenous, their relative returns cannot be computed but must be specified as data. A vector of primary factor input coefficients $\mathbf{f}^t$ is defined as the sum of labor and capital coefficients, weighted by their respective prices:

$$\mathbf{f}^t = \mathbf{l}^t\hat{\mathbf{w}}^t + \alpha^t\mathbf{b}^t \tag{9.4}$$

The wage and interest rates used to weight the two primary factor inputs must be chosen appropriately for the problem at hand. The problem of assigning relative weights to the two major primary factors appears quite generally in comparative static analysis (see Kendrick 1961:284–285 and Denison 1962:140). The same necessity to weigh exogenous labor and capital inputs persists throughout this study, as these two primary factors must be assigned relative weights in the objective functions for linear programming and other computations that compare real cost of alternative structures.

Computed and observed price changes

Given the necessary value weights for labor and capital in value added, prices with 1947 and 1958 physical input structures were computed:

$$\mathbf{p}^{58} = (\mathbf{l}^{58}\hat{\mathbf{w}}^{47} + 0.03\mathbf{b}^{58})\mathbf{Q}^{58} \tag{9.5}$$

and

$$\mathbf{p}^{47} = (\mathbf{l}^{47}\hat{\mathbf{w}}^{47} + 0.03\mathbf{b}^{47})\mathbf{Q}^{47} \tag{9.6}$$

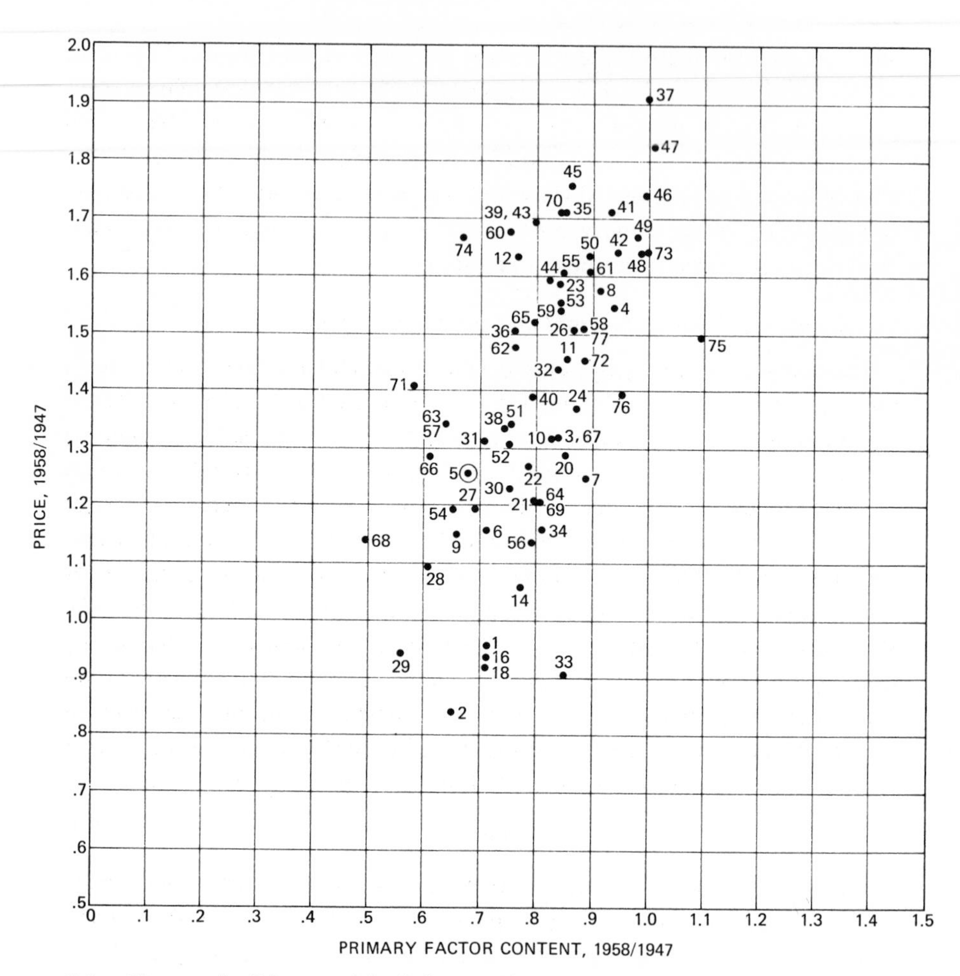

9.1 Changes in Prices and in Primary Factor Contents for 76-Order Sectors, 1958/1947 (for circled points, multiply scales by 2). Industry list, next page.

where $\mathbf{w}$ is a vector of 1947 industry average wage rates; 0.03 is the average long-term interest rate; and replacement coefficients are included in the $\mathbf{A}$ matrices, that is, in $\mathbf{Q}$. Labor and capital are the two primary factors, and $\mathbf{p}^{58}$ and $\mathbf{p}^{47}$ are really measures of total factor content with 1947 factor price weights.

The ratios of prices computed for each sector, p_j^{58}/p_j^{47}, are compared with the corresponding actual price ratio for the two years in figure 9.1. If all the elements of value added were specified with actual prices for both years and

(1) Livestock
(2) Crops
(3) Forestry & fishing
(4) Agric. services
(5) Iron mining
(6) Nonferrous mining
(7) Coal mining
(8) Petroleum mining
(9) Stone & clay mining
(10) Chemical mining
(11) New construction
(12) Maintenance constr.
(14) Food
(15) Tobacco
(16) Textiles
(18) Apparel
(20) Wood & products
(21) Wooden containers
(22) Household furniture
(23) Office furniture
(24) Paper & products
(26) Printing & publishing
(27) Basic chemicals
(28) Synthetic materials
(29) Drugs, soaps, etc.
(30) Paint
(31) Petroleum refining
(32) Rubber prod., etc.
(33) Leather tanning
(34) Shoes
(35) Glass & products
(36) Stone & clay prod.
(37) Iron & steel
(38) Nonferrous metals
(39) Metal containers
(40) Heating, etc.
(41) Stampings, etc.
(42) Hardware, etc.
(43) Engines & turbines
(44) Farm equipment
(45) Constr. & mining eq.
(46) Materials hand. eq.
(47) Metalworking eq.
(48) Special ind. eq.
(49) General ind. eq.
(50) Machine shop prod.
(51) Office & comp. mach.
(52) Service ind. mach.
(53) Electrical apparatus
(54) Household appliances
(55) Light. & wiring eq.
(56) Communication eq.
(57) Electronic compon.
(58) Batteries, etc.
(59) Motor vehicles & eq.
(60) Aircraft
(61) Trains, ships, etc.
(62) Instruments, etc.
(63) Photo. apparatus
(64) Misc. manufactures
(65) Transportation
(66) Telephone
(67) Radio & tv broad.
(68) Utilities
(69) Trade
(70) Finance & insurance
(71) Real estate & rental
(72) Hotels & pers. serv.
(73) Business services
(74) Research & dev.
(75) Auto. repair
(76) Amusements, etc.
(77) Institutions
(80) Noncomp. imports
(81) Bus. travel, etc.
(83) Scrap

---

if observed structural change were guaranteed error free, equation 9.1 would be a tautology; and computed and actual prices would *have* to be equal. However, in the present computation (equations 9.5 and 9.6), only the structural coefficients for labor, capital, and intermediate inputs (**l**, **b**, and **A**) were changed between 1947 and 1958. Changes in wages and interest rates were not introduced; and profits, taxes, and other nonwage income were ignored entirely in the computation. Thus the figure tells, sector by sector, to what extent actual price changes are explained by observed structural change alone.

Certainly, there is a positive association between the observed and computed price changes. Agriculture (1) and (2), utilities (68), textiles (16), apparel (18), and drugs, soaps, and cosmetics (29) are among the sectors with greatest decreases in primary factor content. Their actual prices increase least (most actually decrease) between 1947 and 1958. Iron mining (5), iron and steel (37), and some of the heavy industrial equipment sectors (46), (47),

(49) show little or no decreases in total primary factor requirements and show relatively large price increases. Differences between actual and computed relative price changes are explained by changes in primary factor pricing and missing elements of value added and by errors in the data on structural change. The general price level rose between 1947 and 1958; most relative prices, 1958/1947, were actually greater than 1, but computed relative prices were generally less than 1. Thus, the relationship between computed and actual relative prices has a general price level intercept. Presumably computed prices could be normalized to the same average level as actual prices by specifying the 1947–1958 rise in average prices of primary factors, especially in wage rates.

When the technology of an industry changes, many possible price reactions may be expected, alternatively or in combination. Its product price may change; its wage structure, return on capital, or profits may all be altered. In practice, structural change does not invoke identical pricing responses in all sectors. Industries do not all increase their wage payments proportionally to man-hour requirements, nor are capital payments or profits affected uniformly in all sectors. Since wages constitute by far the largest portion of value added, much of the residual scatter is probably explained by changing skill mixes and changing interindustry wage differentials.

The character of competition within a sector affects not only its wage and price response to a given structural change, but also the rate at which new technologies are introduced and the choice among available structural alternatives. It is tempting to speculate about the role of competitive and oligopolistic elements in explaining some of the dispersion in figure 9.1; that is, to note the fact that agriculture (1) and (2), textiles (16), and clothing (18) prices rose little (decreased) while steel (37), glass (35), engines and turbines (43), construction and mining machinery (45), aircraft (60), and utilities (68) prices rose more, compared with "theoretical" expectations. To what extent do changing profit margins offset the effects of changing techniques on the price of particular sectors? Certainly, industries differ in their price and output reactions over the business cycle. However, without more accurate and explicit information on the changing composition of value added, it would be impossible to distinguish among these various relevant factors. In terms of figure 9.1, how accurately does the price system convey information on changing primary factor requirements from sector to sector? Correspondence is far from perfect; on the whole, prices rose least for those products with greatest structural improvement, and most for those where total factor economies were smallest (as measured here). Price changes are at least rough indicators of changing primary factor content.

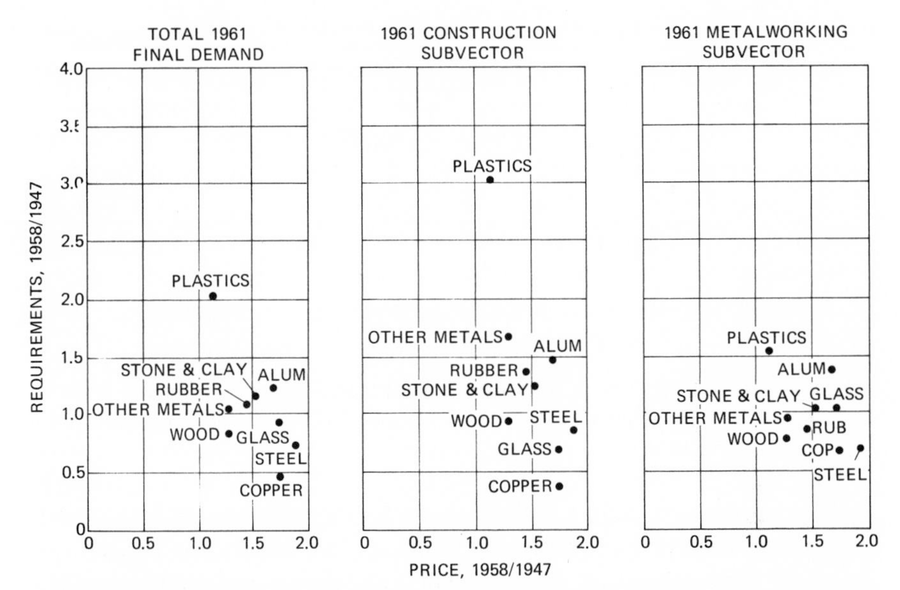

9.2 Changes in Prices and in Requirements for Specific Materials to Deliver Total 1961 Final Demand and Two Specific Subvectors of 1961 Final Demand, 1958/1947.

9.3 Evidence of Quasi Substitution

The next question concerns quasi substitution: Did structures shift so as to make greater use of those inputs whose prices rose least and smaller use of those inputs whose prices rose most over the period 1947–1958? Some evidence is presented in figures 9.2, 9.3, and 9.4. Each figure concerns a particular group of inputs judged to be substitutable on a priori grounds: materials, energy sources, and a group of metalworking components that are important contributors to intermediate, and not just final, markets. The horizontal axes measure relative prices of particular inputs, 1958/1947; the vertical axes measure relative use of these inputs to produce a standardized final demand with the overall input-output structure of 1958, as compared with that of 1947. The figures show changes in intensity of use of specific inputs in the economy with changes in their prices. The first graph in each figure shows changing requirements to deliver the entire 1961 final demand. The next two describe changing requirements for delivering the construction and metalworking (combined machinery, electrical and service equipment, and transportation equipment) subvectors of final demand. In each of these graphs, the vertical axis measures the ratio of direct-plus-indirect requirements for 1958 to those for 1947.

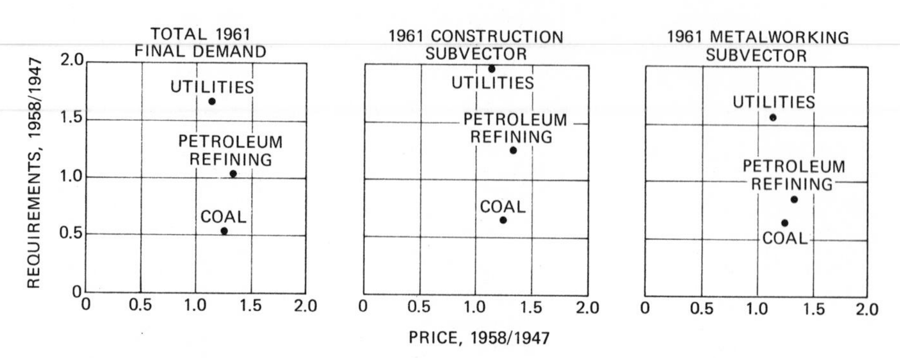

9.3 Changes in Prices and in Requirements for Specific Energy Inputs to Deliver Total 1961 Final Demand and Two Specific Subvectors of 1961 Final Demand, 1958/1947.

Most of the graphs show a slight tendency for those inputs with the smallest price rises to increase in relative importance. This tendency is clearest for materials. Plastic shows the greatest relative increases in use and the smallest increase in price; the greatest decreases in use and the greatest price increases are shown for iron and steel and for copper. Changes in requirements for materials and for metalworking components are different for different subvectors of final demand; energy inputs are general inputs and change more uniformly for different subvectors.

Some question should be raised as to the "direction of causation." Have changes in relative prices of inputs induced substitutions? Or have autonomous increases in the use of some inputs led to falling relative prices through greater economies in their production—economies with fuller utilization of capacities, economies of scale, or the increased relative importance of newer facilities with "best practice" technologies? Undoubtedly both tendencies are simultaneously at work. Some quasi substitution is apparent for each of the individual final demand subvectors. Although figures 9.2 to 9.4 offer some evidence of an inverse relation between relative price change and relative utilization of specific inputs, there is considerable dispersion, and the tendency is hardly striking. Standards for judging whether the amount of substitution, or quasi substitution, was large or small are not established until Chapter 11. There, it becomes clear that quasi substitution is present but that the contributions of changes in intermediate inputs to total factor economy are small relative to those of direct labor.

The cost of using more of one particular input rather than another depends not only on their relative prices, but also on requirements for complementary inputs and on the prices of those other inputs. The advantage of plastic is

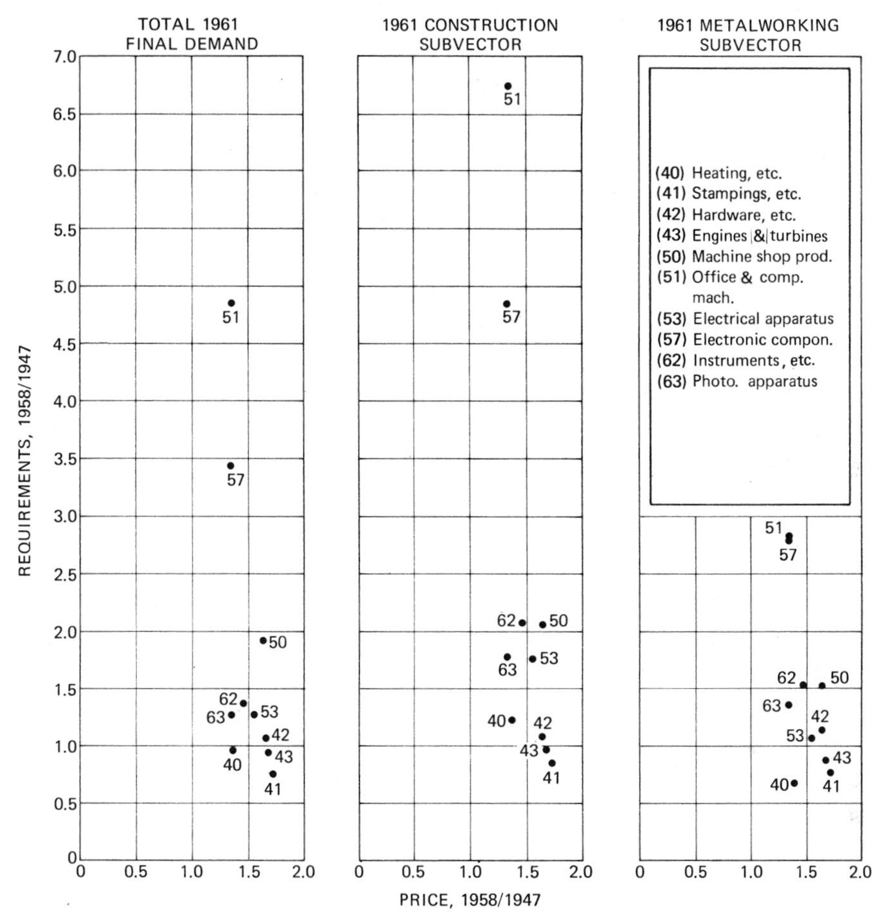

9.4 Changes in Prices and in Requirements for Specific Metalworking Components to Deliver Total 1961 Final Demand and Two Specific Subvectors of 1961 Final Demand, 1958/1947; 76 Order.

not just its price but also the saving of labor that comes with its use. Figures 9.2 to 9.4 show the net effects of changes in constellations of coefficients on specific input requirements, rather than isolated interactions among the specified inputs alone.

Sometimes it is necessary to consider the absolute prices of the inputs in question along with their rates of change. The price of a given input may have to fall below a certain threshold before its substitution for other inputs is warranted. Certainly this is important in understanding the fuel substitution picture. In terms of changing primary factor contents and prices, for example, coal-producing techniques improved more rapidly than those for producing refined petroleum over the period 1947–1958. Yet the relative share of petroleum continued to increase, while that of coal continued to decrease in industrial consumption. Even though coal costs and prices were falling, they had not fallen enough to offset the other advantages—low transport and storage costs and fewer air pollution problems—of using other forms of energy. Landsberg and Schurr (1968: 81–89) point out that, since 1958 further improvements in coal-producing efficiency and new techniques of transporting and handling coal have brightened its competitive prospects.

Improvements in agricultural productivity were even greater than improvements in synthetics production, and relative price changes reflect this. Nevertheless, synthetics gained markedly at the expense of agricultural products in the textile and clothing markets. The apparent price insensitivity observed here may be a result of aggregation, as it is not clear that cotton and wool production gained as rapidly in efficiency as agriculture as a whole. In addition, there are savings in cost to be achieved in processing synthetic, as compared with natural, fibers and familiar qualitative product advantages for which the consumer is willing to pay. Similarly, aluminum and plastic automobile components gain economic advantage not merely because they are becoming cheaper, but also because they are lighter in weight and thus permit economies in other inputs within the framework of overall design considerations. Relative advantages of sets of complex alternatives must be weighed, and the price of a particular input represents only one out of many considerations. This is, after all, what one expects substitution to be like in the real world.

Chapter 10 Old and New Structures as Alternatives: Optimal Combination of 1947 and 1958 Technologies

10.1 General Objectives of Chapters 10 and 11

Two tendencies toward primary factor economy have just been described: (1) reduction in direct requirements and (2) adaptive change. The next step is to establish a clearer idea of their relative importance. Observed 1947 and 1958 structures for each sector are considered as alternatives, and total factor requirements using different combinations of old and new structures are computed and compared. We begin by verifying the superiority of the set of 1958 structures to the 1947 ones, through a simple ex post linear programming analysis: computing the optimal combination of 1947 and 1958 structures. With a few quite plausible exceptions, 1958, rather than 1947, structures are chosen for all industries. The choice of an optimal mix of activities or structures depends, of course, on the specific objective function used as a basis of choice. We go on, then, to ask whether the composition of the optimal activity vector is sensitive to changes in interest rates and wage structure, within a reasonable range. It is not; the superiority of 1958 to 1947 structures stands firm with a shift from 1947 to 1958 wage structure and with hypothetical changes in interest rates from 0 to 15 percent.

Is the advantage of 1958 structure over 1947 in each sector sensitive to structural choice in the others? This question is not answered by routine linear programming techniques. They are too efficient in that they attack the problem of structural choice in all sectors simultaneously. Instead, we set about to consider explicitly some of the inefficient combinations of activities that are eliminated automatically in programming algorithms. We form many hybrid matrices—hypothetical economies with 1947 structures in some sectors and 1958 in others—and compare their efficiencies. Comparisons of total factor saving advantages of introducing individual new techniques separately or simultaneously help to evaluate the importance of adaptive change in the overall picture. Linear programming and sensitivity tests are presented in Chapter 10 and hybrid-matrix computations, in Chapter 11.

10.2 Integrity of Column Structures

How much of the 1947–1958 change in total labor and capital requirements to produce a given final demand can be attributed to observed shifts in direct labor and capital coefficients, and how much to reductions in intermediate inputs and adaptive change? Since changes in direct labor and capital coefficients are large and pervasive, as compared to changes in intermediate coefficients, it is tempting to jump directly to the conclusion that changes in intermediate structure do not matter in a rough appraisal. One could thus seek a quick answer to this question by holding intermediate structure constant and varying the labor and capital coefficients. This procedure is followed in Leontief (1953) although it is not central to the analysis there. In fact, the results of the computation, presented here in table 8.2, can be interpreted as just this kind of approach. It shows that, as labor and capital coefficients changed, total labor and capital savings were similar, although not identical, regardless of which year's intermediate input structure was assumed. Thus, the net effect of changes in intermediate input structure was negligible in the aggregate.

This picture can, in fact, be deceptive for two major reasons. First, economy-wide factor requirements may be stable with respect to intermediate structure, while sectoral requirements are not (see Chapter 8). Second, observed direct economies of primary factors in each sector might not have been possible without the changes in intermediate input structure that accompanied them. Could direct labor coefficients have been reduced without increased inputs of purchased services or the changed division of labor among fabricators? Is it possible to separate increased electricity consumption from automation? What part of the materials substitutions were motivated by cost saving within the materials budget, and what portion by concomitant savings in labor and capital with changing product design? Each sector did indeed operate with the sets of factor proportions observed for given years. Whether it might have been able to do so with other hypothetical sets must either be settled by expert judgment or remain a matter of speculation. A hybrid coefficient *column* that is composed of some coefficients for one year and some for another is not necessarily a workable technological structure.

Thus, it seems important to respect the integrity of observed column structures and not to attempt to alter them piecemeal, except with the support of additional technological analysis. In the computations that follow, the input-output structure of the economy will be varied hypothetically by substituting the column structure of one year for that of another but not by varying individual elements separately. It is meaningful to ask about the impact of using 1947, instead of 1958, factor proportions for producing, say, steel. The interpretation of 1947 intermediate input structure with 1958 labor coefficients is less clear.

One might also argue for recognizing technological interdependence among changes in input structures of different sectors. For example, the input structure of the radio, television, and communications equipment sector in 1958 requires appropriate product mix in the electronic components sector; 1958 structure in the former may call for 1958 structure in the latter. With changing product qualities, the technological feasibility of combining input structures observed for one year in particular sectors with those of another in remaining sectors becomes questionable. The following analysis does not take into account such technological ties among changes in *different* sectors. Essentially, changes in product quality are disregarded. This makes it technologically permissible to mix observed sectoral input structures of different years. In our hybrid matrices, some columns represent the technologies of one year and some of another.

10.3 Optimal Mix of 1947 and 1958 Input Structures

It is generally taken for granted that technological change means economic progress, that the structures observed for a later date are superior to those observed for an earlier one. Now let us test this proposition. Assuming that no information was lost during the period, the input structures of 1947 and 1958 are technological alternatives in 1958. We begin with an ex post programming computation that finds the optimal combination of 1947 and 1958 input structures. This provides a convenient framework for judging to what extent the evolution of technology can be explained in terms of primary factor economies. In this context, sensitivity of technological choice to changes in prices of primary factor inputs is also evaluated. The linear programming formulation is simply to minimize

$$\nu = \mathbf{f}^{47}\mathbf{x}^{47} + \mathbf{f}^{58}\mathbf{x}^{58} \tag{10.1}$$

subject to

$$(\mathbf{I} - \mathbf{A}^{47})\mathbf{x}^{47} + (\mathbf{I} - \mathbf{A}^{58})\mathbf{x}^{58} \geq \mathbf{y}^{58} \tag{10.2}$$

where

ν = total factor requirement, measured in 1947 dollars' worth of combined labor and interest charges

$\mathbf{f}^{47}$ and $\mathbf{f}^{58}$ = vectors of total factor input coefficients, computed in accordance with equation 9.4 and based on 1947 and 1958 man-year coefficients, 1947 wage structure, 1947 and 1958 capital coefficients, and interest charges of 3 percent

$\mathbf{y}^{58}$ = 1958 final demand

$\mathbf{x}^{47}$ and $\mathbf{x}^{58}$ = vectors of output produced with 1947 and 1958 technologies, respectively

$\mathbf{A}^{47}$ and $\mathbf{A}^{58}$ = 1947 and 1958 coefficient matrices, including replacement coefficients

Since total factor input enters as a single primary factor, the optimal solution associates a nonzero activity level with either the 1947 or the 1958 input structure, but not both, for each industry. The level and composition of assumed final demand does not affect the choice of optimal activities (see Samuelson 1951).

The following fourteen sectors (76 order) are those where 1947 structures were chosen in the linear programming computation.

(4) Agricultural services
(5) Iron mining
(8) Petroleum mining
(33) Leather tanning
(37) Iron and steel
(41) Stampings, screw machine products, and fasteners
(42) Hardware, plating, valves, wire products
(46) Materials handling equipment
(47) Metalworking equipment
(48) Special industry equipment
(49) General industrial equipment
(73) Business services
(75) Automobile repair
(76) Amusements and recreation

Table 10.1 is a comparison of total labor and interest charges using the optimal combination, with requirements using only 1958 and only 1947 activities. With 1958 technology in all sectors, the economy was capable of delivering 1958 final demand with a 22 percent lower total factor cost than with 1947 technology in all sectors. Only a 2 percent additional saving would have been achieved by retaining 1947 input structures for the fourteen sectors.

The list of industries where 1947 technologies were chosen is of special interest. It identifies sectors where structural change actually detracted from

Table 10.1 Total Labor Cost and Total Interest Charges to Deliver 1958 Final Demand with 1947, 1958, and the Optimal Combination of 1947 and 1958 Structures (millions of 1947 dollars)

	Input structures			Differences	
	1947	1958	Optimal mix		
	(1)	(2)	(3)	(1) − (2)	(2) − (3)
Total labor cost	$176,685	$136,030	$134,185	$40,655	$1,845
Total interest cost	17,114	14,339	13,805	2,775	534
Total cost	$193,799	$150,369	$147,990	$43,430	$2,379

the overall productivity of primary factors. Compare the list of sectors preferring 1947 technologies with the list of industries showing increasing direct-plus-indirect labor requirements between 1947 and 1958 in figure 8.2. Of the fourteen industries cited, only three—iron mining, materials handling equipment, and automobile repair—showed actual increases in labor required per unit of final demand.* This fact helps clarify the meaning of the linear programming results. Changes in direct-plus-indirect factor requirements per unit of final demand, discussed in Chapter 8, measure improvement, in the system as a whole, in delivering each particular final demand item. The linear programming computation shows that the system would have delivered a fixed bill of final demand (and, actually, any bill of final demand) with even less primary factor input if 1947 technology had been retained instead of that of 1958, in the particular sectors cited. The linear programing computation is, in fact, based on total factor economies, while figure 8.2 concerns labor alone. However, Section 10.4 will show that the optimal choice of structures is hardly changed when capital inputs are disregarded.

For some sectors, the choice of 1947 technology makes apparent good sense. First, in industries that depend directly on scarce natural resources, the "old" technology may not be a real alternative. Take iron mining: exhaustion of the best Mesabi iron-ore mines made it progressively more difficult to extract a given amount of iron between 1947 and 1958. One would expect, therefore, to find 1958 structure inferior to 1947 for this sector. By 1958, compensatory innovations, particularly beneficiation of ores, had been introduced in reaction to this specific deterioration of the nation's resource

* The material in Chapter 8 is presented in terms of the 38-order, rather than the disaggregated 76-order, classification used here. However, the computations for Chapter 8 were performed at 76 order as well, providing the basis for the present comparison.

position. While these innovations were useful, they were apparently insufficient to offset the basic loss. A similar situation existed in petroleum mining, where improved discovery and extraction techniques seemed not quite able to compensate for the need to drill deeper wells. There is some doubt as to the exact balance between changing techniques and resource conditions here. Landsberg and Schurr (1968:91–94) and Schurr and Netschert (1960:370–380) discuss the problem of drilling depths in some detail. In any case, it seemed more realistic to fix 1958 structures as the only feasible alternatives in the mining sectors. The linear programming computation was rerun without the option to use 1947 structures in mining. This limitation did not affect the choice of optimal technologies in other sectors, although it did produce a small increase in total factor requirements to produce the 1958 bill of final demand.

The superiority of 1947 technology for other sectors should not always be taken literally. Consider steel: although new labor-, fuel-, and capital-saving techniques became available for steelmaking during the 1950s very little new capacity employing the new techniques was introduced before 1958 (see McGraw-Hill 1960:93–102). Thus, direct improvements in steelmaking productivity were very small between 1947 and 1958. Two factors tip the apparent balance in favor of 1947 structure. The first is the slightly higher ratio of scrap to ore consumption in the 1947 table. In preliminary versions of this linear programming computation, scrap was treated as a zero-cost by-product. Under that assumption, a process using more scrap, relative to pig iron, would naturally register a cost advantage over a process using less. In the final version, reported here, the purchase cost of scrap was taken into account. This change did not significantly alter the relative advantage of 1958 and 1947 structures. The second, probably overriding consideration was an upgrading in the iron and steel sector's product mix, not wholly taken into account by the 1958/1947 price deflator.

Similar explanations apply for most of the other fourteen sectors cited earlier. Two early metalworking sectors—stampings, screw machine products, and fasteners (41), and hardware, plating, valves, and wire products (42)—and heavy machinery sectors—materials handling equipment (46), and other industrial equipment (47), (48), (49)—registered only minor direct gains in labor or capital productivity over the period. At the same time, their near-diagonal purchases—purchases of components from other closely related metalworking sectors—and general inputs were increasing. The net effect is apparent superiority of the 1947 structures. From all that has been said thus far, it should be clear that these were not among our most dynamic sectors. However, to characterize their structural change as "deterioration" is probably going too far. More conservatively, apparent progress was not sufficient to counterbalance statistical discrepancies and upgrading of the product mix. Note that 1947 structures are favored over 1958, both for iron and steel itself and for many of the major steel-intensive metalworkers. Here is further argument for explaining relative decline of the material, steel, in terms of sluggish progress in fabrication methods as well as in the production of steel itself.

Along similar lines, apparent superiority of 1947 structures for some service sectors undoubtedly depends on qualitative change in their outputs. Leather tanning (33) is a declining industry whose structure changed little between 1947 and 1958. A larger diagonal element in the second year accounts for the apparent structural deterioration, and this difference may well be an accounting discrepancy.

10.4 Sensitivity of Structural Choice to Changes in Wages and Interest Rates

The outcome of any optimizing computation depends on the criterion of optimality, that is, on the objective function. In the linear programming exercise described in Section 10.3, labor and capital charges were combined with particular wage and interest rate weights. The interest rate, in particular, was chosen arbitrarily since it is difficult to judge capital charges from published information (see Section 9.2). However, there is reason to suspect that variations in interest charges, within any reasonable range, have not been an important influence on choice of techniques. A few simple sensitivity tests are useful to show the extent to which the advantage of new over old input structures depends on the specific wage and interest rates assumed.

Structural choice with varying interest rates

The technique used to investigate sensitivity was straightforward. The linear programming system described in Section 10.3 was computed eight times, with interest rates varying from 0 to 15 percent. The results are reassuring. There is hardly any difference in the composition of the optimal vector as interest rates are varied within this range. The list of fourteen sectors where 1947 structure was chosen was based on an interest rate of 0.03 for 1947 and 1958. When the rate is doubled paper and products (24) joins the list. At interest rates of 0.10 for both years, 1947 technology is no longer favored for petroleum mining (8). At 0.15, the list is still the same as it was at 0.03, except for the deletion of sector (8) and the addition of sector (24). Reducing interest rates to 0 shifts favor to 1958 structure for only two sectors: stampings, screw machine products, and fasteners (41) and amusements and recreation (76).

Structural choice with 1947 and 1958 wage structures

With no assurance that available 1947 and 1958 wage information was comparable, the wage coefficient part of total factor input for 1958 was estimated as the product of 1947 wage coefficients and a 1958/1947 index of man-hour requirements per unit of output (see Chapter 8). This is equivalent to assuming that wage differentials among sectors, and skill compositions within sectors, remained fixed over the period 1947–1958. Neither assumption is at all realistic. It is important to ask how alternative assumptions about skill composition and wage structure would affect the optimal mix of 1947 and 1958 input structures.

The linear programming problem was recomputed with 1958, instead of 1947, wage structure. The wage coefficients for 1958 were deflated to the 1947 wage level with a single, across-the-board wage deflator. Then 1947 labor coefficients were estimated by applying each sector's 1947-1958 man-hour index to its "deflated" 1958 wage coefficient. This yielded 1947 and 1958 adjusted man-year coefficients with 1958 wage weights; an interest rate of 3 percent was assumed, and a variant with 15 percent interest rates was also computed. The change from 1947 to 1958 wage structure weights in the objective function did not change the composition of the optimal vector for any set of interest rate assumptions.

Ideally, one would wish to try the computation using 1947 wage structure for 1947, and 1958 wage structure for 1958. This would introduce some implicit allowance for changes in skill intensity in each sector. This could not reasonably be done, since the treatment of unpaid family workers and supplements were not reconciled in the wage vectors for the two years.

Significance of the programming and sensitivity tests

With minor exceptions, the findings of Section 10.3 stand firm with respect to the variations in the objective function just considered. Structures of 1958 are superior to those of 1947 for most sectors; and the superiority of the 1958 structures is not challenged by changes in the interest rate, within a reasonable range. Nor, in general, does the apparent advantage of newer structures rest on special, unrealistic assumptions about interindustry wage differentials. There is no denying the importance of skill requirements in the changing industrial scene. If we assume that all 1958 labor inputs are more skill intensive than those for 1947, the advantage of 1958 over 1947 technology will be narrowed but not eliminated. Capital, too, is presumably upgraded over time. These qualifications should certainly be pursued as information becomes available, but they are not likely to vitiate the present findings. The advantages that are so clear when measured crudely, in terms of undifferentiated man-hours, are not likely to evaporate when the data are refined.

What is the significance of 1958 structural predominance in the optimal vector? Disregarding the layering of old and new structures (to be discussed below), one could argue as follows: Had 1947 and 1958 structures been technological alternatives in 1947, 1958 structures should have been adopted in 1947. They were not adopted because they were not known in 1947. Our findings are presumptive evidence that 1947–1958 differences result from bona fide technological change rather than from simple substitution. The brief excursion into sensitivity analysis reinforces this impression. Structures of 1958 retain their superiority to those of 1947 over a wide range of changes in the relative price of labor to capital. Structural choice was not balanced on a knife edge and *not* sensitive to changes in relative costs of labor and capital, within plausible limits. Moderate changes in wage and interest rates would have changed profit margins, but they would not have given cause for regrets to entrepreneurs responsible for choosing 1958 over 1947 structures. Of course, it is still quite possible that different interest rates and wage structures would have led to different input configurations from those observed either in 1947 or in 1958. Chances are that wage and interest rates work more directly on timing and rates of adoption of a given range of techniques than on kinds of new techniques to be favored.

Structures of 1958 and 1947 are, in fact, averages of structures for different technological layers—for older and newer techniques used side by side in both years. Differences between observed average structures indicate the directions, but not the magnitudes, of differences between older and newer layers. In general, the advantage of the newest structures over the old in 1958 will be even greater than observed differences between "average" structures for the two years (see Chapter 12). However, the sensitivity tests suggest that the advantage of new over older structures is not a matter of "fine tuning."

It is central to our understanding of technological change to find out, in general, how finely tuned technological choices really are. From the business point of view, there are good reasons why fine tuning is out of place. Technological commitment is long term. With heavy investments in equipment and personnel experience, it would be risky to switch to a new technology whose advantage might vanish with small changes in the prices of inputs. To be practical, new techniques should have a high probability of long-term advantage, regardless of short-term fluctuations in primary factor or other input prices. Thus, a new structure must be justifiable in terms of a fair range of input price conditions. In pondering a new technique, it is safe to assume that wages will not fall, that interest rates will be less than 15 percent, and that certain trends affect the costs of intermediate goods. Plastics will become cheaper, copper and petroleum more expensive. Choices that require much more specific knowledge of the future may not seem worth the gamble.

This point of view is not a special facet of business conservatism and inertia. In a broader economic context, this kind of policy is rational. At any given time, there will be some new techniques that are not yet economic (but that may become so if relative wage rates go still higher), and there will be some applications where automation is still too expensive. Thus, Melman (1956:47–57) shows that British factor prices only began to warrant the adoption of certain major laborsaving techniques in the 1950s. American factor prices were at that time well beyond the critical ratio that justified the same changes. Certainly, there were other new technologies available in the United States that were only marginally justified. Given access to the requisite information, one could list structural alternatives in descending order, down to those that would be just marginally economic at current factor prices. These sensitive marginal alternatives never appear at all in our 1947–1958 comparisons. There are two plausible explanations of their absence: 1947 and 1958 are so far apart that the sensitivity of year-to-year changes to factor prices is obscured. What we observe are average, not marginal, differences. A second interpretation, however, is probably more important. Since most change requires investment, there is a limit to the rate at which an economy can incorporate new techniques. Thus, there is always a backlog of structural improvements, ordered in descending priority, to be introduced as resources permit. High on the list are those that are economic for *any* relative factor prices beyond some critical ratio. Lower down on the list will be alternatives that are barely justified with current factor prices. These will be more sensitive to price changes. And even below that, will be some that are still uneconomic, although they may some day prove worthwhile if current price trends continue. Discovery is constantly adding to the choice. The present evidence seems to say that the lower regions of the list are seldom reached. With resources available for growth and changeover, under current conditions, there is always a waiting list of potential changes whose advantage is unequivocal. Their advantage is not sensitive to small changes in input prices. In other words, the system dictates a high cutoff point. Therefore, the new techniques that are actually spreading at any given time do not include all the alternatives that might be economic by comparative cost criteria alone. Some are eliminated by investment constraints that are not subsumed in the market rate of interest.

Chapter 11 Interdependence and Independence of Structural Advantage

11.1 Interdependence of Structural Choice

Chapter 10 touched upon the sensitivity of structural advantage to primary factor prices; this chapter considers how the costs of intermediate inputs affect structural choice. Since sectors produce each others' inputs, the total saving of primary factors effected by structural change in one sector depends on the structures prevailing in the rest of the system. How much labor and capital does the economy save with an improvement in efficiency of producing aluminum? This depends on how much aluminum is used as an input in construction, machinery production and so on. Because there is interdependence, the sum of factor economies due to one-at-a-time structural changes is not generally equal to the effect of simultaneous introduction of the same changes. If aluminum is substituted for steel in automobiles at the same time as efficiency of aluminum production rises, primary factor savings will be greater than the sum saved by introducing both changes separately. Leontief (1953) analyzes changes in total output vectors to deliver a fixed final demand, when structural changes are introduced separately and simultaneously. Effects of separate and simultaneous structural changes on total factor content will be considered below.

Since the advantage of a new input structure in a particular sector depends on techniques used simultaneously in other sectors, new structures may be judged advantageous or not, depending on expected structures of other industries. Thus, the process of structural choice can be quite complex. A new steel technology that uses more oxygen may be economic only if combined with techniques that deliver cheap tonnage oxygen in the chemicals industry. Chemical costs, in turn, depend on the structures of the chemical sector's suppliers. Hence, detailed anticipatory knowledge of new developments throughout the economy might be required for any technicoeconomic choice. General familiarity with the course of technological discovery and skilled judgment as to rates of adoption of new methods may be necessary. Like many other areas of economic decision—price policy, investment—technical choice can require guesswork about parallel decisions throughout the system.

Here, as in Chapter 10, the range of structural alternatives will be limited to observed 1947 and 1958 structures. We survey only a very restricted segment of the total range of technical possibilities. How did the advantage of actual 1958 over 1947 structure in each sector depend on the changing structures of other sectors? To what extent is the set of actual 1947–1958 changes an interdependent package, and to what extent is each change justified separately, that is, independently of the other changes? In theory, the economic effect of any given structural change depends on developments in all other sectors. Of course, the sensitivity of choice depends on the particular values of the structural parameters and on the changes contemplated. Actually, over the period 1947–1958, interindustrial linkages were of minor importance to the advantages of new structures. There is good reason to believe that this will have been quite a usual state of affairs. It turns out that for most sectors, 1958 structure remains superior to 1947 assuming either 1947 or 1958 input structures for all other sectors. This means that decisions to adopt 1958 input structures could have been made unilaterally for each sector, disregarding changes elsewhere in the system.

Structural advantage could often be judged unilaterally because reductions in direct primary factor requirements overshadowed changes in intermediate input structure in most industries. Direct labor coefficients, the largest cost element, were falling markedly everywhere. The fact that labor coefficients were all moving in the same direction served to moderate relative changes in the total factor content of intermediate inputs. For example, in production of coal, oil, and gas, labor productivities rose at rates that differed but were not radically different. Under these conditions, fuel proportions had less influence on total factor requirements than they would have with more disparate productivity changes. Substitution of intermediate for direct primary inputs and adaptation—the tendency to favor intermediate inputs whose relative costs were declining—played a discernible, but secondary, role. The total advantage of each new structure did, then, depend on elements elsewhere in the system, but the effects of this interdependence were not crucial to structural choice.

11.2 Trigger Effects

The strategy for studying the interdependence of structural change in different sectors is as follows: First, we ask whether the 1958 structure of each sector is advantageous if introduced in an all-1947 context, that is, with 1947 structures prevailing in all other sectors. This means forming 76 hybrid matrices, each with 1947 structure for all but a single sector. If total factor requirements are smaller with a hybrid matrix than with all-1947 structures, then 1958 structure is superior to 1947 for the sector singled out. Similarly, 1958-based hybrids are formed by substituting 1947 columns into the 1958 matrix one at a time. The 1958 structure of a sector is superior in a 1958 context if total factor content, with its 1958-based hybrid, exceeds that with all-1958 structures. Suppose the 1958 structure of a sector appears inferior in a 1947 context but superior in a 1958; then the shift in its advantage must depend on the change from 1947 to 1958 structure in one or more other sectors. The structural advantages of these sectors are linked. Change in this sector can be triggered by changes in one or more others.

Advantages of change in 1947 and 1958 contexts

In mathematical terms, the advantage of 1958 input structure in any given industry is measured by the difference:

$$\pi_j^{47} = \nu^{47} - \nu^{47,58_j} = [\mathbf{f}^{47}\mathbf{Q}^{47} - \mathbf{f}^{47,58_j}\mathbf{Q}^{47,58_j}]\mathbf{y}$$

or

$$\pi_j^{47} = \mathbf{f}^{47}\mathbf{x}^{47} - \mathbf{f}^{47,58_j}\mathbf{x}^{47,58_j} \tag{11.1}$$

where

π_j^{47} = primary factor saving effected by introducing 1958 structure for sector j into the 1947 matrix (note that π_j^{47} might have been written alternatively as $\pi^{47,58_j}$)

ν^{47} = total primary factor requirement to produce final demand $\mathbf{y}$ with 1947 input structures

$\nu^{47,58_j}$ = total primary factor requirement to produce $\mathbf{y}$ with 1947 input structures in all sectors, but 1958 structure in j

$\mathbf{f}^{47}$ = vector of 1947 primary factor (labor plus capital) input coefficients

$\mathbf{f}^{47,58_j}$ = vector of 1947 primary factor coefficients with a 1958 coefficient substituted for the 1947 value in sector j

$\mathbf{Q}^{47}$ and $\mathbf{Q}^{47,58_j}$ = respective inverse matrices

$\mathbf{x}^{47}$ and $\mathbf{x}^{47,58_j}$ = vectors of total output requirements to deliver $\mathbf{y}$ with 1947 and hybrid 1947–1958 structure, respectively

For interindustrial comparisons, π_j^{47} is expressed in relative terms:

$$\psi_j^{47} = \frac{2\pi_j^{47}}{\nu^{47} + \nu^{47,58_j}} \tag{11.2}$$

In the present computation, **y** was set equal to 1958 final demand. The 1958 structure for industry j is superior to that of 1947 in a 1947 context, if π_j^{47} (or ψ_j^{47}) > 0. Similarly, the advantage of a 1958 structure over the corresponding 1947 structure in a 1958 context is

$$\pi_j^{58} = \nu^{58,47_j} - \nu^{58} = \mathbf{f}^{58,47_j}\mathbf{x}^{58,47_j} - \mathbf{f}^{58}\mathbf{x}^{58}$$

or

$$\psi_j^{58} = \frac{2\pi_j^{58}}{\nu^{58,47_j} + \nu^{58}} \tag{11.3}$$

A 1958 structure for sector j is superior to that of 1947 in a 1958 context, if π_j^{58} (or ψ_j^{58}) > 0. Trigger effects can occur when $\pi_j^{47} < 0$ and $\pi_j^{58} > 0$.

First π_j^{47} and π_j^{58} were computed for each 76-order sector j. Estimates of π_j^{47} and π_j^{58} involved computing total factor requirements with 152 (2×76) hybrid matrices, each differing from the base-year matrix by a single column. Modifications of the base-year inverse with the change of a single column structure were made in accordance with the method outlined in Hadley (1962:42–49).* For most sectors, π_j^{47} is positive: change to 1958 industry structure brings a net saving in primary factors even in a 1947 structural context. The (sixteen) industries for which this is not the case are listed below:

(4) Agricultural services
(5) Iron mining
(8) Petroleum mining
(32) Rubber and plastic products★
(37) Iron and steel
(41) Stampings, screw machine products, and fasteners
(42) Hardware, plating, valves, and wire products
(46) Materials handling equipment
(47) Metalworking equipment
(48) Special industry equipment
(49) General industrial equipment
(72) Hotels, personal, and repair services★
(73) Business services
(75) Automobile repair
(76) Amusements and recreation

For these industries, 1958 structure was inferior to that of 1947 in a 1947 context. Similarly, for most sectors, π_j^{58} is positive: a net increase in factor

* C. William Benz wrote a special computer program for performing these and related computations described in this chapter.

requirements results when single 1947 structures are substituted into a 1958 context. Except for the starred sectors (32) and (72), 1958 structure was also superior to that of 1947 in a 1958 context. For the starred sectors, 1947 technology was superior in a 1947 context, and 1958 structure was superior in a 1958 context. The linear programming computation in Chapter 10 also chose 1947 structures for the other fourteen sectors.

Sectors (32)—rubber and plastic products—and (72)—hotels, personal, and repair services—are the ones where 1958 structure had a factor-saving advantage in the 1958, but not in the 1947, context. In these two sectors, structural change did not "stand on its own feet" but required the support of changes in other sectors. The next step was to search for those other sectors—ones whose 1947–1958 changes tipped the balance in favor of the 1958 technology for industries (32) and (72).

Specific linkages

The search called for a different set of hybrids, matrices with 1947 structures for all except *two* sectors. The 1958 structure was introduced for a starred sector and then for each other sector in the economy, the latter one at a time:

$$\pi_j^{47_k} = \nu^{47} - \nu^{47,58_{k,j}} \tag{11.5}$$

was computed, where k is a starred sector and j varies from 1 to n. First k was set equal to 32, and $\pi_j^{47,32}$ was computed for all j. Then k was set equal to 72 and $\pi_j^{47_{72}}$ was computed for all j. Comparison of π_j^{47} (factor saving achieved by introducing 1958 structure for j alone), π_k^{47} (factor saving achieved by introducing new structure for k alone), and π_{jk}^{47} (factor saving achieved by introducing k and j together) revealed the important linkages between changes in k and changes in other sectors. In the case of rubber and plastic products (32), the linkage with changes in synthetic materials (28) was clearly the most important. The computed factor savings of introducing the 1958 structures for industries (32) and (28) into the 1947 matrix, one at a time and simultaneously, were as follows:

$$\pi_{32}^{47} = \$-44 \text{ million}$$
$$\pi_{28}^{47} = \$\ 370 \text{ million}$$
$$\pi_{28}^{47_{32}} = \$\ 442 \text{ million}$$

Introducing 1958 structure for industry (32) alone results in a net *increase* in total factor requirements. Introducing 1958 structure for industry (28) alone produces a net decrease in factor requirements of $370 million. When 1958 structures are introduced for both industries (32) and (28) together, the factor saving is $442 million.

Industry (28) shows large increases in direct factor productivity over the period. Because (32) used much more synthetic materials with 1958 than with 1947 structure, the introduction of 1958 structure for (32) permits the economy to take greater advantage of improvements in efficiency of (28). It is this feature that supplies the economic justification for introducing the 1958 structure for industry (32). The 1958 structure for industry (32) is better than that of 1947 in a 1958 context because it uses more of an input, (28), whose production gained greatly in efficiency.

The ultimate advantage of 1958 structure for hotels, personal, and repair services (72) in a 1958 context is not justified by introducing 1958 structure in any single related sector. It becomes economic only if introduced jointly with 1958 structures for several supplying industries, principally trade (69), service equipment (54), and electronic components (57). The following are the relevant factor-saving advantages of introducing 1958 structures in a 1947 context for these sectors, separately and jointly:

One-at-a-time Savings	*Two-at-a-time Savings*
$\pi_{72}^{47} = \$ -62$ million	
$\pi_{69}^{47} = \$5{,}839$ million	$\pi_{69}^{47\,72} = \$5{,}861$ million
$\pi_{57}^{47} = \$\ 238$ million	$\pi_{57}^{47\,72} = \$\ 271$ million
$\pi_{54}^{47} = \$\ 514$ million	$\pi_{54}^{47\,72} = \$\ 534$ million

While 1958 structure for industry (72) is not advantageous if compared to 1947 structure in a 1947 context, it *is* advantageous in a 1958 context. Industry (72) uses larger amounts of trade (69), service equipment (54), and electronic components (57) in 1958 than in 1947; these latter sectors increase markedly in efficiency over the 1947–1958 period. The increased used of these three inputs gives the 1958 structure an advantage over that of 1947 but only in a context where the structures of production for trade, service equipment, and electronic components themselves have been improved. In these sectors, adaptation was important; trigger effects can occur only when adaptive change plays a major role. The total effect of introducing 1958 structures for sectors (72), (69), (57), and (54) all at once was not actually computed, since the linkages are clear from the two-at-a-time computation.

11.3 Economic Significance of Linkages

There was not sufficient information for judging a priori whether 1947 and 1958 structures were really alternative in 1947 or whether 1958 structures embody elements of technology that were not available until after 1947. The above analysis gives a basis for distinguishing ex post between new technical alternatives and triggered substitutions. Suppose the advantage of a given 1958 coefficient vector over its 1947 counterpart was clear in a 1947 context. Then, had it been available in 1947, it should have been chosen as the preferable economic alternative. Most 1958 structures were preferable to 1947 in a 1947 context. The fact that 1958 structures were not chosen in 1947 is presumptive evidence that they were not among the technological alternatives available then. For the "equivocal" industries, (32) and (72), the evidence is not conclusive: 1947 and 1958 structures might both have been technical alternatives in 1947. Their shifts to 1958 structure could have been simple substitutions triggered by the adoption of new structures in other sectors. Had the computation pinpointed no trigger effects at all, one might infer that all 1958 structures were in fact new technical alternatives.

Is it likely that more trigger effects would be revealed with finer disaggregation? At a very fine level of detail, some structural changes were clearly impossible without changes in other sectors: miniaturization of electronic equipment rests on the development of the transistor or of microcircuits; developments in food processing and agriculture presuppose the chemical synthesis of specific preservatives, hormones, insecticides. Many of these qualitative changes are lost in intertemporal comparisons of ordinary input-output tables, even when they are disaggregated. Other specific developments were feasible and did not require important qualitative change in inputs, but they would have been uneconomic without complementary developments in other sectors. As was already noted, basic oxygen steelmaking could not establish an economic advantage over earlier methods before the development of methods for producing cheap tonnage oxygen. Disaggregation should make it easier to identify this latter type of linkage, but it, too, may be obscured in establishment statistics (see Section 11.5).

11.4 Evidence of Adaptation

The fact that so few trigger effects were identified means that adaptive change was not critical to the advantage of 1958 coefficient vectors in most sectors. This impression is reinforced when actual amounts of factor saving, computed with various hybrid matrices, are compared: figure 11.1 shows the percentage change in total factor requirements to produce a fixed final demand as new (1958) structures are introduced one at a time. The vertical axis measures ψ_j^{47}, relative factor saving when 1958 input structure for sector j alone is introduced into the 1947 matrix. The horizontal axis measures

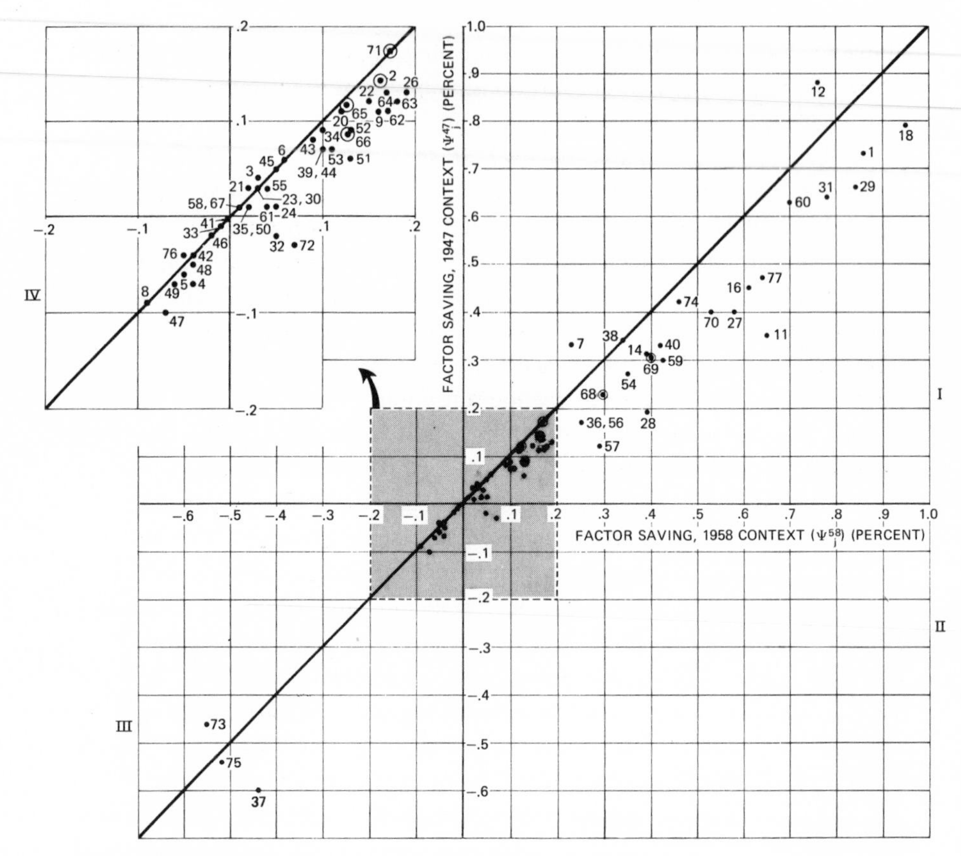

11.1 Percentage Savings in Total Primary Factor Requirements to Deliver 1958 Final Demand with One-At-A-Time Substitutions of 1958 Structures into a 1947 Context and 1947 Structures into a 1958 Context; 76 Order (for circled points, multiply scales by 10). Industry list, next page.

ψ_j^{58}, relative increase in total factor requirements when 1947 structure for industry j alone is introduced into the 1958 matrix. Points in quadrant I, in figure 11.1, represent sectors whose 1958 structures were superior to 1947 in either context; points in quadrant III, sectors whose 1958 structures were inferior in either context. The fourth quadrant contains sectors (32) and (72), already discussed, whose 1958 structures prove superior only in the 1958 context. Note that the factor-saving advantages of 1958 structure in either context are not large for these equivocal sectors. Figure 11.1 brings out a number of important features of structural change.

First, the rates of factor saving contributed by structural change in different sectors varied considerably. Sectors with large values of ψ_j^{47} or ψ_j^{58} show

(1) Livestock
(2) Crops
(3) Forestry & fishing
(4) Agric. services
(5) Iron mining
(6) Nonferrous mining
(7) Coal mining
(8) Petroleum mining
(9) Stone & clay mining
(10) Chemical mining
(11) New construction
(12) Maintenance constr.
(14) Food
(15) Tobacco
(16) Textiles
(18) Apparel
(20) Wood & products
(21) Wooden containers
(22) Household furniture
(23) Office furniture
(24) Paper & products
(26) Printing & publishing
(27) Basic chemicals
(28) Synthetic materials
(29) Drugs, soaps, etc.
(30) Paint
(31) Petroleum refining
(32) Rubber prod., etc.
(33) Leather tanning
(34) Shoes
(35) Glass & products
(36) Stone & clay prod.
(37) Iron & steel
(38) Nonferrous metals
(39) Metal containers
(40) Heating, etc.
(41) Stampings, etc.
(42) Hardware, etc.
(43) Engines & turbines
(44) Farm equipment
(45) Constr. & mining eq.
(46) Materials hand. eq.
(47) Metalworking eq.
(48) Special ind. eq.
(49) General ind. eq.
(50) Machine shop prod.
(51) Office & comp. mach.
(52) Service ind. mach.
(53) Electrical apparatus
(54) Household appliances
(55) Light. & wiring eq.
(56) Communication eq.
(57) Electronic compon.
(58) Batteries, etc.
(59) Motor vehicles & eq.
(60) Aircraft
(61) Trains, ships, etc.
(62) Instruments, etc.
(63) Photo. apparatus
(64) Misc. manufactures
(65) Transportation
(66) Telephone
(67) Radio & tv broad.
(68) Utilities
(69) Trade
(70) Finance & insurance
(71) Real estate & rental
(72) Hotels & pers. serv.
(73) Business services
(74) Research & dev.
(75) Auto. repair
(76) Amusements, etc.
(77) Institutions
(80) Noncomp. imports
(81) Bus. travel, etc.
(83) Scrap

---

greatest relative factor-saving advantage for 1958 over 1947. Aircraft (60), agriculture (1) and (2), trade (69), real estate and rental (71), transportation (65), and utilities (68) are outstanding in this respect.

Second, the ranking of 1958 structures with respect to their rates of factor saving is hardly changed by the shift from the 1947 to the 1958 context. In general, the order of magnitude of factor saving is the same in the two contexts.

Third, for most sectors, ψ_j^{58} exceeds ψ_j^{47}: the relative advantage of individual 1958 over 1947 structures is greater in a 1958 than in a 1947 context. This is evidence of adaptation. The 1958 structures take advantage of differential rates of advance in other sectors, substituting intermediate inputs with markedly improving structures for inputs whose structures improve more sluggishly. However, differences between ψ_j^{58} and ψ_j^{47} are relatively small. In other words, adaptive change is a small component of total factor saving.

After the triggered sectors, electronic components (57), office and computing machines (51), and synthetic materials (28) show the most adaptive change.

11.5 Dominance of Direct Factor Saving in Structural Change

Intuitively, it is easy to see that structural changes stand on their own feet because direct factor saving is large relative to changes in intermediate input requirements. A simple computation permits us to verify this interpretation more concretely. Algebraically, π_j^{47} and π_j^{58}, total factor saving, are divided into two components—one for direct factor saving and one for indirect factor saving. The direct is large relative to the indirect for all but a few sectors. In mathematical terms:

$$\pi_j^{47} = \mathbf{f}^{47}\mathbf{x}^{47} - \mathbf{f}^{47,58_j}\mathbf{x}^{47,58_j} \tag{11.1}$$

$$\mathbf{f}^{47,58_j} = f_1^{47}, f_2^{47}, \ldots, f_j^{47} - \Delta f_j, \ldots, f_n^{47} \tag{11.6}$$

where

$$\Delta f_j = f_j^{47} - f_j^{58} \tag{11.7}$$

thus

$$\pi_j^{47} = \mathbf{f}^{47}(\mathbf{x}^{47} - \mathbf{x}^{47,58_j}) + \Delta f_j x_j^{47,58_j} \tag{11.8}$$

Similarly,

$$\pi_j^{58} = \mathbf{f}^{58}(\mathbf{x}^{58,47_j} - \mathbf{x}^{58}) - \Delta f_j x_j^{58,47_j} \tag{11.9}$$

On the side following the equal sign of equations 11.8 and 11.9, the second terms (those containing Δf_j) measure changes in direct primary factor requirements. The first terms measure indirect changes in primary factor requirements, those associated with changes in intermediate input requirements. The second terms are large, and they overshadow the first. Figure 11.2 is a comparison of direct with total factor saving; that is, of $\Delta f_j x_j^{47,58_j}$ with π_j^{47}, for each j. For most industries, both π_j^{47} and $\Delta f_j x_j^{47,58_j}$ are positive. To avoid complicating the figure, only these sectors are represented in figure 11.2. Industries where π_j^{47} exceeds $\Delta f_j x_j^{47,58_j}$, that is, those represented by points above the 45-degree line, are those where 1958 structure brought indirect, as well as direct, factor saving in a 1947 context. Points falling below the 45-degree line represent sectors where indirect factor saving was negative, that is, where it tended to offset direct factor saving. The great majority of values of $\Delta f_j x_j^{47,58_j}$ are within a range of π_j^{47}, $\pm$ 50 percent.

Since annual labor costs far outweigh annual interest charges in all sectors, these conclusions about the dominance of direct primary factor costs are essentially generalizations about the importance of direct labor charges. Labor and capital charges are assumed to have fixed proportions in this treatment, and Δf_j can be expressed as a sum:

$$\Delta f_j = w_j \Delta l_j + \alpha \Delta b_j$$

where $w_j \Delta l_j$ and $\alpha \Delta b_j$ are changes in wage and interest coefficients. Figure 11.3 compares direct labor saving alone with total factor savings; that is, it compares $w_j \Delta l_j x_j^{47,58_j}$ with π_j^{47}. Note the similarity between figures 11.2 and 11.3. In general, $w_j \Delta l_j x_j^{47,58_j}$ is itself within a range of π_j^{47}, $\pm$ 50 percent.

The clear-cut tendency for direct labor costs to dominate structural change stems from two broad and related developments in the American economy. The first was cited in Section 8.8. Labor is a large element of cost. With rising living standards (and higher training requirements), its price tends to rise systematically relative to the prices of most capital goods and, to a lesser extent, relative to intermediate inputs. Thus, efforts are concentrated on economizing labor, and labor coefficients fall in relation to other input coefficients. On the whole, intermediate input coefficients tend to rise very slowly if at all because substitutability of intermediate inputs for labor is limited and because intermediate input prices follow wages fairly closely. Where, as in the case of business services, intermediate inputs proved substitutable for direct labor, their coefficients did rise over the period studied.

Decisions to substitute one intermediate input for another—to use techniques favoring a particular intermediate input—require the anticipation of developments in other sectors of the economy. Because such decisions are, in principle, interdependent, they are potentially more complex than decisions to economize direct labor. What about the difficult problem of anticipating and adapting to structural changes elsewhere in the system? If technical adaptation could be made instantaneously and without long-term commitment, the price mechanism might provide a framework for rational decisions in this area. Given the inevitable lags, with new structure often presupposing new investment, current prices alone do not supply sufficient information.

Specialization of firms and establishments in itself helps limit the difficulties posed by this problem. Indeed, the linkage of technological changes in two activities is often the rationale for combining them under the same management. As a result, careful analysis of the organization of production might well show a systematic tendency for productive units to internalize or integrate activities whose changes are interdependent. Only a few speculations about the "make-or-buy" decision will be offered here, but investigations in

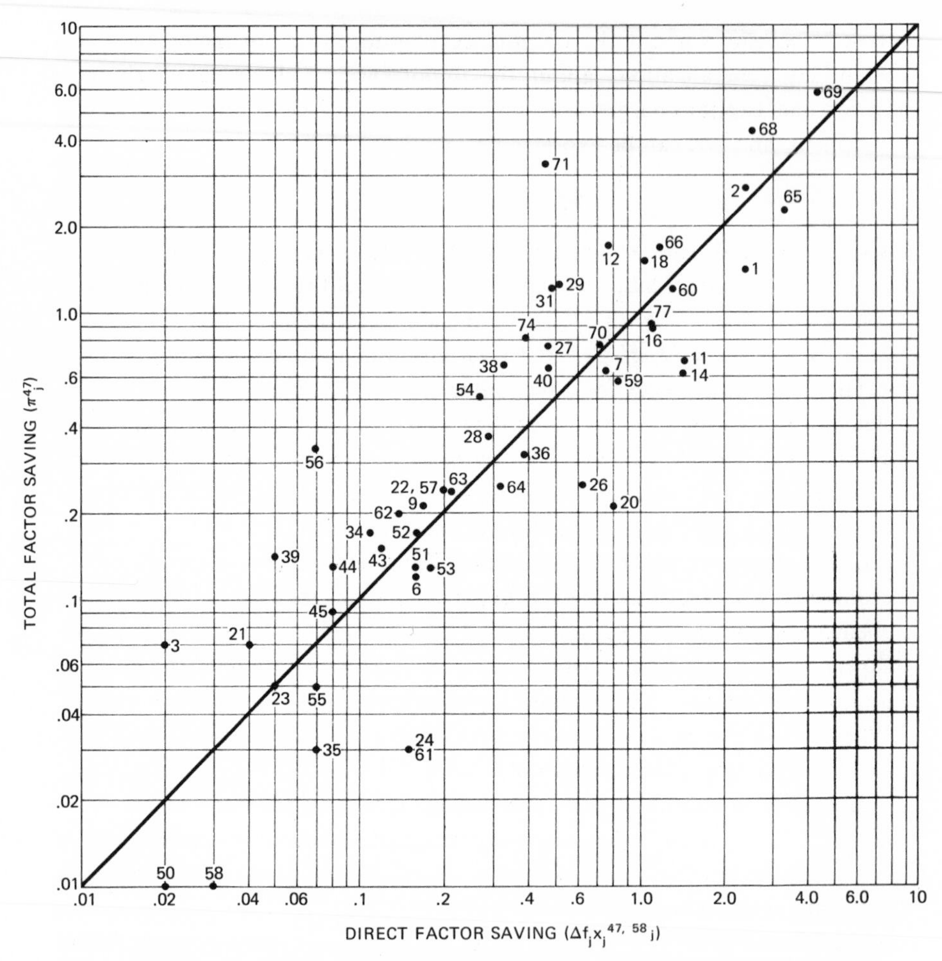

11.2 Direct and Total Savings of Primary Factors with One-At-A-Time Introduction of 1958 Structures into a 1947 Structural Context; 76 Order (dollars times 10^6 in 1947 prices). Points with zero or negative values for direct factor saving or total factor saving are not shown on this graph; see text. Industry list, next page.

this area are essential to a deeper understanding of technical change. There are familiar economic reasons for combining technologically related processes in a single establishment. Specialized components must dovetail in the finished product; adaptation to changes in the design of products or components is easiest with direct communication and coordinated decisions. In an era when many firms are geared to change, specialization of productive units must itself be so oriented. Establishments will tend to combine activities that change together. For example, there has recently been an acceleration of technical

(1) Livestock
(2) Crops
(3) Forestry & fishing
(4) Agric. services
(5) Iron mining
(6) Nonferrous mining
(7) Coal mining
(8) Petroleum mining
(9) Stone & clay mining
(10) Chemical mining
(11) New construction
(12) Maintenance constr.
(14) Food
(15) Tobacco
(16) Textiles
(18) Apparel
(20) Wood & products
(21) Wooden containers
(22) Household furniture
(23) Office furniture
(24) Paper & products
(26) Printing & publishing
(27) Basic chemicals
(28) Synthetic materials
(29) Drugs, soaps, etc.
(30) Paint
(31) Petroleum refining
(32) Rubber prod., etc.
(33) Leather tanning
(34) Shoes
(35) Glass & products
(36) Stone & clay prod.
(37) Iron & steel
(38) Nonferrous metals
(39) Metal containers
(40) Heating, etc.
(41) Stampings, etc.
(42) Hardware, etc.
(43) Engines & turbines
(44) Farm equipment
(45) Constr. & mining eq.
(46) Materials hand. eq.
(47) Metalworking eq.
(48) Special ind. eq.
(49) General ind. eq.
(50) Machine shop prod.
(51) Office & comp. mach.
(52) Service ind. mach.
(53) Electrical apparatus
(54) Household appliances
(55) Light. & wiring eq.
(56) Communication eq.
(57) Electronic compon.
(58) Batteries, etc.
(59) Motor vehicles & eq.
(60) Aircraft
(61) Trains, ships, etc.
(62) Instruments, etc.
(63) Photo. apparatus
(64) Misc. manufactures
(65) Transportation
(66) Telephone
(67) Radio & tv broad.
(68) Utilities
(69) Trade
(70) Finance & insurance
(71) Real estate & rental
(72) Hotels & pers. serv.
(73) Business services
(74) Research & dev.
(75) Auto. repair
(76) Amusements, etc.
(77) Institutions
(80) Noncomp. imports
(81) Bus. travel, etc.
(83) Scrap

---

progress in textiles, accompanied by increased vertical integration and organizational mergers of textile, fiber, and chemical producers. In many sectors, marked economies are offered by an integrated "systems approach" to a sequence of related productive activities. These afford further incentive to coordinate changes in different activities by bringing them under the same roof or, at least, the same planning jurisdiction.

Chapter 4 showed that the degree of specialization of industries measured by the relative volume of intermediate transactions increases only very slowly with the size of the economy. Opportunities for finer division of labor may well be offset by the need to coordinate technical changes in related activities. Change, and therefore growth, would be difficult in an overspecialized economy.

To the extent that processes whose changes are linked do tend to be integrated, the changes affect internal flows that are not reported in establishment

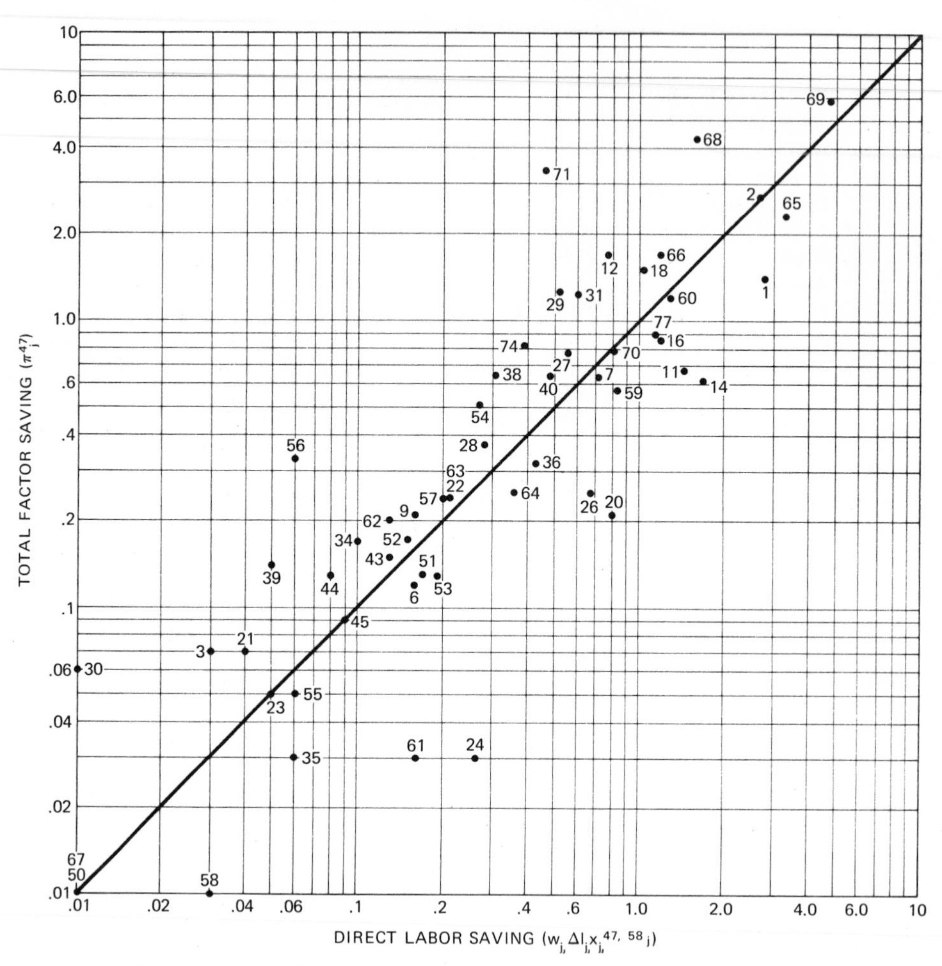

11.3 Direct Labor Savings and Total Savings of Primary Factors with One-At-A-Time Introduction of 1958 Structures into a 1947 Structural Context: 76 Order (dollars times 10^6 in 1947 prices). Points with zero or negative values for direct labor saving or total factor saving are not shown on this graph; see text. Industry list, next page.

statistics. Thus, evidence of change is concentrated at the beginning of the productive sequence, in direct requirements of primary inputs. In part, intermediate input structure remains relatively stable because production is organized so that many rapidly changing "intermediate inputs" are not reported as transactions at all. Adaptive change appears small because establishments combine those activities that are most sensitively attuned to

(1) Livestock
(2) Crops
(3) Forestry & fishing
(4) Agric. services
(5) Iron mining
(6) Nonferrous mining
(7) Coal mining
(8) Petroleum mining
(9) Stone & clay mining
(10) Chemical mining
(11) New construction
(12) Maintenance constr.
(14) Food
(15) Tobacco
(16) Textiles
(18) Apparel
(20) Wood & products
(21) Wooden containers
(22) Household furniture
(23) Office furniture
(24) Paper & products
(26) Printing & publishing
(27) Basic chemicals
(28) Synthetic materials
(29) Drugs, soaps, etc.
(30) Paint
(31) Petroleum refining
(32) Rubber prod., etc.
(33) Leather tanning
(34) Shoes
(35) Glass & products
(36) Stone & clay prod.
(37) Iron & steel
(38) Nonferrous metals
(39) Metal containers
(40) Heating, etc.
(41) Stampings, etc.
(42) Hardware, etc.
(43) Engines & turbines
(44) Farm equipment
(45) Constr. & mining eq.
(46) Materials hand. eq.
(47) Metalworking eq.
(48) Special ind. eq.
(49) General ind. eq.
(50) Machine shop prod.
(51) Office & comp. mach.
(52) Service ind. mach.
(53) Electrical apparatus
(54) Household appliances
(55) Light. & wiring eq.
(56) Communication eq.
(57) Electronic compon.
(58) Batteries, etc.
(59) Motor vehicles & eq.
(60) Aircraft
(61) Trains, ships, etc.
(62) Instruments, etc.
(63) Photo. apparatus
(64) Misc. manufactures
(65) Transportation
(66) Telephone
(67) Radio & tv broad.
(68) Utilities
(69) Trade
(70) Finance & insurance
(71) Real estate & rental
(72) Hotels & pers. serv.
(73) Business services
(74) Research & dev.
(75) Auto. repair
(76) Amusements, etc.
(77) Institutions
(80) Noncomp. imports
(81) Bus. travel, etc.
(83) Scrap

---

each other. When two activities are combined in a single establishment, the statistical reporting system gives no evidence of their interdependence.

The above explanation offers another clue to our central paradox: the stability of input-output structures in the face of general fragmentary evidence of very rapid economic change. This stability is not all due to aggregation or statistical problems in measuring qualitative change. In addition, a great deal of change is systematically concentrated in internal operations. Under these conditions, changes in direct labor and capital requirements provide some indirect indication of how significant these internal, but unreported, changes really are.

Chapter 12 A Linear Programming System Analyzing Embodied Technological Change

*12.1 Background and Objectives**

The structure of each sector is an average of structures for many product and process layers. Thus far, we have concentrated on describing and evaluating changes in average structures only. Information on average structures suffices for analyzing directions of change and for establishing the overwhelming importance of changes in direct primary factor coefficients. Naturally, some sectors advance more rapidly than others; thus far, differential rates of progress in different sectors have been viewed as data.

This chapter introduces a system that explains the rates of structural change in individual sectors in terms of changing relative importance of old and new layers. The approach is highly simplified. The structure of each sector is disaggregated into just two layers: a new or "best-practice" stratum, whose coefficients represent the kinds of technology used in new capacity; and an old stratum, comprising the rest. All structural change is assumed to be "embodied": A sector can increase the importance of old- or new-technology activities only by investing in new capital goods, that is, by gross additions to capacity. Limited decumulation of initial capacity is permitted. The gross amount of new investment in the whole economy is specified for a given period, and optimal levels of new and old activities in each sector are determined by linear programming techniques.

The idea that current input structures depend on the specific qualities of capital goods goes far back in the economic literature. Schumpeter (1954: 1031) contrasted the wide range of technological alternatives at the planning stage with the narrower one, once commitments to specialized equipment were made. Phelps (1963) christened economic models with such ex ante flexibility and ex post rigidity "putty-clay models." While embodiment is often discussed in the economic literature, meaningful evidence on the validity of this assumption is still very sparse. Exploratory attempts at direct verification range over the last twenty years (see Carter 1953 and 1960;

* A preliminary report on the research described in this chapter is given in Carter (1969).

Komiya 1962; Salter 1960). Theoretical interest began to snowball with the publication of Solow (1960). Again, presumably for lack of information, as well as for elegance, our most sophisticated thinkers have concentrated on aggregative formulations. Denison (1964) shows that embodiment is practically unimportant in explaining the overall growth rate for the United States, because the average age of the capital stock of the nation is normally very stable. Phelps (1962:567) points out that, in a long-run aggregative growth model, "the limiting long-run growth rate depends on the rate of technical progress, not the type of progress." However, as Denison suggests, embodiment may be of greater interest for analyzing the allocation of investment than for exploring the average rate of growth.

Embodiment is still just an assumption. Technical evidence shows that technology is capital embodied in electric power generation, but the case is not quite so clear in, say, the manufacture of household appliances. Different equipment is required to mold plastic than to stamp metals. Automation of processes to supplant hand operations requires new inspection and control devices and, often, newly designed process equipment as well. A systems approach to process rationalization is most effective in the design of a whole new plant. However, not every change in input structure requires a complete new plant, and some changes require no investment at all. Old furnaces can be converted to the use of different fuels at minor expense; new alloys can be made with old equipment; personnel policy can increase labor productivity with little or no change in capital goods. It is difficult to weigh and synthesize such fragmentary information into an overall judgment, even for a single sector. In the interests of realism, the assumption that all technical change is capital embodied should certainly be modified as more specific evidence is gathered. Unfortunately, direct surveys of many technical processes are required for this, and they are time-consuming and expensive.

12.2 Linear Programming System

To allow for embodied change, the linear programming model of Chapter 10 is modified in three major respects.

1. The 1947 and 1958 structures are no longer considered as alternatives. Instead, a best-practice structure, representing the new technological layer, is introduced as an alternative to average old-technology input structure in each sector. Procedures for estimating best-practice input structures are described in Section 12.3. It would be better to disaggregate the old structure as well and to view best-practice technique in competition with several older

structural layers, but available information does not permit this yet. Old-technology structure, therefore, is still treated as a single homogeneous layer.

2. Capital is required for the creation of any new capacity, whether for net expansion or for replacement. Capital coefficients govern the expansion of capacity with either old or new techniques. A ceiling, equal to the total volume of investment actually observed in a given time interval, is imposed on the total amount of gross investment, that is, on the total of all expenditures on plant and equipment in the entire economy. Fixing total investment limits the amount of replacement that can take place and, thus, the rate of transition from older to newer technology.

3. The problem is defined as one of delivering a given *increment* to final demand by means of *changes* in the levels of various activities throughout the economy. These changes may increase the levels of production with new input structures, and they may increase or decrease the levels of production with older input structures.

Let

$\mathbf{A}^n$ = a matrix of new-technology input coefficients

$\mathbf{A}^t$ = matrix of "average" input-output structures; that is, of conventional input-output coefficients observed for some base period (replacement is to be determined by the model and hence is not included in flow coefficients)

$\Delta\mathbf{x}^n$ = vector of increments to output levels associated with new-technology input structures

$\Delta\mathbf{x}^t$ = vector of increments to output with old-technology input structures

$\Delta\mathbf{y}$ = vector of increments to final demand

The basic balance equation is

$$(\mathbf{I} - \mathbf{A}^n)\Delta\mathbf{x}^n + (\mathbf{I} - \mathbf{A}^t)\Delta\mathbf{x}^t \geq \Delta\mathbf{y} \tag{12.1}$$

The vector $\Delta\mathbf{x}^t$ is composed of positive elements, representing additions to capacity with old-technology structure, and negative elements, representing retirements of old capacity. (Retirements will include a minimal allowance for wear and tear; they are introduced below.) Distinguishing increases in old-technology capacity from retirements, we partition the vector $\Delta\mathbf{x}^t$ into two parts: $\Delta\mathbf{x}^{t+}$, containing all the positive elements, and $\Delta\mathbf{x}^{t-}$, containing all the negative elements. For convenience in computation, we change the signs of all the elements of $\Delta\mathbf{x}^{t-}$ to plus. Thus

$$\Delta\mathbf{x}^{t+} - \Delta\mathbf{x}^{t-} = \Delta\mathbf{x}^t$$

On the assumption that new-technology capacity will not be retired during the time interval considered, $\Delta\mathbf{x}^n$ is not partitioned. Thus, all elements of $\Delta\mathbf{x}^n$ must be positive or zero. Now, equation 12.1 can be given as

$$(\mathbf{I} - \mathbf{A}^n)\Delta\mathbf{x}^n + (\mathbf{I} - \mathbf{A}^t)\Delta\mathbf{x}^{t+} - (\mathbf{I} - \mathbf{A}^t)\Delta\mathbf{x}^{t-} \geq \Delta\mathbf{y} \qquad (12.2)$$

Increases in capacity with either old or new input structure are permitted, subject to the limit on total investment in the economy. Equation 12.3 specifies this constraint:

$$\mathbf{b}^n\Delta\mathbf{x}^n + \mathbf{b}^t\Delta\mathbf{x}^{t+} \leq \kappa \qquad (12.3)$$

where $\mathbf{b}^n$ and $\mathbf{b}^t$ are vectors of capital coefficients associated with new and old techniques respectively, and κ is the given total amount of gross new investment available to the system over the period studied. For each interval, the total investment ceiling is actual expenditure on plant and equipment in the American economy, summed from the base year through the year before the terminal year of the interval. Retirement of capacity is assumed not to release capacity for other sectors, and that is why no terms involving $\Delta\mathbf{x}^{t-}$ appear in equation 12.3.

The objective function is

$$\mathbf{f}^n\Delta\mathbf{x}^n + \mathbf{f}^t\Delta\mathbf{x}^{t+} - \mathbf{f}^t\Delta\mathbf{x}^{t-} = \text{minimum} \qquad (12.4)$$

where

$\mathbf{f}^t$ = vector of primary factor input coefficients for year t as defined in equation 9.4

$\mathbf{f}^n$ = vector of primary factor input coefficients for best-practice technology

Equations 12.2 to 12.4 are the core of the linear programming model. It chooses an optimal mix of input structures, under conditions of embodied technical change, subject to an overall investment ceiling. Initial capacities are characterized by old (that is, by initial average) technology. Given a specified increment to the bill of goods, $\Delta\mathbf{y}$, and the total lump of investment over the period, the system selects the combination of activity levels with new and old techniques that minimizes total factor costs. Were there no limit on total investment, the system could choose *any* combination of new and old techniques. Depending on the relative advantages of individual new and old structures, each sector would either replace its entire capacity or leave it intact with old structure. Increments to capacity would have matching new or old structures. However, the amount of available investment limits the extent to which this "unlimited" optimum can be achieved. Once capacity requirements for supplying the given increment to final demand are met, investment will be allocated to replace capacities where factor-saving advantages of new techniques are greatest. As suggested in Section 10.4, the system sets up an order of priority for changing over old capacity to new in the various sectors. Those replacements most economic of primary factors come first, and so on. The higher the investment limit, the lower the cutoff point for factor-saving replacement. In a dynamic formulation, of course, the tradeoff between replacement and expansion would have to be considered.

Two additional sets of contraints complete the specification of the system. Initial capacity limits the amount of scrappage of old capacity that can take place in each sector. For most sectors, it is probably more realistic to impose replacement ceilings of less than 100 percent of initial capacity. Initial capacity actually consists of many technological layers, some of them being very similar, if not identical, to best-practice techniques. In reality, then, the factor saving achieved by replacing old with new capacity, in most sectors, decreases as more and more capacity is replaced. In principle, of course, this condition calls for further disaggregation of the initial input structure of each sector. Lacking the data for this, we imposed ceilings on the scrappage rates of all sectors. These are introduced in the form of an additional set of constraints,

$$\Delta \mathbf{x}^{t-} \leq \gamma \bar{\mathbf{x}}^{t} \tag{12.5}$$

where $\bar{\mathbf{x}}^{t}$ represents initial capacities and γ is a constant (γ was varied experimentally within the range from 5 percent per year to 100 percent of initial capacity).

Minimal scrappage rates had to be set to allow for physical attrition and retirements incidental to the needs of changing product mix and geographical relocation of capacity, not explicitly introduced in the model. This meant adding another set of constraints,

$$\Delta \mathbf{x}^{t-} \geq \beta \bar{\mathbf{x}}^{t} \tag{12.6}$$

where β is the minimal scrappage rate, generally equivalent to 3 percent per year.

Beyond the specification of equations 12.5 and 12.6, no a priori distinction between investment for expansion and for replacement was made. Where the new technique is advantageous, its capacity is likely to expand. If this expansion of capacity is compensated by reductions in old capacity, it is "replacement." Otherwise, it is "expansion." Whether a sector expands or replaces depends on the demand for its output and, thus, on final demands and activity levels for other sectors. Except as noted in equation 12.6, we do not prejudge whether a particular addition to capacity will turn out to be an expansion or a replacement. It is determined by the simultaneous workings of the entire system.

Ex ante, an individual producer may distinguish between replacement and expansion of his facilities. If he finds no market for the output of a portion of his capacity, however, what he may have intended to be expansion will turn out, ex post, to be replacement. Of course, physical scrappage of old capital goods will insure that a given purchase of new capital goods is "replacement" for the individual establishment. Insofar as old capacity may be reactivated elsewhere in the sector to compensate for the scrappage, however, the system may still experience net expansion of capacity.

12.3 Empirical Implementation

To implement the system described in equations 12.2 through 12.6 empirically, it was necessary first to estimate the new-technology parameters, $\mathbf{A}^n$ and $\mathbf{f}^n$, the input-output flow and the primary factor coefficients for the newer technological layers. Ideally, one would wish to have the best-practice coefficients estimated directly from sample information (as described in Miernyk 1969; Komiya 1960; or Carter 1963) or on the basis of expert judgment (as described in Harvard Economic Research Project 1964; or Carter 1967a). While Miernyk has estimated sets of best practice coefficients for individual regions, there is nothing approaching a complete set of direct estimates of new-technology structures for the United States as a whole. Rough statistical approximations had to be used instead. "Incremental coefficients" were computed as estimates of these new-technology parameters. In effect, incremental coefficients are generated by running the model of embodied technological change "in reverse." They represent what coefficients for new technology *must have been* in order to produce the changes in input structure actually observed over the period 1947–1958, with given sectoral investment and assuming capital embodiment. Of course, this crude method can only give rough approximations of best-practice structures.

The incremental coefficient matrix was derived as follows. Each 1958 flow coefficient a_{ij}^{58} was viewed as a weighted average of a 1947 (a_{ij}^{47}) and a new-technology (a_{ij}^{n}) coefficient.

$$a_{ij}^{58} = a_{ij}^{n} s_j + a_{ij}^{47}(1 - s_j)$$

Each incremental coefficient could then be estimated from observed 1947 and 1958 coefficients and a set of weights s_j.

$$a_{ij}^{n} = \frac{a_{ij}^{58} - a_{ij}^{47}(1 - s_j)}{s_j} \tag{12.7}$$

where s_j is the proportion of 1958 output characterized by new-technology input structure in industry j. It is approximated by the proportion of 1958 capacity installed during the period 1947–1957 inclusive. This, in turn, is roughly equal to the ratio of gross new investment over the period to gross capital assets at the end of the period.

$$s_j = \frac{\sum_{t=47}^{57} e_j^t}{g_j^{58}} \tag{12.8}$$

where e_j^t is sector j's annual expenditures on new plant and equipment, and g_j^{58} is the value of its gross capital assets in 1958. Incremental primary factor coefficients f_j^n were computed similarly to a_{ij}^n.

Annual capital expenditure time series for each sector were from Koenig and Ritz (1967) and Office of Business Economics (1963:9). All other data were part of the general body of material introduced in earlier chapters. Capital expenditure series were not available to match the full 76-order or the 38-order detail, and so it was necessary to aggregate some sectors, particularly those outside of manufacturing. The 41-order sectoring scheme used for these computations is given in tables 12.1 through 12.3. It is more aggregated than the 38-order sectoring in nonmanufacturing areas but almost as detailed as the 76 order in manufacturing.

The 1947 and 1958 matrices are not entirely comparable cell-by-cell, and the rigid assumption of embodiment for all change is not correct. Change may be tied to investment for some elements of a column and not for others. Thus, there were bound to be some terms in the incremental coefficient matrix that did not make good technological sense. Actually, the computed incremental matrix contained few apparent technological "monstrosities." One across-the-board refinement of these estimates was imposed: when negative coefficients appeared in the incremental matrix, they were replaced by positive coefficients arbitrarily valued at 50 percent of the value of the coefficient for 1958. This adjustment affected only twenty-three cells with 1958 coefficients equal to or greater than 0.005 and should introduce no serious biases into the computations. This single set of incremental coefficients, inferred from observed 1947–1958 coefficient changes, represents new technology for the entire period 1947–1961 in our computations. Coefficients representing average technology in the base year are simply observed input-output coefficients for that base year.

The linear programming problem was solved for three different time intervals: 1947–1958, 1947–1961, and 1958–1961. For the first two intervals, 1947 input-output, labor, and capital coefficients represented the initial technology; in the third case, 1958 coefficients represented initial technology. To repeat, new technology was represented for all three cases by incremental coefficients based on observed 1947–1958 changes. The 1958 capital coefficients were assumed to represent capital requirements with incremental technology, too. Increments to final demands, Δy, were computed directly from observed final demand vectors. The gross investment ceiling imposed on the entire economy for each of the three intervals was estimated as the sum of actual gross expenditures on plant and equipment in all sectors from the base year through the year just *before* the terminal year. Exclusion of investment of the terminal year rests on the assumption of an average one-year lag between expenditure on new investment goods and the emergence of additional output as a result of the lag.

Several variants of the basic model were computed for each interval. For 1947–1958 and 1947–1961, minimal decumulation of initial (1947) capacity was set at 3 percent per year. Versions differed in the arbitrary ceiling imposed on the amount of initial capacity that each sector was permitted to replace. Alternative upper limits ranging from 5 percent per year to 100 percent of initial capacity were imposed on the decumulation of each sector. While the observed total volume of investment was well in excess of requirements to cover a 3 percent per year minimum for 1947–1958 and 1947–1961, total investment for the interval 1958–1960 was barely sufficient to cover 2 percent per year replacement. Minimal decumulation of 1958 capacity for each sector was set at 1 percent per year for the 1958–1961 period and maximum decumulation at 7 percent per year.

12.4 Results of the Computations

The results of the computations are summarized in tables 12.1 through 12.3 and in figures 12.1 through 12.6. Each table shows gross additions to and retirements of capacity computed for a single interval under specified assumptions. For example, the first line of table 12.1 contains the computed results for agriculture (1). The entry in the first column says that initial (1947) agricultural output was $4,617 $\times$ 10^7. To deliver 1958 final demands, according to the linear programming computation, under the assumed limits of 3 and 6 percent per year on replacement, the industry would add $3,327 $\times$ 10^7 worth of capacity. Of this, $2,279 would be replacement of old-technology capacity with best-practice capacity. The next two entries to the right tell us that, if the limits on annual replacement were changed to 3 and 10 percent per year, the industry should add only $2,254 $\times$ 10^7 worth of capacity and retire correspondingly less—$1,314—and so on. In effect, the tables show how investment is rationed under alternative limits on replacement. The few sectors where the computation chose old technology over new for expansion

Table 12.1 Computed Increases and Decreases in Activity Levels with Old- and New-Technology Input Structures for 1947–1958, 41 Order (tens of millions of 1947 dollars)

Sector	Base year output $\mathbf{x}^{47}$	$\Delta \mathbf{x}^{47-} \leq 0.49\bar{\mathbf{x}}^{47}$		$\Delta \mathbf{x}^{47-} \leq 0.69\bar{\mathbf{x}}^{47}$		$\Delta \mathbf{x}^{47-} \leq 1.00\bar{\mathbf{x}}^{47}$	
		Increases[a,b] ($\Delta \mathbf{x}^n$ or $\Delta \mathbf{x}^{47+}$)	*Decreases*[c] ($\Delta \mathbf{x}^{47-}$)	*Increases* ($\Delta \mathbf{x}^n$ or $\Delta \mathbf{x}^{47+}$)	*Decreases* ($\Delta \mathbf{x}^{47-}$)	*Increases* ($\Delta \mathbf{x}^n$ or $\Delta \mathbf{x}^{47+}$)	*Decreases* ($\Delta \mathbf{x}^{47-}$)
(1) Agriculture, etc.	4,617	3,327	2,279[e]	2,254	1,314	2,159	1,314
(2) Mining	974	452	294	427	294	380	294
(3) Transportation	2,256	471	682	457	682	449	682
(4) Communications	384	683	178[e]	620	109	604	109
(5) Utilities	691	1,380	198	1,383	198	1,376	198
(6) Trade & services	15,407	15,349	7,387[e]	16,087	8,199[e]	14,532	6,602[e]
(7) Food	4,538	2,826	1,325	2,756	1,325	2,700	1,325
(9) Textiles & products	2,433	2,100	1,284[e]	2,677	1,784[e]	3,579	2,600[e]
(10) Wood & products	554	291	286[e]	379	397[e]	610	579[e]
(11) Wooden containers	59	19	30[e]	30	42[e]	44	61[e]
(12) Furniture	260	226	130[e]	276	181[e]	355	264[e]
(13) Paper & products	640	518	189	503[d]	189	508[e]	189
(14) Printing & publishing	645	530	322[e]	385	186	378	186
(15) Basic chemicals	489	863	247[e]	967	343[e]	1,168	499[e]
(16) Synthetic materials	157	373	79[e]	436	110[e]	546	161[e]
(17) Drugs, soaps, etc.	319	582	163[e]	644	227[e]	733	330[e]
(18) Paint	134	86	68[e]	108	94[e]	148	138[e]
(19) Petroleum refining	800	933	409[e]	1,083	568[e]	1,324	828[e]
(20) Rubber prod., etc.	350	284	107	282	107	277	107
(21) Leather & shoes	375	228	217[e]	319	302[e]	470	440[e]
(22) Glass & products	116	51	34	51	34	52	34
(23) Stone & clay prod.	297	402	153[e]	330	88	312	88
(24) Iron & steel	1,122	279[d]	372	237[d]	372	167[d]	372
(25) Nonferrous metals	565	530	378[e]	679	525[e]	929	765[e]

Table 12.1 (*continued*)

Sector	Base year output $\mathbf{x}^{47}$	$\Delta\mathbf{x}^{47-} \le 0.49\bar{\mathbf{x}}^{47}$		$\Delta\mathbf{x}^{47-} \le 0.69\bar{\mathbf{x}}^{47}$		$\Delta\mathbf{x}^{47-} \le 1.00\bar{\mathbf{x}}^{47}$	
		Increases[a,b] ($\Delta\mathbf{x}^n$ or $\Delta\mathbf{x}^{47+}$)	*Decreases*[c] ($\Delta\mathbf{x}^{47-}$)	*Increases* ($\Delta\mathbf{x}^n$ or $\Delta\mathbf{x}^{47+}$)	*Decreases* ($\Delta\mathbf{x}^{47-}$)	*Increases* ($\Delta\mathbf{x}^n$ or $\Delta\mathbf{x}^{47+}$)	*Decreases* ($\Delta\mathbf{x}^{47-}$)
(26) Metal forming	954	852	478[e]	1,018	665[e]	1,284	969[e]
(27) Engines & turbines	98	97	49[e]	117	68[e]	145	98[e]
(28) Farm equipment	149	88	74[e]	114	102[e]	156	149[e]
(29) Constr. & mining eq.	176	86	87[e]	45	50	43	50
(30) Other industrial eq.	676	259[d]	205	260[d]	205	247[d]	205
(31) Office & comp. mach.	68	170	34[e]	189	47[e]	209	68[e]
(32) Home & service eq.	358	326	179[e]	396	249[e]	496	362[e]
(33) Electrical eq.	372	293	187[e]	369	260[e]	482	378[e]
(34) Communication eq.	270	633	136[e]	706	189[e]	802	275[e]
(35) Batteries, etc.	93	34	27	36	27	41	27
(36) Motor vehicles & eq.	1,234	1,088	845[e]	1,441	1,175[e]	2,015	1,712[e]
(37) Aircraft	149	758	102[e]	812	142[e]	883	207[e]
(38) Trains, ships, etc.	253	108	125[e]	60	72	62	72
(39) Instruments, etc.	137	188	70[e]	220	97[e]	263	142[e]
(40) Photo. apparatus	67	95	34[e]	109	47[e]	129	69[e]
(41) Misc. manufactures	326	281	168[e]	346	233[e]	457	340[e]

[a] $\Delta\mathbf{x}^n$ is a vector of increases in capacity with new technology input structure.
[b] $\Delta\mathbf{x}^{47+}$ is a vector of increases in capacity with 1947 input structure.
[c] $\Delta\mathbf{x}^{47-}$ is a vector of decumulation of capacity with 1947 input structure.
[d] Sectors where Δx_j^{47+} rather than Δx_j^n have nonzero values.
[e] Sectors where decumulation is greater than the 3 percent per year minimum.

Table 12.2 Computed Increases and Decreases in Activity Levels with Old- and New-Technology Input Structures for 1947–1961, 41 Order (tens of millions of 1947 dollars)

Sector	Base year output $\mathbf{x}^{47}$	$\Delta\mathbf{x}^{47-} \leq 0.51\bar{\mathbf{x}}^{47}$		$\Delta\mathbf{x}^{47-} \leq 0.58\bar{\mathbf{x}}^{47}$		$\Delta\mathbf{x}^{47-} \leq 1.00\bar{\mathbf{x}}^{47}$	
		Increases[a,b] ($\Delta\mathbf{x}^{n}$ or $\Delta\mathbf{x}^{47+}$)	*Decreases*[c] ($\Delta\mathbf{x}^{47-}$)	*Increases* ($\Delta\mathbf{x}^{n}$ or $\Delta\mathbf{x}^{47+}$)	*Decreases* ($\Delta\mathbf{x}^{47-}$)	*Increases* ($\Delta\mathbf{x}^{n}$ or $\Delta\mathbf{x}^{47+}$)	*Decreases* ($\Delta\mathbf{x}^{47-}$)
(1) Agriculture, etc.	4,617	3,538	2,365[e]	3,383	2,171[e]	2,655	1,603
(2) Mining	974	646	358	643	358	585	358
(3) Transportation	2,256	721	832	679	832	676	832
(4) Communications	384	866	197[e]	815	133	800	133
(5) Utilities	691	1,868	357[e]	1,757	242	1,742	242
(6) Trade & services	15,407	18,217	7,665[e]	19,173	8,671[e]	18,159	7,681[e]
(7) Food	4,538	4,586	2,368[e]	3,741	1,616	3,643	1,616
(9) Textiles & products	2,433	2,564	1,332[e]	2,777	1,507[e]	4,004	2,600[e]
(10) Wood & products	554	340	297[e]	375	336[e]	626	579[e]
(11) Wooden containers	59	22	31[e]	27	36[e]	48	61[e]
(12) Furniture	260	259	135[e]	278	153[e]	385	264[e]
(13) Paper & products	640	734	231	727	231	714[e]	231
(14) Printing & publishing	645	656	334[e]	698	378[e]	531	227
(15) Basic chemicals	489	1,073	256[e]	1,130	289[e]	1,372	499[e]
(16) Synthetic materials	157	492	82[e]	521	93[e]	660	161[e]
(17) Drugs, soaps, etc.	319	746	169[e]	768	191[e]	890	330[e]
(18) Paint	134	106	70[e]	114	80[e]	165	138[e]
(19) Petroleum refining	800	1,093	424[e]	1,149	480[e]	1,473	828[e]
(20) Rubber prod., etc.	350	400	130	405	130	392	130
(21) Leather & shoes	375	245	226[e]	276	255[e]	478	440[e]
(22) Glass & products	116	77	42	77	42	78	42
(23) Stone & clay prod.	297	480	159[e]	510	180[e]	411	108
(24) Iron & steel	1,122	473[d]	454	452[d]	454	362[d]	454
(25) Nonferrous metals	565	645	392[e]	695	443[e]	1,026	765[e]

Table 12.2 (*continued*)

Sector	Base year output $\mathbf{x}^{47}$	$\Delta\mathbf{x}^{47-} \leq 0.51\bar{\mathbf{x}}^{47}$		$\Delta\mathbf{x}^{47-} \leq 0.58\bar{\mathbf{x}}^{47}$		$\Delta\mathbf{x}^{47-} \leq 1.00\bar{\mathbf{x}}^{47}$	
		Increases[a,b] ($\Delta\mathbf{x}^n$ or $\Delta\mathbf{x}^{47+}$)	*Decreases*[c] ($\Delta\mathbf{x}^{47-}$)	*Increases* ($\Delta\mathbf{x}^n$ or $\Delta\mathbf{x}^{47+}$)	*Decreases* ($\Delta\mathbf{x}^{47-}$)	*Increases* ($\Delta\mathbf{x}^n$ or $\Delta\mathbf{x}^{47+}$)	*Decreases* ($\Delta\mathbf{x}^{47-}$)
(26) Metal forming	954	1,054	496[e]	1,110	562[e]	1,462	969[e]
(27) Engines & turbines	98	101	50[e]	108	57[e]	149	98[e]
(28) Farm equipment	149	80	76[e]	91	86[e]	148	149[e]
(29) Constr. & mining eq.	176	93	90[e]	104	102[e]	57	61
(30) Other industrial eq.	676	394[d]	250	393[d]	251	384[d]	250
(31) Office & comp. mach.	68	232	35[e]	242	40[e]	273	68[e]
(32) Home & service eq.	358	418	186[e]	446	210[e]	586	362[e]
(33) Electrical eq.	372	387	194[e]	417	219[e]	572	378[e]
(34) Communication eq.	270	990	141[e]	1,021	159[e]	1,158	275[e]
(35) Batteries, etc.	93	55	33	55	33	60	33
(36) Motor vehicles & eq.	1,234	1,480	877[e]	1,600	992[e]	2,370	1,712[e]
(37) Aircraft	149	858	106[e]	882	120[e]	983	207[e]
(38) Trains, ships, etc.	253	110	130[e]	127	147[e]	74	88
(39) Instruments, etc.	137	259	73[e]	273	82[e]	334	142[e]
(40) Photo. apparatus	67	116	35[e]	122	40[e]	150	69[e]
(41) Misc. manufactures	326	338	174[e]	362	197[e]	507	340[e]

[a] $\Delta\mathbf{x}^n$ is a vector of increases in capacity with new technology input structure.

[b] $\Delta\mathbf{x}^{47+}$ is a vector of increases in capacity with 1947 input structure.

[c] $\Delta\mathbf{x}^{47-}$ is a vector of decumulation of capacity with 1947 input structure.

[d] Sectors where x_j^{47+} rather than x_j^n have nonzero values.

[e] Sectors where decumulation is greater than the 3 percent per year minimum.

Table 12.3 Computed Increases and Decreases in Activity Levels with Old- and New-Technology Input Structures for 1958–1961, 41 Order (tens of millions of 1947 dollars)

Sector	Base year output $\mathbf{x}^{58}$	$\Delta\mathbf{x}^{58-} \leq 0.20\bar{\mathbf{x}}^{58}$	
		Increases[a,b] ($\Delta\mathbf{x}^{n}$ or $\Delta\mathbf{x}^{58+}$)	*Decreases*[c] ($\Delta\mathbf{x}^{58-}$)
(1) Agriculture, etc.	5,599	364	166
(2) Mining	1,120	184	48
(3) Transportation	2,190	149	75
(4) Communications	842	188	25
(5) Utilities	1,782	376	60
(6) Trade & services	23,775	3,259	709
(7) Food	6,029	816	196
(9) Textiles & products	3,106	1,174	711[e]
(10) Wood & products	615	53	22
(11) Wooden containers	37	12	9[e]
(12) Furniture	350	110	82[e]
(13) Paper & products	958	204	34
(14) Printing & publishing	835	140	27
(15) Basic chemicals	986	247	33
(16) Synthetic materials	385	150	13
(17) Drugs, soaps, etc.	698	303	152[e]
(18) Paint	153	22	5
(19) Petroleum refining	1,315	182	42
(20) Rubber prod., etc.	476	114	17
(21) Leather & shoes	364	92	79[e]
(22) Glass & products	125	23[d]	5
(23) Stone & clay prod.	498	93	19
(24) Iron & steel	997	143[d]	44
(25) Nonferrous metals	690	124	30
(26) Metal forming	1,290	212	45
(27) Engines & turbines	130	40	36[e]
(28) Farm equipment	154	31	42[e]
(29) Constr. & mining eq.	175	10	7
(30) Other industrial eq.	719	130[d]	30
(31) Office & comp. mach.	165	69	7
(32) Home & service eq.	459	214	129[e]
(33) Electrical eq.	465	219	130[e]
(34) Communication eq.	701	558	197[e]
(35) Batteries, etc.	101	19	4
(36) Motor vehicles & eq.	1,484	441	78
(37) Aircraft	730	282	172[e]
(38) Trains, ships, etc.	231	7	8
(39) Instruments	231	80	8
(40) Photo. apparatus	115	47	26[e]
(41) Misc. manufactures	439	97	45[e]

[a] $\Delta\mathbf{x}^{n}$ is a vector of increases in capacity with new technology input structure.
[b] $\Delta\mathbf{x}^{58+}$ is a vector of increases in capacity with 1958 input structure.
[c] $\Delta\mathbf{x}^{58-}$ is a vector of decumulation of capacity with 1958 input structure.
[d] Sectors where Δx_j^{58+} rather than Δx_j^{n} have nonzero values.
[e] Sectors where decumulation is greater than the 1 percent per year minimum.

of capacity, that is, where $\Delta \mathbf{x}^{t+}$ was not zero, are marked by a "d". These are sectors where incremental technology, as estimated here, brought net disadvantages as compared with the technology of the base year when used in combination with the structures chosen for other sectors. Those sectors where incremental technology is inferior are generally ones whose 1958 structures were already shown to be disadvantageous in Chapter 10.

Industries in the retirements column that are marked with an "e" are ones that decumulated old capacity by more than the minimum. The optimizing procedure tends, of course, to allocate as much investment as possible to those sectors where new technology has the greatest factor-saving advantage over old. With higher sectoral replacement ceilings, retirements tend to be concentrated in relatively fewer sectors. Therefore, there are fewer such sectors when replacement ceilings are higher. Retirement of old capacity has higher priority in textiles (9), drugs, soaps, and cosmetics (17), and the various electrical and service machinery sectors (31)–(34), among others, than it does in printing and publishing (14), stone and clay products (23), construction and mining equipment (29), and trains, ships, and cycles (38). Note that aggregation poses problems here. With very fine specification of both new and old structures, one could very likely see some high (low) ranking opportunities in areas where the average advantage of new structures is low (high). In reality, there are some profitable innovative opportunities in sectors where new techniques are on the whole unimpressive, and vice versa.

Because investment data for the nonmanufacturing sectors were scarce, sectors (1)–(6) are very heavily aggregated. More than minimal replacement was generally warranted for only two of these sectors, namely agriculture (1) and trade and services (6). When the linear programming system allocates only minimal investment resources to nonmanufacturing sectors, the amount available to manufacturing sectors becomes relatively liberal; and the great majority of all manufacturing sectors can replace the maximum permitted by the sectoral decumulation ceilings between 1947 and 1961. It would be interesting to see whether disaggregation of the nonmanufacturing sectors would divert significant investment resources from the manufacturing sectors in these calculations.

Some question of the consistency of the input-output coefficients, capital coefficients, and investment data still remains. For it appears that observed gross investment is more than sufficient to cover the postulated 3 percent per year lower limit on scrappage in the longer intervals, 1947–1958 and 1947–1961, and to permit additional replacement in many sectors where it was economically warranted. It is not clear why, between 1958 and 1961, observed gross investment was only sufficient to permit 1 percent per year replacement of initial capacity, with a modest amount left over for replacement. Is the apparent "tightness" of investment for 1958–1961 due to data problems; to low final demand in the recession year, 1958; or to the inaccuracy of assuming a one-year lag between investment and capacity, that is, to cumulating investment from 1958 through 1960 to explain additions required in 1961 capacity; or to all three?

Figures 12.1 through 12.6 give more details of how the system performed. Figures 12.1–12.3 are comparisons of computed with actual gross investment, that is, gross additions to capital stock, in each sector. The results shown in these figures were based on decumulation ceilings of 10 percent per year for 1947–1958 and 1947–1961, and 7 percent per year for 1958–1961. The minimal rates of decumulation of capacity were 3 percent per year for the first two intervals, and 1 percent per year for the last.

Figures 12.4b, 12.5b, and 12.6b show total factor requirements as estimated in the computations and actual factor employment in the terminal year in each sector. Figures 12.4a, 12.5a, and 12.6a are comparisons of actual factor employment in the final year with employment required to produce the bill of goods of that last year with intermediate input and primary factor coefficients of the *base year*. Thus, 12.4a, 12.5a, and 12.6a tell what error is introduced into industry factor-cost estimates by reliance on obsolescent input-output and primary factor coefficients. Figures 12.4b, 12.5b, and 12.6b show what error remains when embodied technical change updates these coefficients in our present linear programming model. The charts can speak for themselves. Estimated primary factor requirements based on outdated coefficients are almost always larger than actual factor requirements; hence, the clustering of points below the 45-degree line in the "a" graphs. When embodied technical change is introduced with the linear programming system, there is no very clear tendency to over- or underestimate labor requirements. Points tend to lie about equally above and below the 45-degree line, except possibly for the 1958–1961 comparisons. Here, computed requirements seem slightly to understate actual requirements.

12.5 Conclusions

On the whole, figures 12.1–12.3 show that the estimated allocation of investment among sectors is not unreasonable when lower and upper limits are placed on the scrappage of initial capacity. Over the periods studied, an accelerator and a replacement rate, limited to between 3 and 6 percent per year, give a fair approximation of the allocation of new investment among the

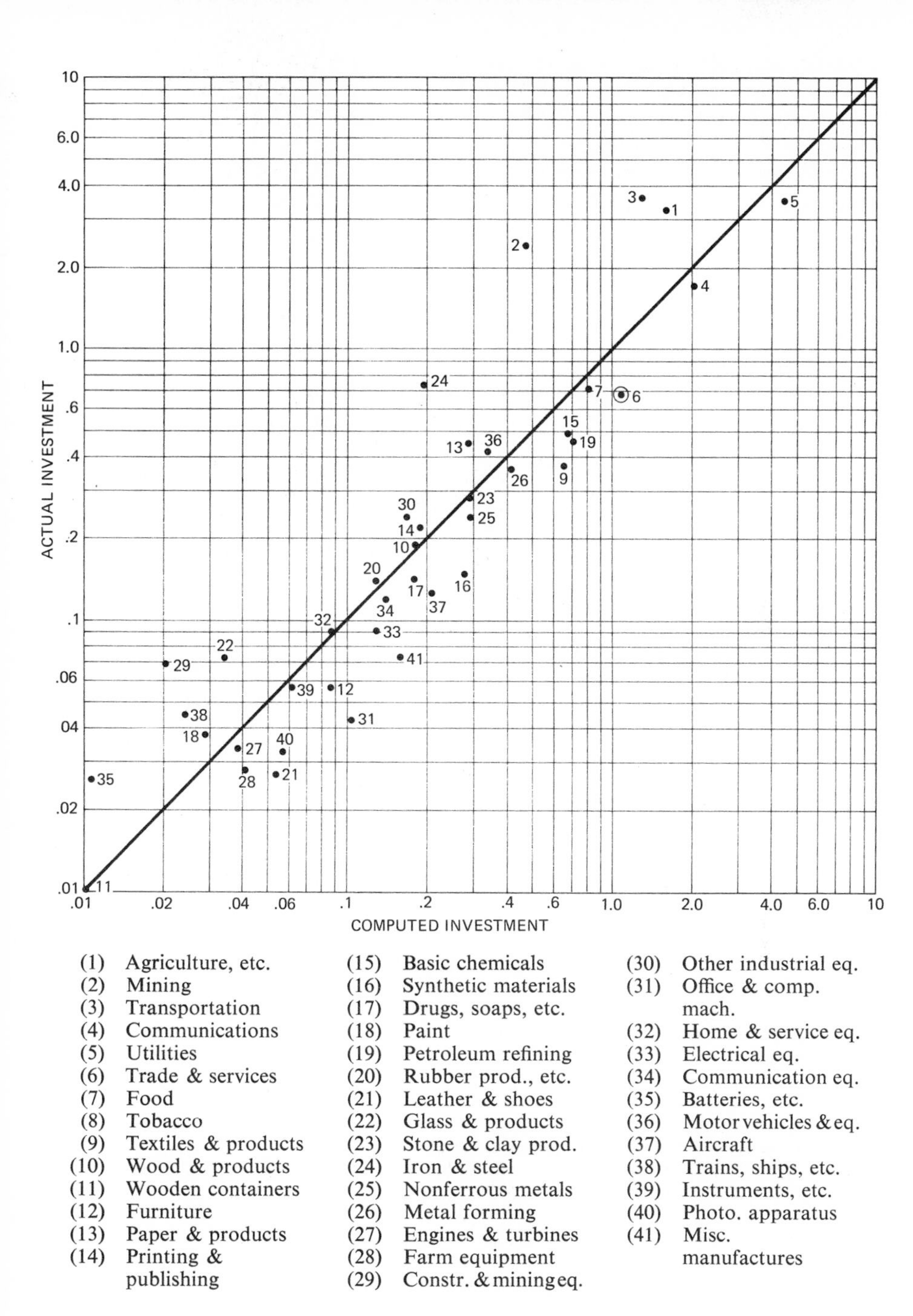

12.1 Computed and Actual Gross Investment Expenditure from 1947 to 1957 (dollars times 10^{10} in 1947 prices). Each point indicates the value of investment expenditure by a 41-order sector computed with the linear programming model using a 10 percent per year ceiling and observed investment expenditure by the same sector (for circled points, multiply scales by 10).

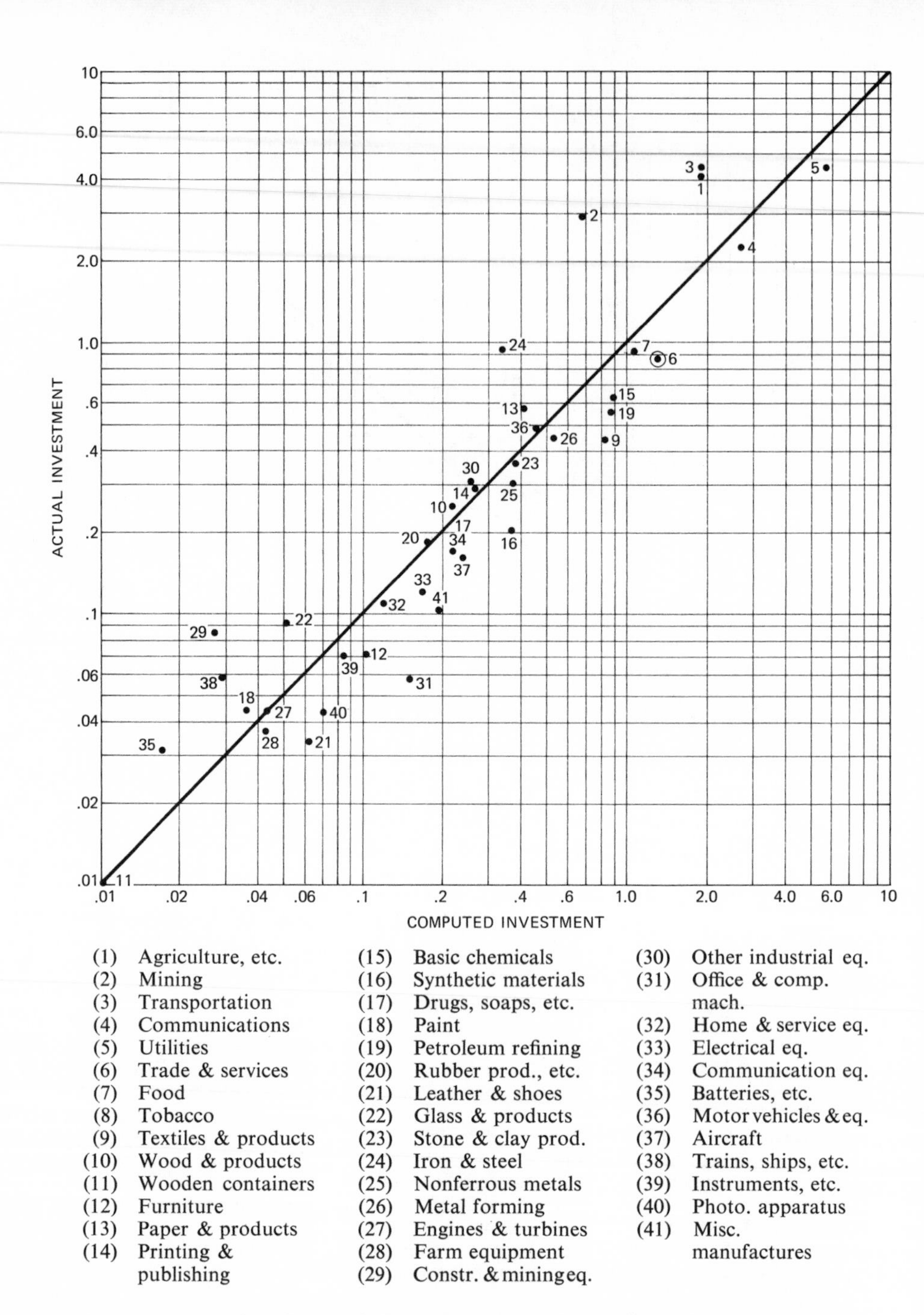

12.2 Computed and Actual Gross Investment Expenditure from 1947 to 1960 (dollars times 10^{10} in 1947 prices). Each point indicates the value of investment expenditure by a 41-order sector computed with the linear programming model using a 10 percent per year ceiling and observed investment expenditure by the same sector (for circled points, multiply scales by 10).

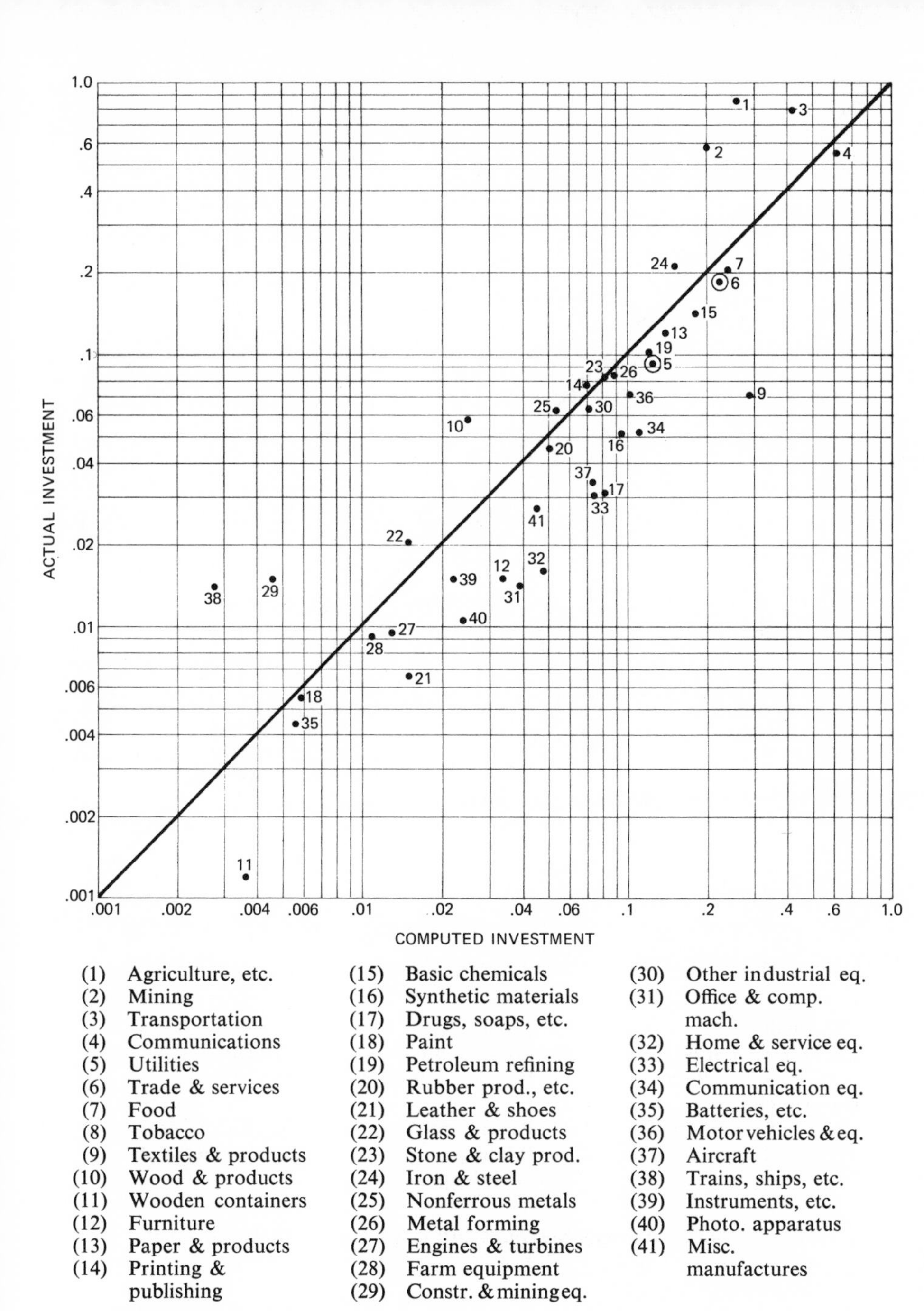

12.3 Computed and Actual Gross Investment Expenditure from 1958 to 1960 (dollars times 10^{10} in 1947 prices). Each point indicates the value of invest ment expenditure by a 41-order sector computed with the linear programming model using a 7 percent per year ceiling and observed investment expenditure by the same sector (for circled points, multiply scales by 10).

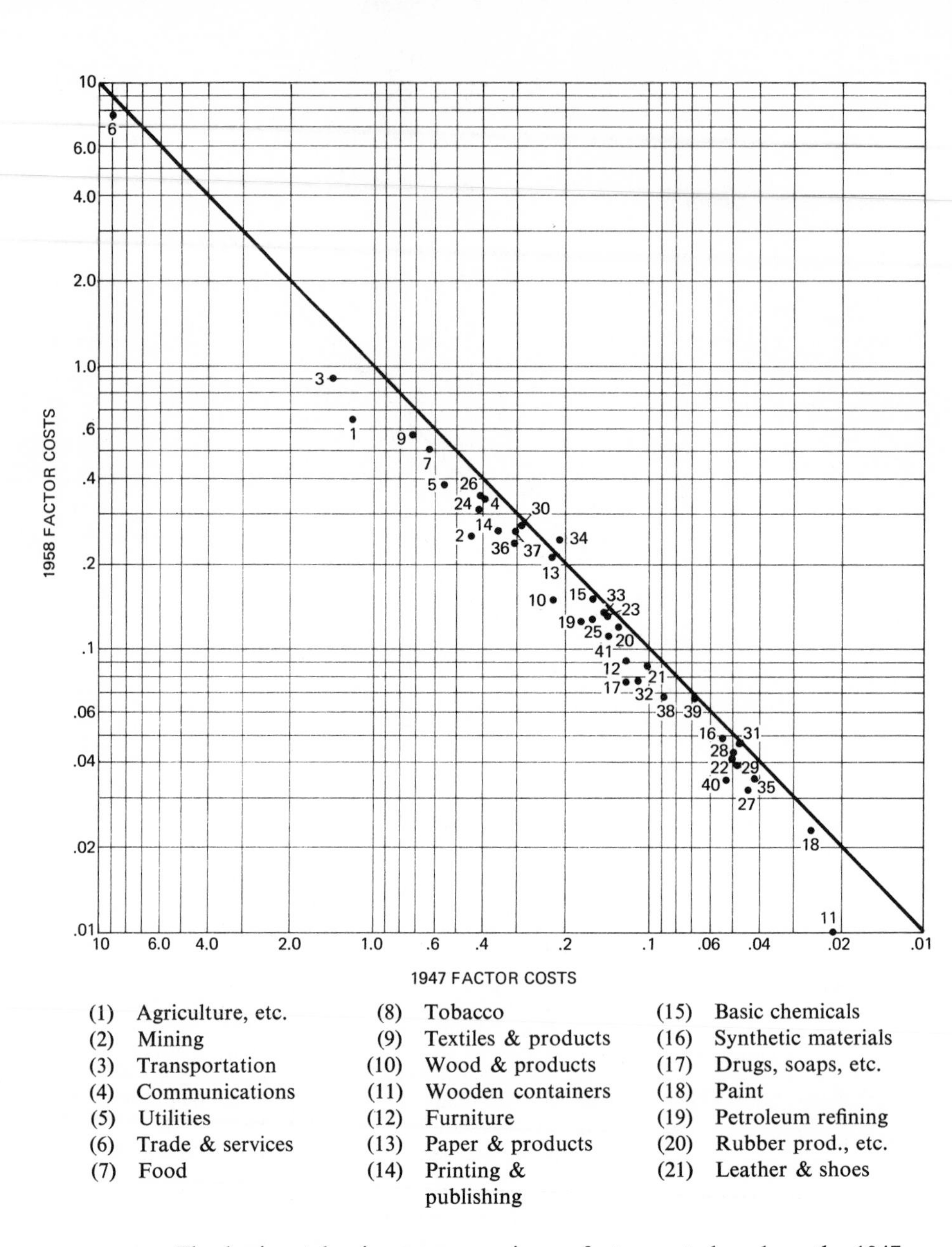

a. The horizontal axis measures primary factor costs based on the 1947 structures.

12.4 Value of Primary Factor Requirements in Each 41-Order Sector to Deliver 1958 Final Demand Based on 1947 and 1958 Technologies and the Linear Programming Model (dollars time 10^{10} in 1947 prices). Each point indicates the value of combined labor and capital charges for a particular sector computed with two different sets of structures; the vertical axes measure observed primary factor costs for 1958.

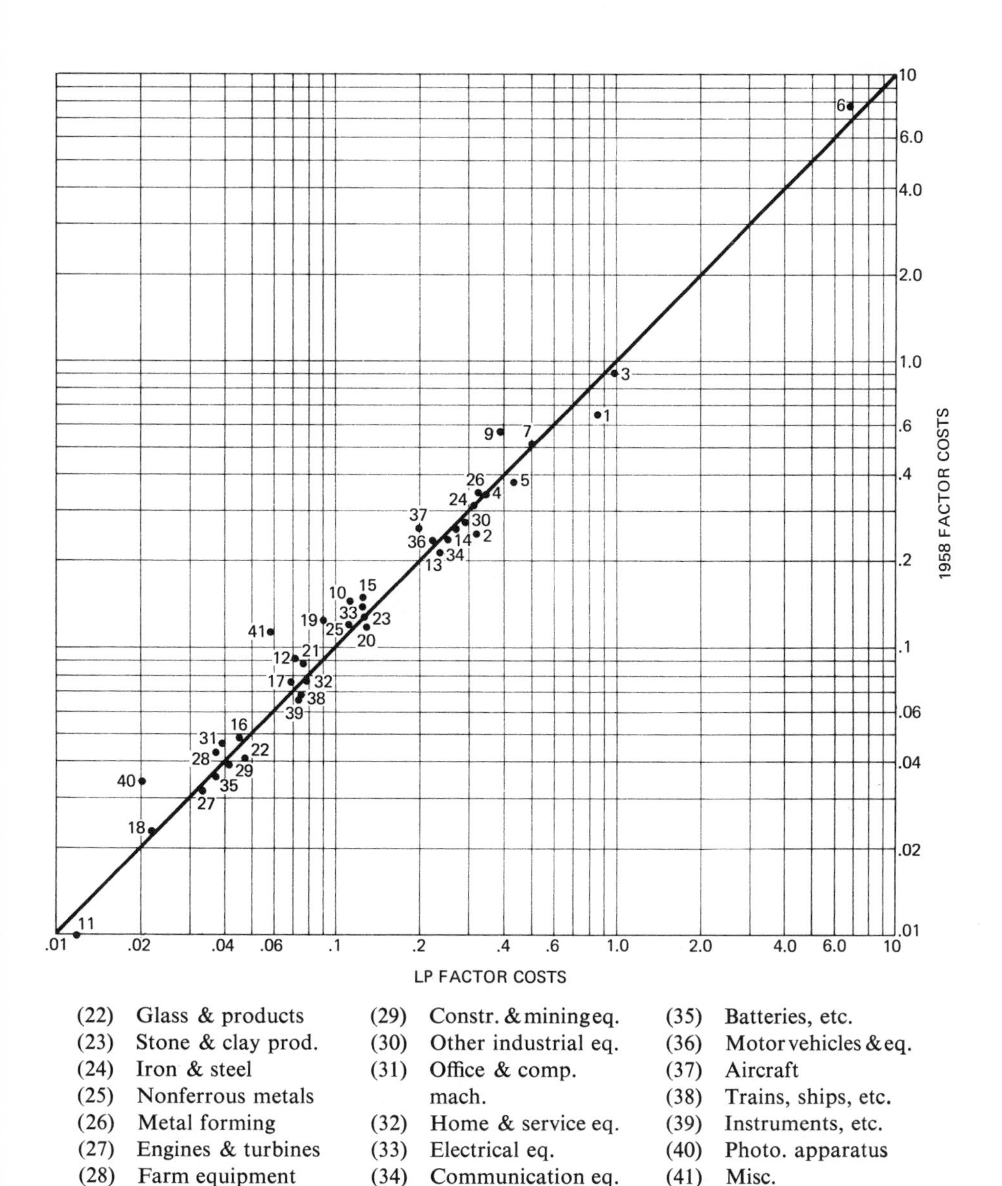

(22)	Glass & products	(29)	Constr. & mining eq.	(35)	Batteries, etc.
(23)	Stone & clay prod.	(30)	Other industrial eq.	(36)	Motor vehicles & eq.
(24)	Iron & steel	(31)	Office & comp. mach.	(37)	Aircraft
(25)	Nonferrous metals			(38)	Trains, ships, etc.
(26)	Metal forming	(32)	Home & service eq.	(39)	Instruments, etc.
(27)	Engines & turbines	(33)	Electrical eq.	(40)	Photo. apparatus
(28)	Farm equipment	(34)	Communication eq.	(41)	Misc. manufactures

b. The horizontal axis measures factor costs computed on the basis of the linear programming model.

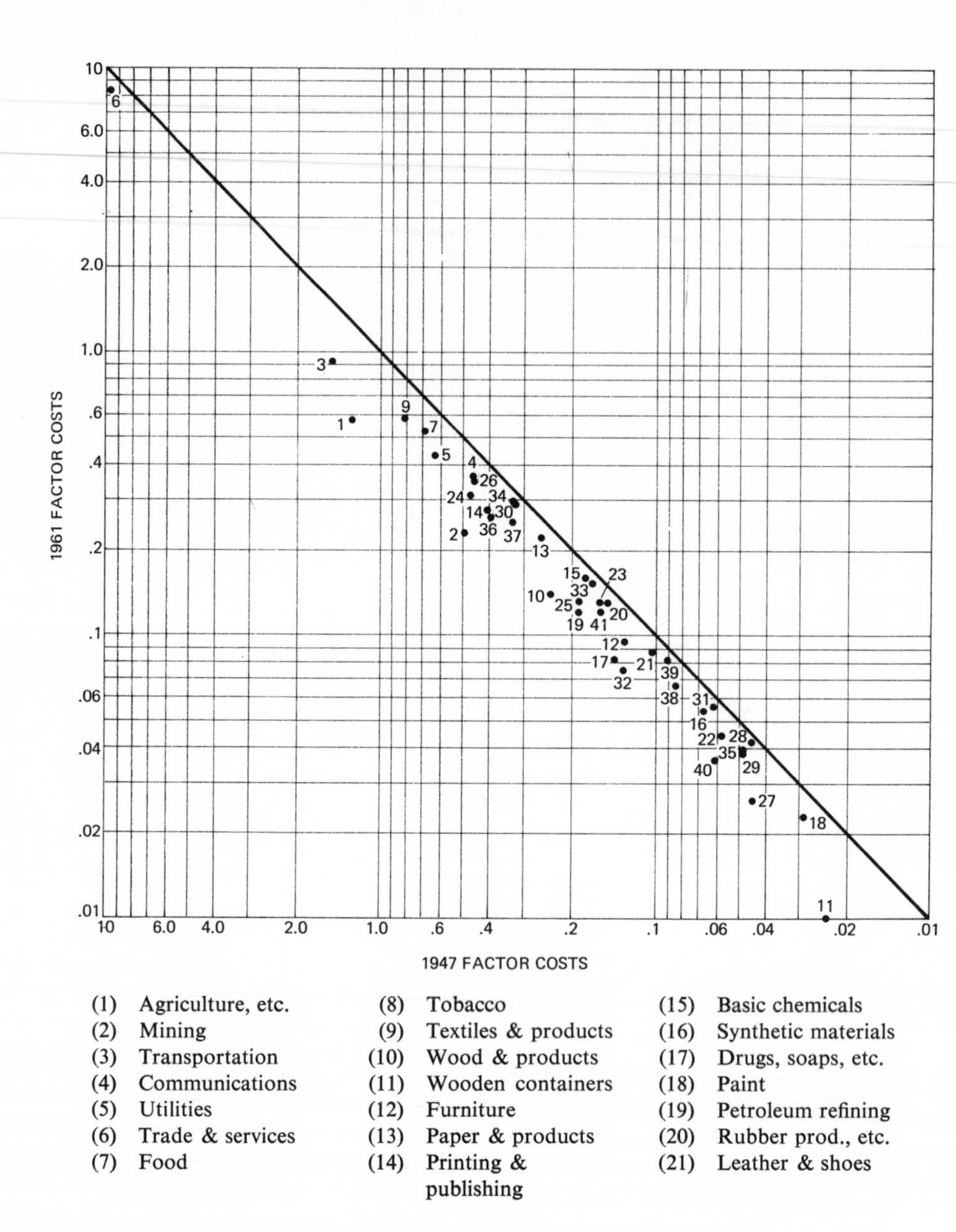

a. The horizontal axis measures primary factor costs based on the 1947 structures.

12.5 Value of Primary Factor Requirements in Each 41-Order Sector to Deliver 1961 Final Demand Based on 1947 and 1961 Technologies and the Linear Programming Model (dollars times 10^{10} in 1947 prices). Each point indicates the value of combined labor and capital charges for a particular sector computed with two different sets of structures; the vertical axes measure observed primary factor costs for 1961.

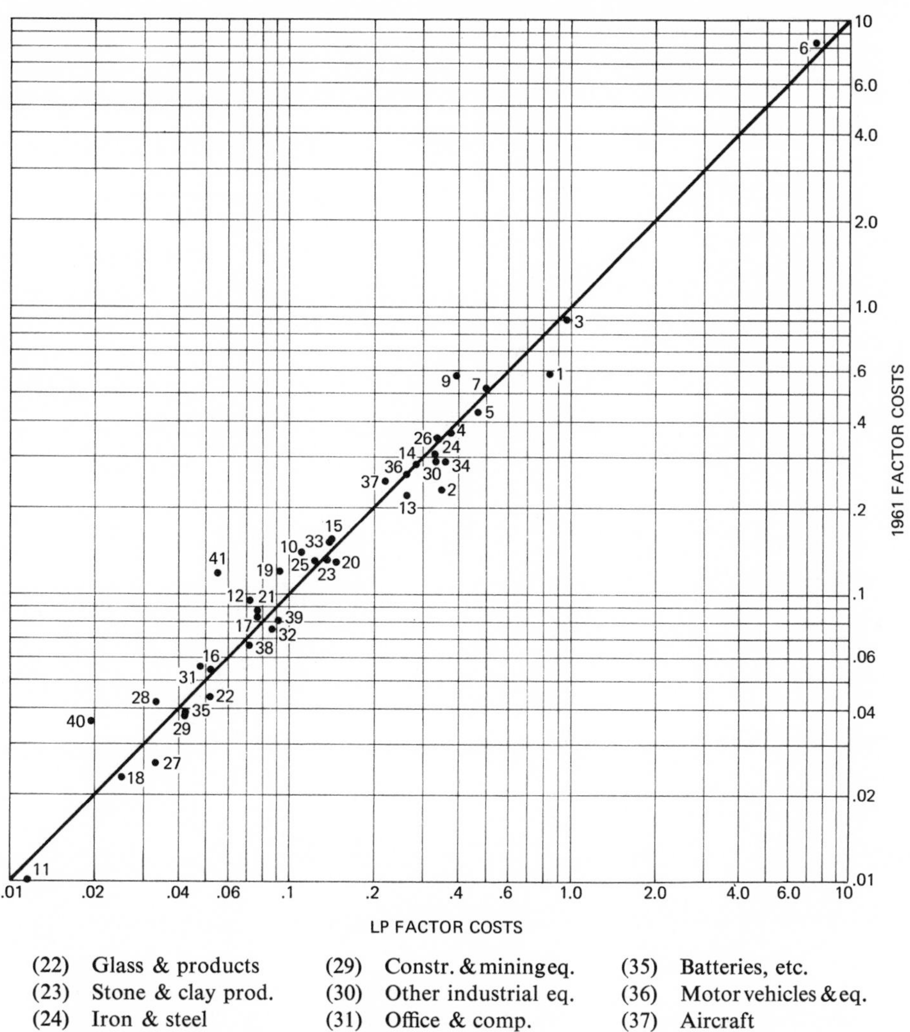

(22) Glass & products
(23) Stone & clay prod.
(24) Iron & steel
(25) Nonferrous metals
(26) Metal forming
(27) Engines & turbines
(28) Farm equipment
(29) Constr. & mining eq.
(30) Other industrial eq.
(31) Office & comp. mach.
(32) Home & service eq.
(33) Electrical eq.
(34) Communication eq.
(35) Batteries, etc.
(36) Motor vehicles & eq.
(37) Aircraft
(38) Trains, ships, etc.
(39) Instruments, etc.
(40) Photo. apparatus
(41) Misc. manufactures

b. The horizontal axis measures factor costs computed on the basis of the linear programming model.

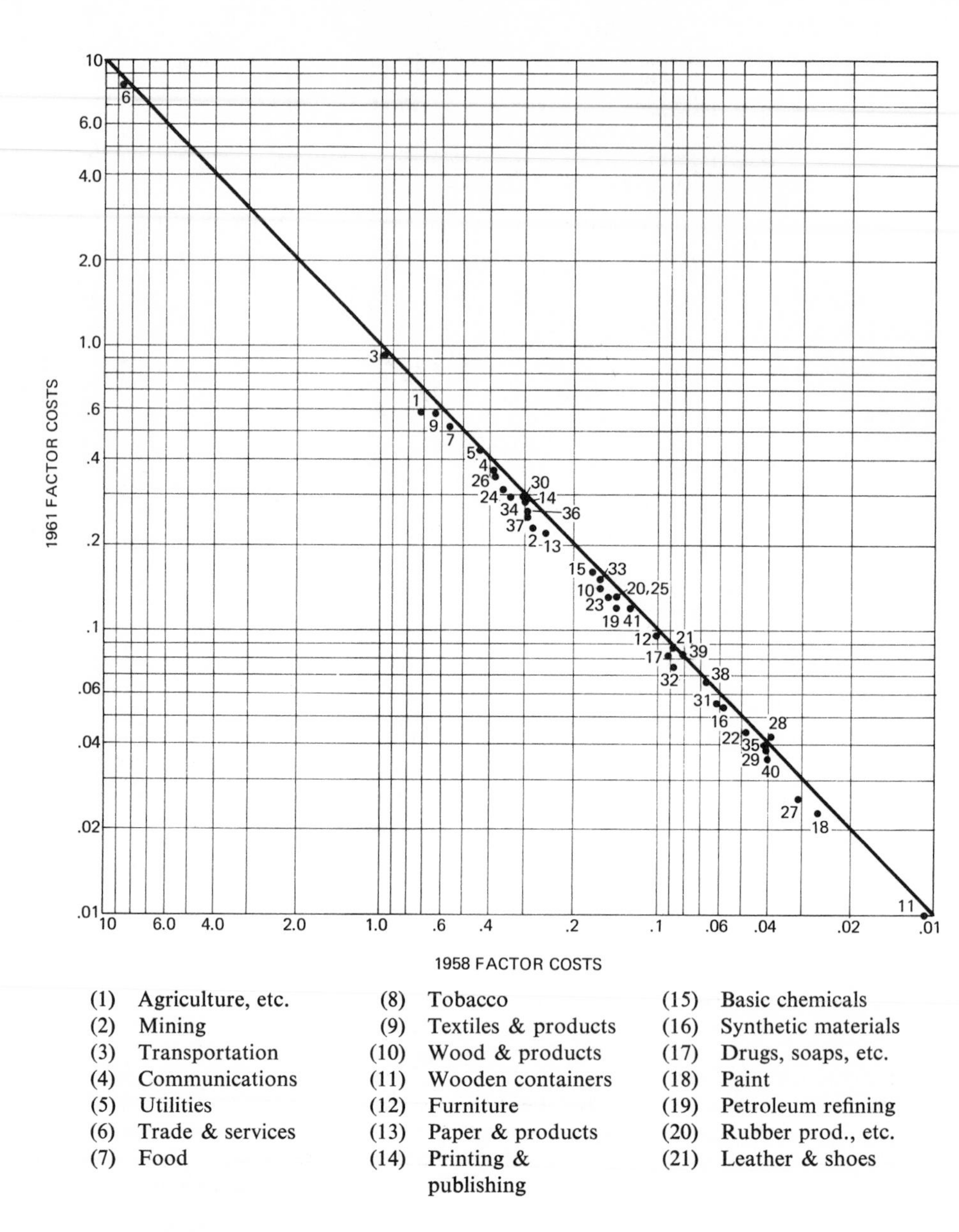

a. The horizontal axis measures primary factor costs based on the 1958 structures.

12.6 Value of Primary Factor Requirements in Each 41-Order Sector to Deliver 1961 Final Demand Based on 1958 and 1961 Technologies and the Linear Programming Model (dollars times 10^{10} in 1947 prices). Each point indicates the value of combined labor and capital charges for a particular sector computed with two different sets of structures; the vertical axes measure observed primary factor costs for 1961.

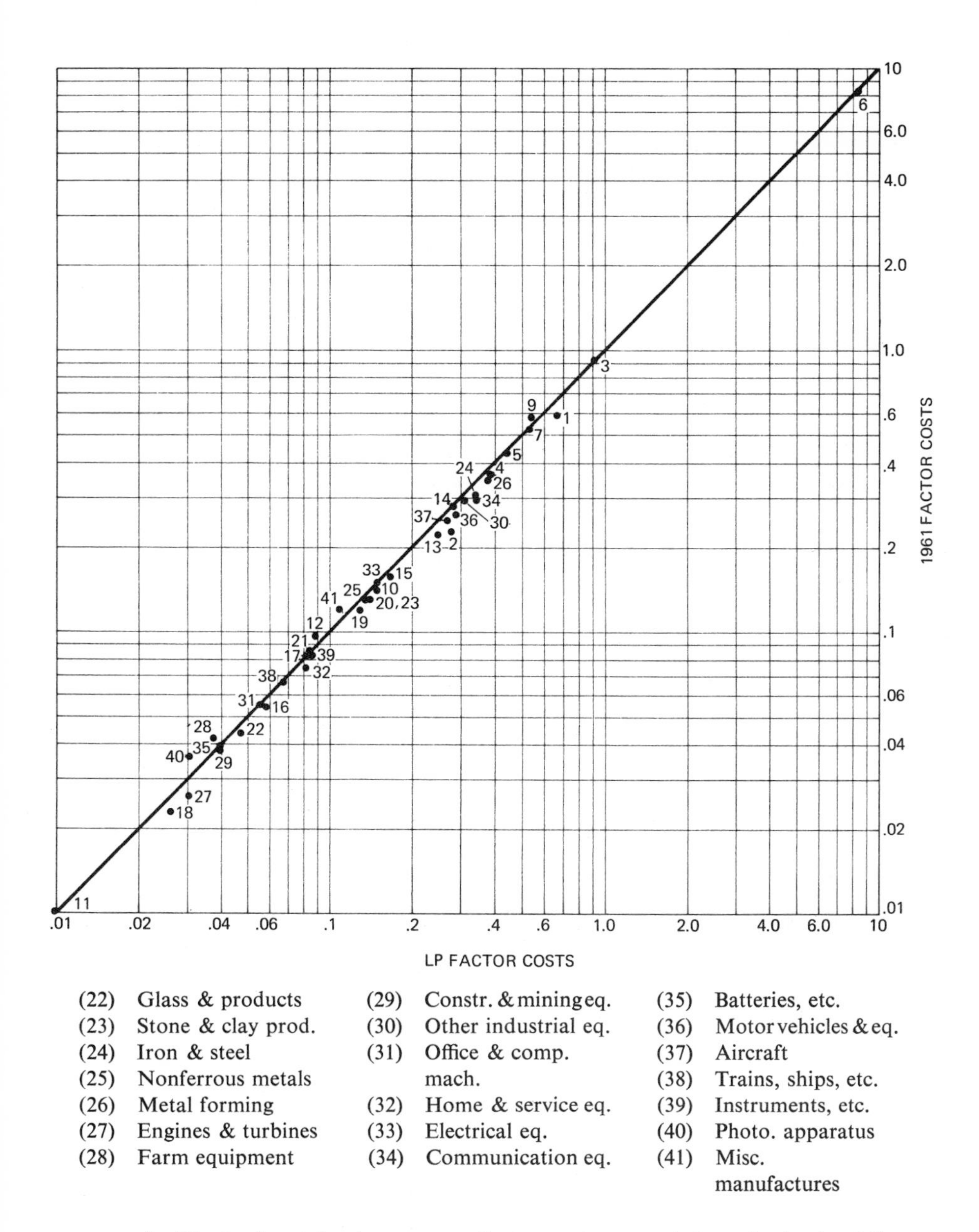

(22) Glass & products
(23) Stone & clay prod.
(24) Iron & steel
(25) Nonferrous metals
(26) Metal forming
(27) Engines & turbines
(28) Farm equipment
(29) Constr. & mining eq.
(30) Other industrial eq.
(31) Office & comp. mach.
(32) Home & service eq.
(33) Electrical eq.
(34) Communication eq.
(35) Batteries, etc.
(36) Motor vehicles & eq.
(37) Aircraft
(38) Trains, ships, etc.
(39) Instruments, etc.
(40) Photo. apparatus
(41) Misc. manufactures

b. The horizontal axis measures factor costs computed on the basis of the linear programming model.

specified sectors. Comparison of input structures with new and old technologies serves to establish a hierarchy of advantages of replacement opportunities among different sectors in this system, but the overall level of investment available over the period is predetermined. Thus, any estimates of new-technology structures that preserve the same ranking of advantages of new over old structures among industries will yield approximately the same distribution of investment. In a more dynamic formulation, investment levels themselves would depend on the development paths of capital-producing industries and their suppliers.

The use of incremental coefficients to represent new technology structures is a special kind of trend extrapolation, where change in coefficients is made proportional to investment rather than to time. The basic weakness of any extrapolation, with present data limitations, is that it rests on only two sets of observations—coefficients for 1947 and the corresponding coefficients for 1958. With such sparse and rough information, there is not much sense in trying to demonstrate that one type of extrapolation gives better results than another. The long-run advantage of the method described here is that it will permit us to make use of better new-technology information as it becomes available. When reliable new-technology coefficients are known, they can be substituted for the makeshift incremental coefficients. In forecasting future input-output structures, data gathering resources should be devoted to the search for reasonable approximations of such new-technology structures. Mutually consistent rates of adoption of the new technologies can then be estimated according to the system just presented.

Chapter 13 Conclusion

In the present study, structural change means changes in input-output coefficients. The analysis covers the period 1939 to 1961 at various levels of detail; it deals most intensively with the interval between 1947 to 1958. The input-output model has "fixed coefficients," a feature that has made its implementation practical on a regular basis. But a fixed-coefficient model does not make all the distinctions that are traditional in the economic analysis of production. Thus, economic theory distinguishes substitution from technological change, long-run from short-run substitutability, invention from innovation, and innovation from diffusion. Invention in itself does not affect input-output coefficients, but innovation and diffusion do. Substitutions, long- and short-run, are not distinguished from the diffusion of new technical alternatives. In sum, changes in input-output structure subsume long- and short-run substitution along with innovation and diffusion of new techniques. Operationally, some of these distinctions are very difficult to make, even in a framework that specifies variable coefficients; it was not possible to make them here.

Although our data did not distinguish all of these aspects of change directly, some conclusions about the nature and importance of substitution and about rates of diffusion of new input combinations do emerge. Chapters 10 and 11 demonstrate that a significant portion of structural change between 1947 and 1958 resulted from the assimilation of new techniques rather than classical substitution. In Chapter 10, the 1947 and 1958 structures were treated as alternatives for each sector, and the optimal mix of 1947 and 1958 structures was computed by linear programming techniques. Within a wide range of primary factor price variations, most 1958 industrial structures were superior to those of 1947. Chapter 11 showed that the factor-saving advantages of 1958 structures in most sectors were not contingent on 1947–1958 changes in the structures of other sectors. Only two trigger effects were detected at 76 order. Otherwise, 1958 input structures were superior to those of 1947, assuming either 1947 or 1958 primary and intermediate factor supply conditions. However, there is evidence of some adaptivity. The factor-saving advantages of new structures in most individual sectors were greater when computed with new than with old structures prevailing in all other sectors. Thus, the evidence in Chapter 11 does not preclude the possibility that some structural changes were induced by changes in other sectors—that changing prices of intermediate goods led to some substitutions among inputs. These substitutions were much smaller than direct primary factor economies, particularly labor economies, in the total factor-saving picture.

Direct labor saving is the most striking feature of structural change. Labor coefficients are very large compared to those for most other inputs specified; they decrease over time in virtually all sectors. Viewed in the context of a closed system, labor saving itself becomes a form of adaptive change. While the input structures of most intermediate and capital items improve steadily, the input structure for labor "deteriorates," that is, the standard of living and investment in human capital get higher and higher. This provides strong incentive to substitute other intermediate and capital inputs for labor when the opportunity arises. Economists have only a very incomplete understanding of how changing cost conditions affect the new technologies that will be devised. However, it is clear that a large proportion of new techniques that were adopted economized direct labor.

The dominance of *direct* labor saving in total factor economies is less readily explained. Industrial specialization patterns seem to be important. Choice of techniques means committing the firm to particular processes for some future period. Industrial organization tends toward "economic homeostasis": the advantage of substituting one intermediate input for another depends on expected developments in other sectors, and such developments are difficult to anticipate. To minimize the difficulties of coordination, firms tend to internalize those activities whose changes must dovetail. Under modern conditions, the course of wages is easier to gauge. Changes that economize direct labor are favored because they are more readily evaluated with today's information on wages and capital goods prices.

Economic homeostasis may also explain the stability of intermediate transactions in relation to gross output. In a growing economy, one might expect to find increasing opportunities for division of labor and thus more and more indirectness; actually, the proportion of intermediate transactions grows very little over the period studied. Again, we look for a linkage between organization and change. Division of labor might lead to greater efficiency in production at any given time, but a fine specialization of tasks imposes some rigidity and presents obstacles to rapid change. Thus, our economy may sacrifice some of the potential economies of specialization, as well as of "learning," for the dynamic requirements of flexibility and coordination of technological progress.

While the total amount of indirectness is very stable, there are significant shifts in patterns of industrial specialization. Most sectors increase their dependence on the general sectors—producers of services, communications, energy, transportation, and trade. Relatively more coordinating activity is required as the economy grows. However, some increases in services may represent transfer of function from product-oriented to specialized service establishments, rather than net growth in service activity.

Increased requirements for general inputs are counterbalanced by decreases in other sets of coefficients. The relative contributions of materials producers decline across the board. Improvements in materials and in the technologies of using them have permitted increasing diversification in this area. Durable goods are made of lighter and cheaper materials, and their designs are more compact.

Along with construction, the metalworking sectors deliver the great bulk of investment goods for expansion of capacity, for replacement, and for maintenance and repair. Structural change comes with shifting methods of production and with change in the character of capital goods to be delivered. On the production side, numerical control and automatic assembly techniques were just beginning to be introduced at the end of the period studied. Because of increasing complexity and sophistication of capital goods, however, changes in metalworking specialization patterns occurred during the period. The result was a marked growth in requirements from electrical, electronics, and instrumentation sectors and a decline in requirements from general metalworkers.

For a period of alleged technological revolution, structural change seems a bit sluggish. There were a few areas of very rapid change—labor, aircraft, instruments, coal mining—but in most sectors change was very gradual. Diffusion of new developments takes time. The model of embodied technological change described in Chapter 12 shows how structural change can be constrained by the rate of gross investment. Even with rapid development of advantageous new alternatives, the level of investment now prevailing in the United States can permit only moderate rates of structural improvement. In addition, our methodology hides some changes that might otherwise be observed. Compared with economy-wide production functions, input-output analysis provides a very detailed view of economic structure. However, by the standards of microtheory, input-output sectors at 38 or 76 order are still broad aggregates. Technological change has many more than 38 or 76 or even 5000 dimensions.

Aggregation problems are most serious in the present treatment of primary factors. Capital-to-capacity ratios expressed in deflated value units change little over time, declining moderately for most sectors. However, there is clear evidence of changing qualitative composition of capital over the period. This change is obscured when capital items are summed and expressed in deflated value units. Similar problems arise in the measurement of labor inputs in terms of undifferentiated man-hours. Changes in skill structure and human capital requirements are hidden in the summary measures. Certainly discernment and understanding of real change in the area of primary inputs would be improved by greater detail and better specification.

At any level of detail, it is all too easy to complain about aggregation and natural to call for more and better information. Analysis at an intermediate level of detail has provided some meaningful insights into the relation of structural change to the economic process. Still more detailed answers would come with more detailed data. But it is misleading and dangerous to rely too heavily upon progressive disaggregation for deeper understanding. Larger input-output tables are needed, particularly for specific purposes of business and planning; and very detailed information is being generated with increased business and government applications. This is all to the good. Still, despite computers, it is prudent to consider informational needs selectively and to think of what we need to know in broader terms.

Certainly, many areas of this study need to be refined and verified by more information and by wider experience. But some of the more interesting and pressing questions call not merely for disaggregation but also for more fundamental innovations in the analysis. The model of embodied change described in Chapter 12 could serve only to illustrate the potential contribution of a general equilibrium approach in this area. Deeper insight requires new kinds of information about new processes and discards. More important, the model itself is really too simple to give any but very rough answers. An explicitly dynamic formulation would allow more sophisticated treatment of the diffusion process. It might be more meaningful to pinpoint quality change in the capital stock through the growth of a few specific detailed elements than to rely on a more cumbersome overall disaggregation.

The problem of industrial specialization has received relatively little attention in modern economics. Nelson (1960), Ames and Rosenberg (1963), and Stigler (1951) are a few interesting exceptions. This study suggests that division of labor is a key element in understanding change. Perhaps our principal contribution will be to redirect attention to this old subject.

Appendices
References
Index

APPENDIX A SENSITIVITY OF MEASURES OF CHANGING INDUSTRIAL SPECIALIZATION TO THE ASSUMED STRUCTURE OF FINAL DEMAND

In Part I of this study, changing industrial specialization is described in terms of changes in intermediate output requirements to deliver a given final demand as input-output structures change over time. Table A1 shows how estimated rates of change in intermediate output requirements are affected by the assumed structure of final demand. Four actual bills of final demand, those of 1939, 1947, 1958, and 1961, were singled out. For each bill of final demand, vectors of total intermediate output requirements $\mathbf{z}^t$, with input-output structures of 1939, 1947, and 1958, were estimated. The average rate of change in requirements of intermediate output for each sector was then computed. Least-squares time trends were fitted to the logarithms of estimated requirements. The slopes of these trends measure the rates of change in requirements for each specific input as input-output structures change over time.

Each column in table A1 shows the slope computed on the basis of a different fixed final demand. Thus, the first entry in the first column tells the annual percentage change in requirements for agricultural products to deliver 1939 final demand, as input-output structures vary over time. The first entry in the second column shows the corresponding rate computed on the basis of 1947, rather than 1939, final demand.

In general, computed rates of change in requirements do not vary much with the different assumed final demand weights. Changing the final demand weights reversed the direction of change for only two 38-order sectors, petroleum mining (5) and nonmetallic mining (6). Positive and negative rates of change computed for those sectors remained close to zero. The evidence in table A1 suggests that the general features of changing industrial specialization would not appear to be very different if earlier final demand weights were substituted for those of 1961 in the basic computations.

Table A1 Average Annual Rates of Change in Intermediate Output Requirements with Changing Input-Output Structures, Assuming Four Alternative Bills of Final Demand, 38 Order (percent per year)

Sector	Final demand of year			
	1939	1947	1958	1961
(1) Agriculture, etc.	−1.3	−1.4	−1.4	−1.5
(2) Iron mining	−1.8	−1.7	−2.0	−1.9
(3) Nonferrous mining	−1.7	−2.0	−2.0	−2.0
(4) Coal mining	−3.8	−4.1	−4.2	−4.1
(5) Petroleum mining	0.1	0.0	−0.1	−0.1
(6) Nonmetallic mining	0.1	−0.1	−0.1	−0.1
(7) Construction	−3.1	−3.2	−3.3	−3.3
(8) Food	1.0	0.9	0.8	0.7
(10) Textiles & products	0.3	0.3	0.1	0.0
(11) Wood & products	−1.0	−1.0	−1.1	−1.1
(12) Paper & publishing	0.3	0.0	0.0	0.1
(13) Chemicals	2.0	2.0	1.8	1.8
(14) Petroleum refining	2.0	1.8	1.6	1.6
(15) Rubber prod., etc.	1.4	1.1	1.2	1.2
(16) Leather & shoes	−0.6	−0.8	−1.4	−1.6
(17) Stone, clay, & glass	1.8	1.7	1.7	1.6
(18) Iron & steel	−1.7	−1.6	−1.8	−1.8
(19) Nonferrous metals	−0.1	−0.3	−0.4	−0.5
(20) Metal forming	1.2	0.9	1.0	1.0
(21) Nonelectrical eq.	2.6	2.8	2.9	3.0
(22) Engines & turbines	2.6	2.9	3.1	3.1
(23) Electrical eq., etc.	2.5	2.3	2.6	2.5
(24) Motor vehicles	0.6	0.6	0.6	0.5
(25) Aircraft	8.6	8.5	6.4	6.5
(26) Trains, ships, etc.	−0.3	−0.3	−0.4	−0.4
(27) Instruments, etc.	3.9	3.4	3.4	3.3
(28) Misc. manufactures	−0.7	−0.8	−0.9	−0.9
(29) Transportation	−0.3	−0.4	−0.5	−0.5
(30) Communications	3.7	3.4	3.3	3.3
(31) Utilities	3.6	3.4	3.4	3.4
(32) Trade	2.7	2.5	2.4	2.4
(33) Finance & insurance	1.3	1.0	0.9	0.9
(34) Real estate & rental	0.6	0.3	0.1	0.1
(35) Business serv., etc.	2.3	1.9	1.9	1.9
(36) Auto. repair	2.9	3.0	2.7	2.7
(37) Institutions, etc.	1.3	1.6	1.6	1.6
(38) Scrap	0.8	0.6	0.7	0.6

APPENDIX B CLASSIFICATION SCHEMES

The basis of the classification scheme for this study was the 83-order sectoring plan of the 1958 input-output table. Its detailed four-digit SIC composition is described in Office of Business Economics (1965). The 76-order classification used in this study differs from the 83-order in only a few sectors:

1. Ordnance and accessories (13) is endogenous in the 83-order table but part of final demand at 76 order.
2. The activities of government enterprises—federal government enterprises (78) and state and local government enterprises (79)—are divided between their private counterparts and government final demand in the 76-order classification.
3. Office supplies (82), a "dummy" sector in the 83-order classification, was abolished for the 76-order. At 76 order, goods formerly dubbed "office supplies" are delivered directly by the sectors that produce them to intermediate or final consumers.
4. Textile-producing sectors—broad and narrow fabrics, yarn and thread mills (16), miscellaneous textile goods and floor coverings (17), and miscellaneous fabricated textile products (19)—distinguished separately at 83 order, are combined into textiles (16) at 76 order.
5. Paper and allied products except containers (24) and paperboard containers and boxes (25), distinguished at 83 order, are combined at 76 order into paper and products (24). All of these changes were made in order to improve the comparability of the 1947 and 1958 tables. Some of the original sector names were changed to clarify their content to the casual reader or to shorten them. The 76-order sectors are listed in table B1. Except for items marked by a "c," each sector has the same definition as its 83-order counterpart with the same sector number.

In the course of the analysis, it became clear that apparent structural changes in the tobacco industry (15) were out of line with changes in other sectors and inconsistent with general qualitative evidence about developments in the tobacco sector itself. The publication of the 1963 Census of Manufactures included major revisions of 1958 census statistics for the tobacco industry, confirming the impression that the 1958 structural information was questionable. Therefore, findings for the tobacco sector (15) are not cited in many of the detailed tables and figures.

The 38-order classification provides a bridge between the 76-order classification of the 1947 and 1958 tables and the 98-order scheme of the 1939 study. No "splits" are required to go from 76 to 38 order, that is, 38-order sectors for 1947 and 1958 are derived by aggregating whole 76-order sectors. There are three minor exceptions. Water services, part of electric, gas, water and sanitary utilities (68), are excluded entirely from economic activity at 38 order to permit comparison of the 38-order electric and gas utilities sector (31) over the period 1939–1947–1958. The 76-order dummy sector—business travel, entertainment, and gifts (81)—is in final demand in the 38-order classification. Noncompetitive imports (80), which is a row at 76 order, is included with value added at 38 order. The correspondence between the 38- and 76-order classifications is indicated in the first two columns of table B1.

The last column of table B1 enumerates the sectors of the original 98-order, 1939 sectoring plan. The 38- and 98-order classifications do not dovetail perfectly, and many 98-order categories had to be split among two or more 38-order sectors. Where a 98-order sector is split among more than one 38-order category, its name is repeated next to each 38-order sector to which part of it is assigned. A "b" after its name signifies that more than half of the value of the given 98-order sector is assigned to the 38-order sector. An "a" indicates that less than half of a 98-order sector's value is assigned to the 38-order sector.

Table B1 covers endogenous sectors only. The following list enumerates activities included in final demand at 38 and 76 order in all three years:

Personal consumption expenditures
Gross private capital formation
Net inventory change
Exports
Competitive imports (negative)
Federal government purchases
State and local government purchases
Ordnance and accessories
Federal government enterprises (with no private counterparts)
State and local government enterprises (with no private counterparts)
Business travel, entertainment, and gifts (at 38 order only)

The industrial composition of the subvectors of final demand is specified in table B2.

Table B3 contains the alignment of the 41-order classification used in Chapter 12 and the 76-order classification.

Table B1 Alignment of 38-, 76-, and 98-Order Sectoring Plans

38 Order	76 Order	98 Order
(1) Agriculture, forestry, & fishing	(1) Livestock	(1) Field crops
	(2) Crops	(2) Vegetables
	(3) Forestry & fishing	(3) Fruits & nuts
	(4) Agricultural services	(4) Horticultural specialities
		(5) Forest products
		(6) Dairy products
		(7) Poultry & poultry products
		(8) Livestock & livestock products
		(9) Fisheries
(2) Iron mining	(5) Iron mining	(22) Iron mining
		(23) Nonferrous metal mining[a]
(3) Nonferrous mining	(6) Nonferrous mining	(39) Nonferrous metal mining[b]
(4) Coal mining	(7) Coal mining	(47) Anthracite coal
		(48) Bituminous coal
(5) Petroleum mining	(8) Petroleum mining	(45) Petroleum & natural gas
(6) Nonmetallic mining	(9) Stone & clay mining	(43) Nonmetallic mining
	(10) Chemical mining	
(7) Construction	(11) New construction	(68) New construction
	(12) Maintenance construction	(69) Maintenance construction
(8) Food	(14) Food	(10) Flour & grist mill products
		(11) Canning, preserving
		(12) Bread & bakery products
		(13) Sugar refining
		(14) Starch & glucose products

Table B1 (*continued*)

38 Order	76 Order	98 Order
		(15) Alcoholic beverages
		(16) Nonalcoholic beverages
		(18) Slaughtering & meat packing
		(19) Manufactured dairy products
		(20) Edible fats and oils
		(21) Other food products
		(53) Chemicals[a]
		(67) Industries, n.e.c.[a]
(9) Tobacco	(15) Tobacco	(17) Tobacco manufactures
(10) Textiles & products	(16) Textiles[c]	(58) Cotton yarn & cloth
	(18) Apparel	(59) Silk and rayon products
		(60) Woolen & worsted manufactures
		(61) Clothing
		(62) Other textile products[b]
(11) Wood & products	(20) Wood & products	(54) Lumber & timber products
	(21) Wooden containers	(55) Furniture and other manufactures of wood[b]
	(22) Household furniture	
	(23) Office furniture	
(12) Paper & publishing	(24) Paper & products[c]	(56) Wood, pulp, paper, & paper products
	(26) Printing & publishing	(57) Printing & publishing
(13) Chemicals	(27) Basic chemicals	(53) Chemicals[b]
	(28) Synthetic materials	(49) Coke & manufactured solid fuels[a]
	(29) Drugs, soaps, cosmetics	(28) Explosives[b]
	(30) Paint	

Table B1 (*continued*)

38 Order	76 Order	98 Order
(14) Petroleum refining	(31) Petroleum refining	(46) Petroleum refining (44) Nonmetallic mineral manufactures[a] (49) Coke & manufactured solid fuels[a]
(15) Rubber & plastic products	(32) Rubber & plastic products	(66) Rubber products (53) Chemicals[a]
(16) Leather & shoes	(33) Leather tanning (34) Shoes & other leather products	(63) Leather (64) Leather shoes (65) Leather products, n.e.c.
(17) Stone, clay, & glass	(35) Glass & products (36) Stone & clay products	(44) Nonmetallic minerals manufactures[b]
(18) Iron & steel	(37) Iron & steel	(23) Blast furnace (24) Steel works & rolling mills (25) Iron & steel foundary products (49) Coke & manufactured solid fuels[b] (38) Iron & steel, n.e.c.[a]
(19) Nonferrous metals	(38) Nonferrous metals	(40) Smelting & refining of nonferrous metals (41) Aluminum (42) Nonferrous metal manufactures & alloys[b] (37) Electrical equipment[a] (53) Chemicals[a] (38) Iron & steel, n.e.c.[a]

Table B1 (*continued*)

38 Order	76 Order	98 Order
(20) Metal forming	(39) Metal containers	(38) Iron & steel, n.e.c.[b]
	(40) Heating, plumbing, structural metals	(42) Nonferrous metal manufactures & alloys[a]
	(41) Stampings, screw machine products, & fasteners	(34) Industrial & household equipment[a]
	(42) Hardware, plating, valves, wire products	(67) Industries, n.e.c.[a]
(21) Nonelectrical equipment	(44) Farm equipment	(29) Agricultural machinery
	(45) Construction & mining equipment	(34) Industrial & household equipment[b]
	(46) Materials handling equipment	(35) Machine tools
	(47) Metalworking equipment	
	(48) Special industry equipment	
	(49) General industrial equipment	
	(50) Machine shop products	
(22) Engines & turbines	(43) Engines & turbines	(30) Engines & turbines
(23) Electrical & service equipment	(51) Office & computing machines	(36) Merchandising & service machines[b]
	(52) Service industry machines	(37) Electrical equipment[b]
	(53) Electrical apparatus & motors	(34) Industrial & household equipment[a]
	(54) Household appliances	
	(55) Lighting & wiring equipment	
	(56) Radio, tv, & communication equipment	
	(57) Electronic components	
	(58) Batteries, x-ray, & engine electric equipment	
(24) Motor vehicles & equipment	(59) Motor vehicles & equipment	(31) Automobiles
		(34) Industrial & household equipment[a]

Table B1 *(continued)*

38 Order	76 Order	98 Order
(25) Aircraft	(60) Aircraft	(32) Aircraft
(26) Trains, ships, & cycles	(61) Trains, ships, & cycles	(26) Shipbuilding (33) Transportation equipment
(27) Instruments & cameras	(62) Instruments & clocks (63) Optical & photographic apparatus	(34) Industrial & household equipment[a] (67) Industries, n.e.c.[a]
(28) Miscellaneous manufactures	(64) Miscellaneous manufactures	(67) Industries, n.e.c.[b] (62) Other textile products[a] (36) Merchandising & service machines[a] (55) Furniture & other manufactures of wood[a]
(29) Transportation & storage	(65) Transportation & storage	(70) Transportation (71) Coastwise & inland water transportation (72) Transoceanic transportation (73) Steam railroad transportation (80) Services allied to transportation
(30) Communications	(66) Telephone & telegraph (67) Radio & tv broadcasting	(51) Communications
(31) Electric & gas utilities	(68) Electric, gas, water, & sanitary utilities[d]	(50) Manufactured gas (52) Electric public utilities
(32) Trade	(69) Trade	(74) Trade
(33) Finance & insurance	(70) Finance & insurance	(76) Banking (77) Insurance
(34) Real estate & rental	(71) Real estate & rental	(84) Commercial renting (85) Home renting

Table B1 (*continued*)

38 Order	76 Order	98 Order
(35) Business & personal services & hotels	(72) Hotels, personal, & repair services (73) Business services	(78) Business services—other than advertising[b] (79) Business services—advertising (82) Repair services—other than automotive (87) Laundries, drycleaners (88) Personal services (86) Hotels
(36) Automobile repair	(75) Automobile repair	(81) Automotive repair & services
(37) Institutions, research & entertainment	(74) Research & development (76) Amusements & recreation (77) Medical & educational institutions	(83) Rental agencies (89) Professional entertainment (90) Motion picture theatres (91) Amusement places (67) Industries, n.e.c.[a] (78) Business services—other than advertising[a] (94) Households[a] (98) Research & development
(38) Scrap	(83) Scrap (80) Noncompetitive imports[d] (81) Business travel, entertainment, & gifts[c]	(97) Scrap

[a] Less than half of the value of the 98-order sector is assigned to the 38-order sector.
[b] More than half of the value of the 98-order sector is assigned to the 38-order sector.
[c] These activities are not included in endogenous sectors at 38 order (see Appendix B text).
[d] 76-order definitions differ from Office of Business Economics 83-order definitions (see Appendix B text).

Table B2 Industrial Composition of Subvectors of Final Demand, 38 and 76 Order

Subvector	38 Order	76 Order
(1) Food & tobacco	(1) Agriculture, forestry, & fishing	(1) Livestock
	(8) Food	(2) Crops
	(9) Tobacco	(3) Forestry & fishing
		(4) Agricultural services
		(14) Food
		(15) Tobacco
(2) Textiles & clothing	(10) Textiles & products	(16) Textiles
	(16) Leather & shoes	(18) Apparel
		(33) Leather tanning
		(34) Shoes
(3) Paper & chemicals	(12) Paper & publishing	(24) Paper & products
	(13) Chemicals	(26) Printing & publishing
		(27) Basic chemicals
		(28) Synthetic materials
		(29) Drugs, soaps, etc.
		(30) Paint
(4) Construction	(7) Construction	(11) New construction
		(12) Maintenance construction
(5) Machinery	(21) Nonelectrical equipment	(43) Engines & turbines
	(22) Engines & turbines	(44) Farm equipment
		(45) Construction & mining equipment
		(46) Materials handling equipment
		(47) Metalworking equipment
		(48) Special industry equipment
		(49) General industrial equipment
		(50) Machine shop products

Table B2 (*continued*)

Subvector	38 Order	76 Order
(6) Electrical & service equipment	(23) Electrical & service equipment	(51) Office & computing machines
		(52) Service industry machines
		(53) Electrical apparatus & motors
		(54) Household appliances
		(55) Light & wiring equipment
		(56) Radio, tv, & communication equipment
		(57) Electronic components
		(58) Batteries, x-ray, & engine electrical equipment
(7) Transportation equipment	(24) Motor vehicles & equipment	(59) Motor vehicles & equipment
	(25) Aircraft	(60) Aircraft
	(26) Trains, ships, & cycles	(61) Trains, ships, & cycles
(8) Services & transportation	(29) Transportation & storage	(65) Transportation & storage
	(30) Communications	(66) Telephone & telegraph
		(67) Radio & tv broadcasting
	(31) Electric & gas utilities	(68) Electric, gas, water, & sanitary utilities
	(32) Trade	(69) Trade
	(33) Finance & insurance	(70) Finance & insurance
	(34) Real estate & rental	(71) Real estate & rental
	(35) Business & personal services & hotels	(72) Hotels, personal, & repair services
		(73) Business services
	(36) Automobile repair	(75) Automobile repair
	(37) Institutions, research, & entertainment	(76) Amusements & recreation
(9) Furniture		(22) Household furniture
		(23) Office furniture

Table B3 Alignment of 41- and 76-Order Sectoring Plans

41 Order	76 Order
(1) Agriculture, forestry, & fishing	(1) Livestock
	(2) Crops
	(3) Forestry & fishing
	(4) Agricultural services
(2) Mining	(5) Iron mining
	(6) Nonferrous mining
	(7) Coal mining
	(8) Petroleum mining
	(9) Stone & clay mining
	(10) Chemical mining
(3) Transportation & storage	(65) Transportation
(4) Communications	(66) Telephone & telegraph
	(67) Radio & tv broadcasting
(5) Electric, gas, water, & sanitary utilities	(68) Electric, gas, water, & sanitary utilities
(6) Trade & services	(11) New construction
	(12) Maintenance construction
	(69) Trade
	(70) Finance & insurance
	(71) Real estate & rental
	(72) Hotels & personnel, & repair services
	(73) Business services
	(74) Research & development
	(75) Automobile repair
	(76) Amusements & recreation
	(77) Medical & educational institutions
	(80) Noncompetitive imports
	(81) Business travel, entertainment, & gifts
	(83) Scrap
(7) Food	(14) Food
(8) Tobacco	(15) Tobacco
(9) Textiles & products	(16) Textiles
	(18) Apparel
(10) Wood & products	(20) Wood & products
(11) Wooden containers	(21) Wooden containers
(12) Furniture	(22) Household furniture
	(23) Office furniture
(13) Paper & products	(24) Paper & products
(14) Printing & publishing	(26) Printing & publishing
(15) Basic chemicals	(27) Basic chemicals
(16) Synthetic materials	(28) Synthetic materials

Table B3 (*continued*)

41 Order	76 Order
(17) Drugs, soaps, etc.	(29) Drugs, soaps, etc.
(18) Paint	(30) Paint
(19) Petroleum refining	(31) Petroleum refining
(20) Rubber & plastic products	(32) Rubber & plastic products
(21) Leather & shoes	(33) Leather tanning (34) Shoes
(22) Glass & products	(35) Glass & products
(23) Stone & clay products	(36) Stone & clay products
(24) Iron & steel	(37) Iron & steel
(25) Nonferrous metals	(38) Nonferrous metals
(26) Metal forming	(39) Metal containers (40) Heating, plumbing, & structural metals (41) Stampings, screw machine products, & fasteners (42) Hardware, plating, valves, wire products
(27) Engines & turbines	(43) Engines & turbines
(28) Farm equipment	(44) Farm equipment
(29) Construction & mining equipment	(45) Construction & mining equipment
(30) Other industrial equipment	(46) Materials handling equipment (47) Metalworking equipment (48) Special industry equipment (49) General industrial equipment (50) Machine shop products
(31) Office & computing machines	(51) Office & computing machines
(32) Home & service equipment	(52) Service industry machines (54) Household appliances
(33) Electrical equipment	(53) Electrical apparatus & motors (55) Lighting & wiring equipment
(34) Communication equipment	(56) Radio, tv, & communication equipment (57) Electronic components
(35) Batteries, x-ray, & engine electrical equipment	(58) Batteries, x-ray, & engine electrical equipment
(36) Motor vehicles & equipment	(59) Motor vehicles & equipment
(37) Aircraft	(60) Aircraft
(38) Trains, ships, & cycles	(61) Trains, ships, & cycles
(39) Instruments & clocks	(62) Instruments & clocks
(40) Optical & photographic apparatus	(63) Optical & photographic apparatus
(41) Miscellaneous manufactures	(64) Miscellaneous manufactures

APPENDIX C 38-ORDER DATA

Appendix C contains 38-order data central to the analysis of Chapters 4 through 8. These data, which are discussed in Chapter 2, include final demand vectors and input-output, labor, and capital coefficients.

Table C1 Input-Output Coefficients for 1958, 38 Order (dollars per ten thousand dollars of output in 1947 prices)

Sector	Sector												
	(1)	(2)	(3)	(4)	(5)	(6)	(7)	(8)	(9)	(10)	(11)	(12)	(13)
(1) Agriculture, etc.	3,231						39	3,986	1,769	514	504		19
(2) Iron mining		866	122			3							17
(3) Nonferrous mining		761	2,217			6							25
(4) Coal mining	1	125	12	1,818		15		6	3	5	3	35	35
(5) Petroleum mining					212	5							6
(6) Nonmetallic mining	15		12	6		252	138	2				21	147
(7) Construction	67	13	8	7	4	8	1	24		3	11	34	10
(8) Food	726						5	1,839	70	14	33	55	256
(9) Tobacco									1,961				
(10) Textiles & products	26	6	27	10	4	2	1	34	3	4,104	309	68	32
(11) Wood & products	15	161	12	73	7	1	696	14	19	7	2,346	329	18
(12) Paper & publishing	8	5	9	33	10	99	63	175	259	84	180	2,984	279
(13) Chemicals	188	303	337	144	74	150	238	86	260	514	216	319	2,218
(14) Petroleum refining	126	243	71	103	64	227	233	35	5	9	67	66	309
(15) Rubber prod., etc.	22	13	28	71	35	130	57	17	15	45	151	79	87
(16) Leather & shoes	1								1	21	8	1	
(17) Stone, clay, & glass	3	18	47	19	5	497	693	52		6	106	24	60
(18) Iron & steel		332	285	51	3	97	271			1	117	1	76
(19) Nonferrous metals		38	46	56	10	14	192	4	10	1	34	12	79
(20) Metal forming	12	49	17	83	64	8	1,084	177	23	8	254	61	118
(21) Nonelectrical eq.	23	419	195	292	121	397	97	2	1	14	37	35	49
(22) Engines & turbines		7	3		14	1							
(23) Electrical eq., etc.	4	65	40	29	51	24	280	4	1	1	28	15	12
(24) Motor vehicles & eq.	6	11	7	17	9	28					3	1	
(25) Aircraft											1	7	
(26) Trains, ships, etc.	3	69		40		2					6		
(27) Instruments, etc.		6	4	1	1	3	27			3	6	26	18
(28) Misc. manufactures	1	1		16	1	4	23	4	13	112	38	23	12
(29) Transportation	102	492	117	66	94	121	326	301	114	102	358	263	212
(30) Communications	19	40	17	7	3	18	21	21	4	18	36	87	32
(31) Utilities	29	422	257	274	92	293	25	48	9	51	64	114	140
(32) Trade	345	381	203	329	126	326	1,031	292	130	249	480	355	239
(33) Finance & insurance	53	106	105	68	97	69	72	34	15	33	51	64	53
(34) Real estate & rental	386	1,292	360	194	1,425	193	34	32	11	55	76	196	75
(35) Business serv., etc.	103	130	61	39	340	74	303	191	430	67	89	294	410
(36) Auto. repair	13			3	16	1	41	31	3	3	52	9	8
(37) Institutions, etc.	19	17	8	8	10	8	9	7	8	7	10	11	18
(38) Scrap		16	7	25	183	30	19			17	3	160	2

Table C1 (*continued*)

Sector	(14)	(15)	(16)	(17)	(18)	(19)	(20)	(21)	(22)	(23)	(24)	(25)	(26)
(1) Agriculture, etc.			155	9				2					
(2) Iron mining	1			10	442	21				1			
(3) Nonferrous mining				5	4	1,007	1			1			
(4) Coal mining	6	18	5	89	433	15	3	4	15	3	8	3	10
(5) Petroleum mining	4,592												
(6) Nonmetallic mining	34	20	2	656	54	8	2	8					1
(7) Construction	12	8	1	4	79	3	7	14	3	8	29	19	4
(8) Food	14	3	1,084	10	9	1		1					
(9) Tobacco													
(10) Textiles & products	3	1,493	415	40	21	48	31	27	20	24	213	28	53
(11) Wood & products	1	20	59	93	19	12	85	33	4	99	9	45	474
(12) Paper & publishing	52	161	184	530	78	49	139	41	104	147	66	37	39
(13) Chemicals	384	2,302	209	493	176	273	142	47	35	170	100	48	256
(14) Petroleum refining	720	26	9	110	115	45	60	66	52	20	24	32	55
(15) Rubber prod., etc.	4	302	372	96	43	18	62	141	58	159	301	76	98
(16) Leather & shoes		21	3,265	1			3	8	2	4	5		9
(17) Stone, clay, & glass	20	83	29	1,119	180	41	88	90	84	123	138	46	169
(18) Iron & steel	1	15		25	2,076	108	1,948	1,048	973	300	718	302	979
(19) Nonferrous metals	1	21	3	23	248	3,272	702	358	415	489	133	359	215
(20) Metal forming	144	139	46	125	306	172	475	447	245	384	646	295	779
(21) Nonelectrical eq.	2	47	4	31	238	120	311	1,089	1,226	153	232	442	382
(22) Engines & turbines					1		25	158	900	32	35	16	293
(23) Electrical eq., etc.	5	37	13	45	62	179	199	424	380	1,342	310	610	538
(24) Motor vehicles & eq.				2	21	13	82	172	250	36	2,997	67	150
(25) Aircraft		27	1	1	1		15	63	97	37	4	1,930	56
(26) Trains, ships, etc.		3			9	5	35	32	89	13	9	2	649
(27) Instruments, etc.	1	19	24	7	4	4	52	28	10	110	44	197	15
(28) Misc. manufactures	5	51	33	25	10	17	27	30	21	19	11	23	33
(29) Transportation	494	213	111	561	656	198	180	145	144	119	198	107	196
(30) Communications	14	34	25	41	53	32	38	90	34	39	25	60	35
(31) Utilities	150	111	35	382	389	272	85	79	53	53	62	68	76
(32) Trade	100	372	223	367	518	342	413	495	345	444	355	240	582
(33) Finance & insurance	43	45	43	84	79	48	64	68	55	33	32	28	45
(34) Real estate & rental	72	99	55	79	46	37	63	109	46	81	28	63	47
(35) Business serv., etc.	184	220	170	152	116	75	126	163	129	230	270	36	100
(36) Auto. repair	10	3	3	29	4	6	14	11	3	3	4	2	10
(37) Institutions, etc.	11	10	8	13	23	13	11	11	67	10	15	20	11
(38) Scrap	2	10		81	593	515	37	26	41	4	6	8	40

Table C1 (*continued*)

Sector	Sector											
	(27)	(28)	(29)	(30)	(31)	(32)	(33)	(34)	(35)	(36)	(37)	(38)
(1) Agriculture, etc.	14	25	17			13		501			12	
(2) Iron mining								1				
(3) Nonferrous mining	2				1			1				
(4) Coal mining	6	1	11		356		3	2	11	17		
(5) Petroleum mining					372			19				
(6) Nonmetallic mining		1	1					2				
(7) Construction	2	22	357	220	222	61	48	833	15	126	233	
(8) Food	33	30	39			53		10	6		73	15
(9) Tobacco		1										31
(10) Textiles & products	150	372	21	19	1	14	40	8	204	81	56	1,246
(11) Wood & products	35	192	10		1	18		5	6		1	
(12) Paper & publishing	246	603	48	124	8	113	273	15	1,791	22	164	1,079
(13) Chemicals	302	435	35	2	6	25	9	20	129	111	396	
(14) Petroleum refining	19	28	542	16	92	70	47	62	80	40	31	141
(15) Rubber prod., etc.	117	337	79	6	2	21	24	7	41	385	30	71
(16) Leather & shoes	22	161	2			2	1	1	4		7	21
(17) Stone, clay, & glass	138	55	3		3	18		4	15	174	2	
(18) Iron & steel	116	160	9		29	1		3			1	1,294
(19) Nonferrous metals	407	428	17	23	4	2		2	7		4	649
(20) Metal forming	269	261	15	2	23	20		4	8	126	17	952
(21) Nonelectrical eq.	193	38	16			15		12	46	121	37	948
(22) Engines & turbines			22		1	1		1	26		14	25
(23) Electrical eq., etc.	738	167	52	188	4	27	4	11	324	200	277	1,027
(24) Motor vehicles & eq.	111	11	26			17		3	1	1,335	52	1,086
(25) Aircraft	102	7	44			7		2			435	224
(26) Trains, ships, etc.	11	22	86	4		1	4	1	9	14	9	36
(27) Instruments, etc.	656	13	8			8	1	2	132	22	146	335
(28) Misc. manufactures	62	472	20	22	2	15	24	3	177	4	57	217
(29) Transportation	118	131	617	16	193	39	106	60	55	97	40	602
(30) Communications	46	52	99	138	9	101	210	41	720	82	82	
(31) Utilities	42	44	49	54	1,757	205	69	45	150	238	176	
(32) Trade	416	549	364	63	145	159	131	166	323	1,002	174	
(33) Finance & insurance	42	55	209	50	17	123	2,039	285	122	215	103	
(34) Real estate & rental	105	120	310	182	9	416	841	194	410	379	577	
(35) Business serv., etc.	262	185	141	170	57	398	402	229	358	177	274	
(36) Auto. repair	3	12	256	13	4	70	36	18	67	169	15	
(37) Institutions, etc.	13	9	18	266	3	16	61	21	14	9	578	
(38) Scrap		19				14			20	140		

Table C2 Input-Output Coefficients for 1947, 38 Order (dollars per ten thousand dollars of output in 1947 prices)

Sector	Sector (1)	(2)	(3)	(4)	(5)	(6)	(7)	(8)	(9)	(10)	(11)	(12)	(13)
(1) Agriculture, etc.	3,193						32	4,472	2,011	919	506	5	85
(2) Iron mining		354											6
(3) Nonferrous mining			1,833			32							26
(4) Coal mining	1	66	29	1,590		42	1	13	6	15	9	50	95
(5) Petroleum mining					130		3						25
(6) Nonmetallic mining	9			4		369	92	1				14	158
(7) Construction	123	14	12	13	18	16	2	19	4	17	19	34	26
(8) Food	550			2			4	1,465	38	61	34	48	990
(9) Tobacco									2,510				1
(10) Textiles & products	24				16	22	6	49		3,866	359	85	49
(11) Wood & products	33	142	271	168	14	11	852	23	48	17	1,828	225	55
(12) Paper & publishing	1			14	78	201	59	156	240	92	131	2,941	447
(13) Chemicals	138	184	231	147	148	273	215	76	95	388	172	221	1,892
(14) Petroleum refining	112	65	95	49	66	206	204	25	12	29	99	91	234
(15) Rubber prod., etc.	30	2	2	4	20	32	26	13	8	25	78	44	61
(16) Leather & shoes					3	2			1	24	18	6	6
(17) Stone, clay, & glass	6	26	17	33	45	14	574	56	8	5	75	16	109
(18) Iron & steel	5	197	122	87	45	214	356	5	13	11	222	31	152
(19) Nonferrous metals	2	31	20	91	21	34	265	3	21	3	50	25	149
(20) Metal forming	16	15	11	77	124	41	965	128	54	10	230	34	170
(21) Nonelectrical eq.	16	146	119	312	99	92	95	5	31	17	100	68	31
(22) Engines & turbines				38	24	29		3			15	4	
(23) Electrical eq., etc.	5	2	1	35	68	73	185	5	6	4	36	26	38
(24) Motor vehicles & eq.	16	6	5	45	22	81	7	4	7	5	32	11	15
(25) Aircraft										4			
(26) Trains, ships, etc.	2	9	5	13	2	3			1	1	2	1	2
(27) Instruments, etc.				5	6	5	7		3	2	42	42	20
(28) Misc. manufactures	1				3	2	15	2	1	88	13	3	11
(29) Transportation	217	93	101	33	49	103	437	244	112	124	473	363	367
(30) Communications		37	15	10	25	25	18	12	7	16	28	47	31
(31) Utilities	12	278	400	158	26	240	10	38	6	45	41	71	97
(32) Trade	322	158	136	84	154	279	901	158	155	304	328	286	246
(33) Finance & insurance	61	76	81	62	101	164	79	28	26	37	114	60	82
(34) Real estate & rental	480	1,009	278	175	1,181	244	30	25	15	66	80	208	82
(35) Business serv., etc.	25	25	16	24	56	44	310	131	263	82	135	140	426
(36) Auto. repair	45	24	23	15	29	240	117	28	4	4	63	12	14
(37) Institutions, etc.	12	8	6	3	9	9	14	9	9	12	14	18	18
(38) Scrap							11	22		9		189	11

Table C2 (*continued*)

Sector	Sector (14)	(15)	(16)	(17)	(18)	(19)	(20)	(21)	(22)	(23)	(24)	(25)	(26)
(1) Agriculture, etc.		38	131	1									
(2) Iron mining				9	380	11				2			
(3) Nonferrous mining	1			15	20	1,072				3			
(4) Coal mining	24	35	16	171	441	22	11	11	17	11	12	6	12
(5) Petroleum mining	4,952												
(6) Nonmetallic mining	12	11		626	42	8	1	3	1	2			
(7) Construction	18	33	39	60	54	31	28	28	30	23	25	41	44
(8) Food	29	3	1,183	5	9								
(9) Tobacco													
(10) Textiles & products	2	1,393	260	115	3	53	16	17	3	30	139	37	56
(11) Wood & products	20	18	80	83	32	70	121	57	35	174	47	57	252
(12) Paper & publishing	202	267	245	561	27	89	172	85	66	174	26	172	75
(13) Chemicals	235	1,112	357	279	115	148	152	69	32	208	124	101	108
(14) Petroleum refining	944	73	36	121	180	90	46	45	53	46	29	64	40
(15) Rubber prod., etc.	20	150	168	55	3	28	44	205	102	193	496	72	88
(16) Leather & shoes	2	18	3,101	15		3	27	17	2	5	12	3	12
(17) Stone, clay, & glass	48	55	32	808	133	53	71	51	112	181	124	30	63
(18) Iron & steel	48	74	33	157	2,146	353	1,987	1,222	1,372	540	928	602	1,156
(19) Nonferrous metals	9	11	11	36	499	2,774	574	246	290	629	228	489	143
(20) Metal forming	141	136	75	82	51	106	595	529	513	606	689	414	559
(21) Nonelectrical eq.		29		44	58	70	288	823	1,793	256	381	283	472
(22) Engines & turbines	11	17		13	19	4	11	237	392	36	72		275
(23) Electrical eq., etc.	14	62	14	93	20	130	267	451	522	1,130	162	139	395
(24) Motor vehicles & eq.	15	20	17	26	3	14	56	76	47	31	2,679	165	366
(25) Aircraft							1	1	1		4	1,226	49
(26) Trains, ships, etc.	1	2	2	3	2	2	3	1	1	2	11	67	767
(27) Instruments, etc.	5	12	6	11	1	12	47	24	18	50	45	145	18
(28) Misc. manufactures	2	105	63	4	5	2	4	6	5	21	2	105	30
(29) Transportation	668	234	195	654	705	273	196	160	168	163	189	145	191
(30) Communications	15	32	21	32	19	18	33	31	28	35	16	57	23
(31) Utilities	68	93	26	278	180	139	76	63	42	54	40	80	55
(32) Trade	131	342	297	324	219	367	313	273	201	339	53	292	334
(33) Finance & insurance	87	80	60	114	54	55	66	58	184	69	33	76	75
(34) Real estate & rental	127	93	80	76	48	66	59	57	134	92	24	127	68
(35) Business serv., etc.	135	129	168	101	52	58	99	93	66	140	88	114	71
(36) Auto. repair	11	12	10	40	4	9	14	12	6	10	2	13	10
(37) Institutions, etc.	12	16	14	19	14	12	17	16	15	17	12	41	14
(38) Scrap		20		28	644	1,010	25	20	131	3			

Table C2 (*continued*)

Sector	Sector (27)	(28)	(29)	(30)	(31)	(32)	(33)	(34)	(35)	(36)	(37)	(38)
(1) Agriculture, etc.	17	46	4			1		695			130	
(2) Iron mining	1							1				
(3) Nonferrous mining	2							1				
(4) Coal mining	9	11	196	1	667	13	2	23	4	5	21	
(5) Petroleum mining			3		327			20				
(6) Nonmetallic mining	3	134	2					2	4			
(7) Construction	30	31	572	469	466	45	57	1,356	32	32	154	
(8) Food	23	109	66		1	15		14	1		153	444
(9) Tobacco								1				124
(10) Textiles & products	230	190	12	3	4	13	19	13	32	53	9	352
(11) Wood & products	161	391	4	8	2	7	51	23	1		6	208
(12) Paper & publishing	503	216	39	149	9	171	379	24	1,572		185	535
(13) Chemicals	311	264	26	27	17	22	29	14	85	120	126	290
(14) Petroleum refining	46	55	381	23	285	78	39	26	44	65	31	22
(15) Rubber prod., etc.	132	408	74	11	2	14	22	7	23	82	5	104
(16) Leather & shoes	64	60	1	1		1	2	1	38		3	200
(17) Stone, clay, & glass	150	42	6	9	26	17	12	4	2	59	2	48
(18) Iron & steel	137	88	77	17	190	1	23	9	5	4	1	2,443
(19) Nonferrous metals	349	661	25	13	1	2	7	2	8	16	1	534
(20) Metal forming	514	218	27	19	6	17	25	5	7	86	4	1,411
(21) Nonelectrical eq.	181	51	27	1	3	6		5	121	102	1	574
(22) Engines & turbines	13	2	6		1			5				60
(23) Electrical eq., etc.	363	150	32	228	33	16	29	12	61	343	12	765
(24) Motor vehicles & eq.	19		54	16	16	39	13	4	11	1,213	1	1,312
(25) Aircraft			17			1		1			241	72
(26) Trains, ships, etc.	1	12	130	1	1	1	1	2	1	8		193
(27) Instruments, etc.	917	1	2	3		1	4	2	44	11	123	108
(28) Misc. manufactures	83	636	6	6	3	15	19	6	218		23	200
(29) Transportation	143	167	483	43	305	95	46	160	48	64	76	
(30) Communications	44	27	34	186	32	72	101	40	361	61	70	
(31) Utilities	42	46	54	27	1,347	110	60	55	139	176	89	
(32) Trade	355	441	248	137	75	262	254	191	199	557	288	
(33) Finance & insurance	68	41	142	59	84	145	2,191	294	71	9	49	
(34) Real estate & rental	94	88	93	179	142	429	345	73	389	420	721	
(35) Business serv., etc.	256	190	62	340	39	451	316	150	333	246	169	2
(36) Auto. repair	10	9	124	31	18	122	18	15	68	134	24	
(37) Institutions, etc.	19	15	7	16	10	12	38	81	22	10	738	
(38) Scrap		9										

Table C3 Input-Output Coefficients for 1939, 38 Order (dollars per ten thousand dollars of output in 1947 prices)

Sector	Sector (1)	(2)	(3)	(4)	(5)	(6)	(7)	(8)	(9)	(10)	(11)	(12)	(13)
(1) Agriculture, etc.	3,630	18	58	27	24	22	111	4,479	3,994	1,085	207	8	252
(2) Iron mining		352				6							11
(3) Nonferrous mining			1,828										13
(4) Coal mining	1	68	34	1,580	1	71	2	16	10	15	18	56	74
(5) Petroleum mining					221		2						10
(6) Nonmetallic mining	4	1	3	3	1	392	162	2	3		2	21	83
(7) Construction	46	10	4	17	101	28		10	6	6	11	16	9
(8) Food	430	17	55	27	23	21	39	1,323	75	62	42	73	629
(9) Tobacco		1	3	1	1	1	2	1	2,997		2		
(10) Textiles & products	43	23	72	34	31	33	45	53	75	3,649	408	106	7
(11) Wood & products	31	143	137	118	8	7	790	15	114	64	2,245	176	13
(12) Paper & publishing	8	20	53	29	23	89	74	137	335	47	53	2,982	186
(13) Chemicals	108	299	231	113	25	135	303	83	143	177	228	162	1,854
(14) Petroleum refining	73	74	43	29	23	104	173	14	13	12	55	42	75
(15) Rubber prod., etc.	18	9	23	11	12	15	23	7	23	18	30	6	21
(16) Leather & shoes	2	5	12	6	5	5	9	3	12	20	31	12	
(17) Stone, clay, & glass	3	43	19	7	17	6	450	48	13	1	55	16	91
(18) Iron & steel	8	193	65	84	15	152	417	9	38	3	205	6	116
(19) Nonferrous metals	4	31	26	24	8	23	127	4	67	1	23	15	236
(20) Metal forming	17	28	34	26	14	33	678	133	69	7	261	12	90
(21) Nonelectrical eq.	18	122	59	64	50	26	92	7	22	17	38	31	7
(22) Engines & turbines	1	1	4	2	2	2	2	1	4		3		
(23) Electrical eq., etc.	5	24	33	16	14	13	226	7	34	6	23	18	6
(24) Motor vehicles & eq.	1	10	10	5	4	4	6	3	11	1	7	1	
(25) Aircraft			2	1	1	1	1		2		1		
(26) Trains, ships, etc.	2	7	5	2	2	2	2	1	4		9		
(27) Instruments, etc.		1	3	1	1	1	4	1	3		2	16	
(28) Misc. manufactures	4	8	26	12	11	10	38	6	27	85	34	26	22
(29) Transportation	75	96	115	63	54	85	354	202	145	62	323	235	169
(30) Communications	5	38	6	10	18	13	11	9	9	11	17	33	13
(31) Utilities	7	284	161	127	28	164	7	35	16	37	33	61	44
(32) Trade	127	218	149	86	69	159	464	183	365	148	212	266	89
(33) Finance & insurance	23	103	57	108	88	70	72	22	32	55	63	45	19
(34) Real estate & rental	251	527	27	44	689	51	125	26	26	55	42	104	30
(35) Business serv., etc.	30	20	41	63	22	16	121	138	1,007	108	65	75	189
(36) Auto. repair	19	28	7	4	5	16	10	5	7		9	1	
(37) Institutions, etc.	8	8	5	3	9	6	11	8	16	10	14	18	14
(38) Scrap										3		148	

Table C3 (*continued*)

Sector	(14)	(15)	(16)	(17)	(18)	(19)	(20)	(21)	(22)	(23)	(24)	(25)	(26)
(1) Agriculture, etc.	26	60	46	15	36	30	59	78	49	25	4	20	79
(2) Iron mining				23	523	15		1		8			
(3) Nonferrous mining					30	1,548		1					1
(4) Coal mining	20	41	15	237	487	29	18	27	30	19	15	3	21
(5) Petroleum mining	5,784							1					1
(6) Nonmetallic mining	22	18	2	813	91	13	3	3	2	2		1	3
(7) Construction	27	11	1	25	44	12	14	16		6	9	31	20
(8) Food	45	59	1,109	20	36	28	56	75	47	24	4	19	75
(9) Tobacco	1	3	2	1	2	1	3	4	2	1		1	4
(10) Textiles & products	35	1,628	212	46	47	56	75	133	62	40	354	44	145
(11) Wood & products	26	20	49	31	34	24	101	62	16	121	56	7	159
(12) Paper & publishing	196	111	99	270	34	57	99	112	50	89	32	26	73
(13) Chemicals	145	282	567	145	137	89	191	108	74	195	137	154	87
(14) Petroleum refining	697	53	13	24	95	32	25	38	77	25	17	244	39
(15) Rubber prod., etc.	14	299	197	6	14	13	32	174	38	133	457	25	128
(16) Leather & shoes	5	47	3,061	4	7	6	13	22	10	6	17	4	16
(17) Stone, clay, & glass	48	30	13	752	104	38	44	51	34	149	99	13	69
(18) Iron & steel	69	38	29	46	1,878	107	2,666	1,174	1,038	346	1,137	443	1,149
(19) Nonferrous metals	9	20	14	10	307	3,418	481	154	279	822	171	451	232
(20) Metal forming	141	34	82	9	78	28	405	478	1,337	557	786	11	607
(21) Nonelectrical eq.	53	29	19	15	206	10	121	353	1,346	45	206	7	294
(22) Engines & turbines	2	4	3	1	2	2	5	68	172	24	16	3	368
(23) Electrical eq., etc.	15	34	25	9	20	18	188	466	186	1,254	291	97	212
(24) Motor vehicles & eq.	5	11	8	3	83	5	346	34	63	262	2,834	300	303
(25) Aircraft	1	2	1		1	1	2	3	3	6	4	1,216	2
(26) Trains, ships, etc.	2	4	3	1	14	2	29	35	14	12	8	2	247
(27) Instruments, etc.	1	3	2	1	2	2	25	13	12	39	16	108	14
(28) Misc. manufactures	12	28	20	12	16	14	40	36	23	28	9	56	170
(29) Transportation	578	304	117	645	621	202	230	274	190	119	124	124	208
(30) Communications	9	18	11	21	14	10	24	37	34	29	10	29	19
(31) Utilities	56	65	25	262	149	95	80	74	29	51	30	26	47
(32) Trade	89	170	156	249	253	122	226	421	304	285	167	186	347
(33) Finance & insurance	24	48	50	80	85	31	55	70	46	48	33	43	60
(34) Real estate & rental	189	51	41	51	44	29	42	61	20	54	13	17	48
(35) Business serv., etc.	202	150	96	93	63	67	87	138	126	143	107	44	78
(36) Auto. repair	8	8	5	2	4	4	10	10	6	3	4	2	10
(37) Institutions, etc.	14	18	11	23	16	13	22	29	40	30	13	64	20
(38) Scrap		5			642	517	10	59			11		23

Table C3 (*continued*)

Sector	Sector (27)	(28)	(29)	(30)	(31)	(32)	(33)	(34)	(35)	(36)	(37)	(38)
(1) Agriculture, etc.	32	265	14	133	67	36	119	691	2	51	142	
(2) Iron mining	1	1						1				
(3) Nonferrous mining	2	1		1	1		1	1				
(4) Coal mining	10	24	347	5	569	21	14	24	17	13	23	
(5) Petroleum mining		1	5	1	238		1	20				
(6) Nonmetallic mining	4	54		6	4	2	5	2		2		
(7) Construction	25	15	1,023	675	864	31	34	1,785	12	7	149	
(8) Food	37	299	73	128	67	51	114	17	4	49	166	864
(9) Tobacco	1	7	1	7	3	2	6	1		3		132
(10) Textiles & products	223	309	14	167	89	58	149	17	46	64	16	3,219
(11) Wood & products	37	175	4	43	28	18	65	34	13	16	8	60
(12) Paper & publishing	345	384	48	185	66	281	282	20	2,137	67	190	6
(13) Chemicals	270	271	41	115	71	58	103	16	169	373	143	379
(14) Petroleum refining	14	77	376	45	177	63	42	20	55	31	30	
(15) Rubber prod., etc.	99	95	124	56	30	30	53	5	55	524	8	
(16) Leather & shoes	71	74	2	27	15	8	24	1	36	10	6	443
(17) Stone, clay, & glass	146	78	13	29	15	9	26	4	1	137	3	30
(18) Iron & steel	97	103	149	85	406	23	75	11	4	32	2	2,325
(19) Nonferrous metals	348	496	35	46	24	12	39	3	4	17	1	874
(20) Metal forming	294	205	44	75	50	34	67	7	24	114	5	539
(21) Nonelectrical eq.	103	76	23	46	24	14	41	6	3	114	1	
(22) Engines & turbines	1	10	6	9	6	2	8	5		3		
(23) Electrical eq., etc.	140	82	20	158	57	21	78	13	34	68	4	48
(24) Motor vehicles & eq.	5	26	3	24	16	6	21	4	5	9	1	830
(25) Aircraft		4	4	4	2	1	3	1		1	52	
(26) Trains, ships, etc.	1	25	159	8	4	2	8	2		3		
(27) Instruments, etc.	1,114	8	2	7	4	2	6	2	26	3	136	
(28) Misc. manufactures	122	282	17	61	33	28	62	5	320	23	37	250
(29) Transportation	125	294	585	228	441	136	190	158	24	129	79	
(30) Communications	32	28	30	210	29	71	69	24	474	75	71	
(31) Utilities	40	60	67	54	1,138	127	68	35	96	54	67	
(32) Trade	235	585	149	323	149	149	282	174	119	747	288	
(33) Finance & insurance	26	66	198	82	91	183	1,207	327	48	40	47	
(34) Real estate & rental	56	168	156	369	186	706	329	119	331	456	811	
(35) Business serv., etc.	261	213	57	216	147	476	264	151	105	357	103	
(36) Auto. repair	5	18	100	23	17	67	40	10	8	24	16	
(37) Institutions, etc.	29	19	9	24	13	15	30	116	21	20	177	
(38) Scrap												

Table C4 Final Demands for 1939, 1947, 1958, and 1961; 38 Order (millions of 1947 dollars)

Sector	1939	1947	1958	1961
(1) Agriculture, etc.	7,819	5,422	9,280	8,410
(2) Iron mining	−118	−94	−262	−236
(3) Nonferrous mining	51	−151	−100	28
(4) Coal mining	658	1,093	426	293
(5) Petroleum mining	26	−93	−731	−654
(6) Nonmetallic mining	−46	−61	−47	−39
(7) Construction	18,241	22,109	39,348	41,719
(8) Food	21,706	33,870	43,257	47,681
(9) Tobacco	974	2,963	3,587	4,083
(10) Textiles & products	8,463	12,803	14,732	16,449
(11) Wood & products	1,417	2,837	2,610	2,630
(12) Paper & publishing	949	1,697	2,153	2,827
(13) Chemicals	2,632	2,906	6,198	7,813
(14) Petroleum refining	2,097	3,150	6,185	6,823
(15) Rubber prod., etc.	757	1,144	1,232	1,514
(16) Leather & shoes	1,933	2,298	2,223	2,239
(17) Stone, clay, & glass	197	592	260	260
(18) Iron & steel	306	984	144	214
(19) Nonferrous metals	288	70	−24	−17
(20) Metal forming	535	1,807	1,401	1,487
(21) Nonelectrical eq.	2,286	6,201	5,619	5,892
(22) Engines & turbines	75	365	625	576
(23) Electrical eq., etc.	2,003	7,322	10,120	13,923
(24) Motor vehicles & eq.	4,346	7,539	8,662	10,995
(25) Aircraft	433	920	4,552	4,976
(26) Trains, ships, etc.	873	1,950	1,718	1,648
(27) Instruments, etc.	472	1,183	1,679	2,195
(28) Misc. manufactures	1,004	2,103	2,552	2,808
(29) Transportation	4,675	12,230	10,493	10,376
(30) Communications	789	1,898	3,690	4,510
(31) Utilities	1,537	2,553	7,760	9,056
(32) Trade	21,166	41,095	58,335	64,324
(33) Finance & insurance	5,410	4,519	6,978	7,841
(34) Real estate & rental	13,466	16,508	29,511	33,470
(35) Business serv., etc.	3,860	7,108	9,040	10,261
(36) Auto. repair	686	1,876	3,100	3,309
(37) Institutions, etc.	6,356	11,160	18,948	21,326
(38) Scrap	−335	−1,252	−741	−121

Table C5 Labor Coefficients for 1939, 1947, 1958, and 1961; 38 Order (man-years per thousand dollars of output in 1947 prices)

Sector	1939	1947	1958	1961
(1) Agriculture, etc.	0.2306	0.2187	0.1089	0.0952
(2) Iron mining	0.1088	0.1119	0.1206	0.0904
(3) Nonferrous mining	0.1127	0.1193	0.0650	0.0575
(4) Coal mining	0.2496	0.1663	0.1028	0.0894
(5) Petroleum mining	0.0357	0.0215	0.0194	0.0164
(6) Nonmetallic mining	0.1591	0.1107	0.0727	0.0620
(7) Construction	0.1901	0.1248	0.1045	0.1008
(8) Food	0.0489	0.0436	0.0318	0.0294
(9) Tobacco	0.0618	0.0305	0.0212	0.0179
(10) Textiles & products	0.1235	0.0970	0.0683	0.0619
(11) Wood & products	0.1338	0.1440	0.1041	0.0975
(12) Paper & publishing	0.1005	0.1020	0.0825	0.0758
(13) Chemicals	0.0550	0.0623	0.0375	0.0314
(14) Petroleum refining	0.0304	0.0289	0.0170	0.0141
(15) Rubber prod., etc.	0.0860	0.0875	0.0724	0.0615
(16) Leather & shoes	0.1155	0.1108	0.0992	0.0999
(17) Stone, clay, & glass	0.1592	0.1257	0.0916	0.0850
(18) Iron & steel	0.1214	0.0906	0.0849	0.0769
(19) Nonferrous metals	0.0558	0.0546	0.0446	0.0408
(20) Metal forming	0.1471	0.1073	0.0844	0.0800
(21) Nonelectrical eq.	0.1510	0.1130	0.1038	0.0815
(22) Engines & turbines	0.1352	0.0956	0.0693	0.0610
(23) Electrical eq., etc.	0.1174	0.1010	0.0786	0.0681
(24) Motor vehicles & eq.	0.0533	0.0560	0.0407	0.0350
(25) Aircraft	0.1234	0.1506	0.1036	0.0921
(26) Trains, ships, etc.	0.1984	0.1157	0.0944	0.0928
(27) Instruments, etc.	0.1272	0.1327	0.0928	0.0874
(28) Misc. manufactures	0.1229	0.1172	0.0864	0.0783
(29) Transportation	0.2560	0.1713	0.1218	0.1096
(30) Communications	0.2317	0.1601	0.0980	0.0834
(31) Utilities	0.1889	0.0700	0.0310	0.0258
(32) Trade	0.2132	0.1924	0.1629	0.1527
(33) Finance & insurance	0.2199	0.1651	0.1410	0.1441
(34) Real estate & rental	0.0304	0.0217	0.0160	0.0148
(35) Business serv., etc.	0.2328	0.2279	0.1837	0.1769
(36) Auto. repair	0.3151	0.1620	0.1597	0.1530
(37) Institutions, etc.	0.2329	0.2262	0.2061	0.2046
(38) Scrap	0.1214	0.0906	0.0849	0.0769

Table C6 Capital Coefficients for 1939, 1947, and 1958; 38 Order (dollar value of stock per dollar of capacity per year in 1947 prices)

Sector	1939	1947	1958
(1) Agriculture, etc.	0.5988	0.6135	0.7019
(2) Iron mining	1.4559	1.3956	1.4963
(3) Nonferrous mining	0.7379	1.6673	1.2944
(4) Coal mining	1.8108	1.2351	0.6902
(5) Petroleum mining	3.3607	1.5877	1.2156
(6) Nonmetallic mining	1.1243	1.3868	0.8942
(7) Construction	0.1058	0.2357	0.1900
(8) Food	0.2162	0.2691	0.2937
(9) Tobacco	0.2530	0.1456	0.1318
(10) Textiles & products	0.4132	0.3080	0.2497
(11) Wood & products	0.4253	0.5150	0.4092
(12) Paper & publishing	0.5630	0.5440	0.6145
(13) Chemicals	0.3384	0.5304	0.5648
(14) Petroleum refining	0.3862	0.5739	0.6644
(15) Rubber prod., etc.	0.7151	0.4300	0.4465
(16) Leather & shoes	0.2932	0.2368	0.1672
(17) Stone, clay, & glass	1.3805	0.9548	0.8439
(18) Iron & steel	1.1217	0.8311	1.0113
(19) Nonferrous metals	0.3852	0.6519	0.4380
(20) Metal forming	0.5565	0.5159	0.4156
(21) Nonelectrical eq.	0.4685	0.6212	0.5172
(22) Engines & turbines	0.5543	0.3178	0.3360
(23) Electrical eq., etc.	0.2824	0.3538	0.2738
(24) Motor vehicles & eq.	0.2178	0.2915	0.2331
(25) Aircraft	0.1244	0.3157	0.2583
(26) Trains, ships, etc.	0.2431	0.4745	0.4044
(27) Instruments, etc.	0.5608	0.5608	0.3552
(28) Misc. manufactures	0.9048	0.3022	0.4602
(29) Transportation	4.2566	3.7115	2.8204
(30) Communications	5.7348	4.3871	3.2835
(31) Utilities	9.7437	6.3805	3.2614
(32) Trade	0.9369	0.9369	0.8705
(33) Finance & insurance	0.2791	0.2938	0.3870
(34) Real estate & rental	0.0630	0.0630	0.0482
(35) Business serv., etc.	0.9213	0.9213	0.7938
(36) Auto. repair	1.5010	1.5010	2.3705
(37) Institutions, etc.	2.4539	2.4539	2.1512
(38) Scrap	0.8311	0.8311	1.0113

APPENDIX D 76-ORDER DATA

Except for the modifications noted in Appendix B on classification, the 76-order table for 1958 is identical to the domestic-base version of the 1958 input-output table published by the Office of Business Economics. That agency now furnishes on request a computer tape of the 1958 table and the 1947 table that were used for this study. Hence, it is unnecessary to undertake the cumbersome task of reproducing them here. Some auxiliary data—replacement coefficients, wage coefficients, capital coefficients, and total factor costs—are presented in the tables that follow. The crude and approximate nature of this information is cited in the text, Chapters 2, 8, 9, and 10.

Table D1 Labor Coefficients for 1947 and 1958, 76 Order (man-years per thousand dollars of output in 1947 prices)

Sector	1947	1958
(1) Livestock	0.2061	0.1004
(2) Crops	0.2276	0.1108
(3) Forestry & fishing	0.1396	0.1312
(4) Agric. services	0.2306	0.2167
(5) Iron mining	0.1119	0.1206
(6) Nonferrous mining	0.1193	0.0650
(7) Coal mining	0.1663	0.1028
(8) Petroleum mining	0.0215	0.0194
(9) Stone & clay mining	0.1207	0.0777
(10) Chemical mining	0.0848	0.0546
(11) New construction	0.1094	0.0946
(12) Maintenance constr.	0.1610	0.1392
(14) Food	0.0436	0.0318
(15) Tobacco	0.0305	0.0212
(16) Textiles	0.0870	0.0539
(18) Apparel	0.1090	0.0824
(20) Wood & products	0.1451	0.1013
(21) Wooden containers	0.1472	0.1200
(22) Household furniture	0.1346	0.1022
(23) Office furniture	0.1287	0.1080
(24) Paper & products	0.0668	0.0590
(26) Printing & publishing	0.1367	0.1081
(27) Basic chemicals	0.0640	0.0409
(28) Synthetic materials	0.0804	0.0371
(29) Drugs, soaps, etc.	0.0505	0.0274
(30) Paint	0.0415	0.0400
(31) Petroleum refining	0.0289	0.0170
(32) Rubber prod., etc.	0.0875	0.0724
(33) Leather tanning	0.0500	0.0425
(34) Shoes	0.1365	0.1195
(35) Glass & products	0.1310	0.1144
(36) Stone & clay prod.	0.1224	0.0858
(37) Iron & steel	0.0906	0.0849
(38) Nonferrous metals	0.0546	0.0446
(39) Metal containers	0.0728	0.0567
(40) Heating, etc.	0.1021	0.0737
(41) Stampings, etc.	0.1239	0.1168
(42) Hardware, etc.	0.0968	0.0872
(43) Engines & turbines	0.0956	0.0693
(44) Farm equipment	0.0895	0.0741
(45) Constr. & mining eq.	0.1000	0.0828

Table D1 (*continued*)

Sector	1947	1958
(46) Materials hand. eq.	0.0999	0.0978
(47) Metalworking eq.	0.1260	0.1197
(48) Special ind. eq.	0.1114	0.1058
(49) General ind. eq.	0.0967	0.0919
(50) Machine shop prod.	0.1571	0.1493
(51) Office & comp. mach.	0.1257	0.0803
(52) Service ind. mach.	0.0783	0.0526
(53) Electrical apparatus	0.1112	0.0926
(54) Household appliances	0.0850	0.0490
(55) Light. & wiring eq.	0.1018	0.0865
(56) Communication eq.	0.0781	0.0754
(57) Electronic compon.	0.1628	0.0902
(58) Batteries, etc.	0.1032	0.0957
(59) Motor vehicles & eq.	0.0560	0.0407
(60) Aircraft	0.1506	0.1036
(61) Trains, ships, etc.	0.1157	0.0944
(62) Instruments, etc.	0.1167	0.0935
(63) Photo. apparatus	0.1600	0.0894
(64) Misc. manufactures	0.1172	0.0864
(65) Transportation	0.1713	0.1218
(66) Telephone	0.1692	0.1015
(67) Radio & tv broad.	0.0847	0.0760
(68) Utilities	0.0700	0.0310
(69) Trade	0.1924	0.1629
(70) Finance & insurance	0.1651	0.1410
(71) Real estate & rental	0.0217	0.0160
(72) Hotels & pers. serv.	0.3574	0.3110
(73) Business services	0.1126	0.1110
(74) Research & dev.	0.1115	0.1115
(75) Auto. repair	0.1620	0.1597
(76) Amusements, etc.	0.1732	0.1548
(77) Institutions	0.2529	0.2314
(81) Bus. travel, etc.	0.1732	0.1548
(83) Scrap	0.0906	0.0849

Table D2 Capital Coefficients for 1947 and 1958, 76 Order (dollar value of fixed capital stock per dollar of capacity per year in 1947 prices)

Sector	1947	1958
(1) Livestock	0.5909	0.6807
(2) Crops	0.6283	0.7134
(3) Forestry & fishing	0.8107	0.9089
(4) Agric. services	0.3361	0.4643
(5) Iron mining	1.3956	1.4963
(6) Nonferrous mining	1.6673	1.2944
(7) Coal mining	1.2351	0.6902
(8) Petroleum mining	1.5877	1.2156
(9) Stone & clay mining	1.4911	0.9063
(10) Chemical mining	1.1176	0.8497
(11) New construction	0.2357	0.1825
(12) Maintenance constr.	0.2357	0.2176
(14) Food	0.2691	0.2937
(15) Tobacco	0.1456	0.1318
(16) Textiles	0.3540	0.3940
(18) Apparel	0.2525	0.1423
(20) Wood & products	0.5955	0.4707
(21) Wooden containers	0.2759	0.2950
(22) Household furniture	0.3603	0.2880
(23) Office furniture	0.3886	0.3699
(24) Paper & products	0.5711	0.7066
(26) Printing & publishing	0.5152	0.4961
(27) Basic chemicals	0.6104	0.7138
(28) Synthetic materials	0.8584	0.6338
(29) Drugs, soaps, etc.	0.3203	0.2721
(30) Paint	0.1840	0.2644
(31) Petroleum refining	0.5739	0.6644
(32) Rubber prod., etc.	0.4300	0.4465
(33) Leather tanning	0.2249	0.2071
(34) Shoes	0.2405	0.1544
(35) Glass & products	1.0803	0.6568
(36) Stone & clay prod.	0.8968	0.8901
(37) Iron & steel	0.8311	1.0113
(38) Nonferrous metals	0.6519	0.4380
(39) Metal containers	0.6857	0.4547
(40) Heating, etc.	0.3964	0.3187
(41) Stampings, etc.	0.4326	0.4869
(42) Hardware, etc.	0.6213	0.4772
(43) Engines & turbines	0.3178	0.3360
(44) Farm equipment	0.6024	0.3623
(45) Constr. & mining eq.	0.4256	0.4614

Table D2 (*continued*)

Sector	1947	1958
(46) Materials hand. eq.	0.5116	0.3782
(47) Metalworking eq.	0.6019	0.6296
(48) Special ind. eq.	0.7632	0.5672
(49) General ind. eq.	0.5065	0.4862
(50) Machine shop prod.	0.8592	0.5596
(51) Office & comp. mach.	0.4716	0.5620
(52) Service ind. mach.	0.3365	0.2355
(53) Electrical apparatus	0.3259	0.3662
(54) Household appliances	0.2859	0.2127
(55) Light. & wiring eq.	0.4301	0.2699
(56) Communication eq.	0.3372	0.1831
(57) Electronic compon.	0.3481	0.2442
(58) Batteries, etc.	0.3405	0.2963
(59) Motor vehicles & eq.	0.2915	0.2331
(60) Aircraft	0.3157	0.2583
(61) Trains, ships, etc.	0.4745	0.4044
(62) Instruments, etc.	0.5667	0.2762
(63) Photo. apparatus	0.5259	0.5172
(64) Misc. manufactures	0.3022	0.4602
(65) Transportation	3.7115	2.8204
(66) Telephone	4.8463	3.7131
(67) Radio & tv broad.	0.5994	0.6236
(68) Utilities	6.3805	3.2614
(69) Trade	0.9369	0.8705
(70) Finance & insurance	0.2938	0.3870
(71) Real estate & rental	0.0630	0.0482
(72) Hotels & pers. serv.	1.5686	1.1413
(73) Business services	0.3403	0.5219
(74) Research & dev.	0.3157	0.2583
(75) Auto. repair	1.5010	2.3705
(76) Amusements, etc.	1.2770	1.4107
(77) Institutions	2.9583	2.3027
(81) Bus. travel, etc.	1.2770	1.4107
(83) Scrap	0.8311	1.0113

Table D3 Total Factor Cost for 1947 and 1958, 76 Order (dollars per dollar of output in 1947 prices)

Sector[a]	1947	1958	Sector[a]	1947	1958
(1)	0.1787	0.0988	(43)	0.3268	0.2401
(2)	0.1900	0.1047	(44)	0.3366	0.2746
(3)	0.5691	0.5393	(45)	0.2592	0.2179
(4)	0.1016	0.1000	(46)	0.2923	0.2825
(5)	0.3756	0.4043	(47)	0.4924	0.4696
(6)	0.3644	0.2102	(48)	0.3746	0.3512
(7)	0.5045	0.3096	(49)	0.3313	0.3149
(8)	0.1758	0.1520	(50)	0.3644	0.3385
(9)	0.4399	0.2817	(51)	0.4157	0.2735
(10)	0.2178	0.1442	(52)	0.2838	0.1910
(11)	0.2988	0.2578	(53)	0.3702	0.3112
(12)	0.3944	0.3415	(54)	0.2410	0.1402
(14)	0.1074	0.0812	(55)	0.2878	0.2415
(15)	0.0557	0.0397	(56)	0.3749	0.3579
(16)	0.2229	0.1467	(57)	0.4444	0.2477
(18)	0.2872	0.2157	(58)	0.3708	0.3432
(20)	0.3298	0.2319	(59)	0.2149	0.1569
(21)	0.3193	0.2623	(60)	0.5200	0.3590
(22)	0.3354	0.2550	(61)	0.3596	0.2939
(23)	0.3146	0.2652	(62)	0.3626	0.2851
(24)	0.2374	0.2145	(63)	0.5124	0.2931
(26)	0.3899	0.3111	(64)	0.3341	0.2534
(27)	0.2113	0.1446	(65)	0.5359	0.3865
(28)	0.2468	0.1211	(66)	0.6471	0.4124
(29)	0.1914	0.1067	(67)	0.1398	0.1280
(30)	0.1503	0.1475	(68)	0.3823	0.1824
(31)	0.1328	0.0881	(69)	0.4414	0.3762
(32)	0.2953	0.2473	(70)	0.3671	0.3176
(33)	0.1723	0.1469	(71)	0.0416	0.0306
(34)	0.3088	0.2685	(72)	0.4693	0.4016
(35)	0.3794	0.3227	(73)	0.2526	0.2547
(36)	0.3473	0.2513	(74)	0.5200	0.3590
(37)	0.3108	0.2982	(75)	0.4769	0.4969
(38)	0.2153	0.1733	(76)	0.3600	0.3300
(39)	0.2135	0.1639	(77)	0.9752	0.8802
(40)	0.2994	0.2171	(81)	0.3600	0.3300
(41)	0.3296	0.3131	(83)	0.3108	0.2982
(42)	0.3607	0.3225			

[a] Sector names are given in tables B1, D1, and D2.

Table D4 Replacement Coefficients for 1958, 76 Order[a] (dollars per ten thousand dollars of output in 1947 prices)

Sector	Sector (1)	(2)	(3)	(4)	(5)	(6)	(7)	(8)
(11) New construction	54	58	41	38	314	206	79	173
(16) Textiles								
(20) Wood & products					1			
(22) Household furniture					1			1
(23) Office furniture								
(24) Paper & products								
(27) Basic chemicals								
(30) Paint								
(32) Rubber prod., etc.					4	2		
(36) Stone & clay prod.								
(37) Iron & steel					4	9	2	
(38) Nonferrous metals								
(39) Metal containers								
(40) Heating, etc.					2	1		78
(41) Stampings, etc.								
(42) Hardware, etc.					1	1		98
(43) Engines & turbines								84
(44) Farm equipment	163	202	166	132	2	3	1	
(45) Constr. & mining eq.	13	16	13	11	256	189	190	296
(46) Materials hand. eq.	11	14	11	9	32	21	12	
(47) Metalworking eq.					26	15	2	
(48) Special ind. eq.								
(49) General ind. eq.					32	14	3	
(50) Machine shop prod.								
(51) Office & comp. mach.					2	1	1	1
(52) Service ind. mach.					1	1		
(53) Electrical apparatus					73	36	12	
(54) Household appliances					1			
(55) Light. & wiring eq.								
(56) Communication eq.								
(57) Electronic compon.								
(58) Batteries, etc.			10					
(59) Motor vehicles & eq.	39	1			54	21		
(60) Aircraft								
(61) Trains, ships, etc.			247		25	21	5	
(62) Instruments, etc.					3	1		18
(63) Photo. apparatus								
(64) Misc. manufactures								
(65) Transportation	4	6	5	4	6	4	2	6
(66) Telephone								
(68) Utilities								
(70) Finance & insurance								
(73) Business services								
(77) Institutions								

Table D4 (*continued*)

Sector	Sector							
	(9)	(10)	(11)	(12)	(14)	(15)	(16)	(18)
(11) New construction	136	101	10	11	14	7	24	13
(16) Textiles								
(20) Wood & products	1	2			1	1	4	4
(22) Household furniture	1							
(23) Office furniture							1	1
(24) Paper & products								
(27) Basic chemicals		1						
(30) Paint		1					1	1
(32) Rubber prod., etc.	3	2	1	1				
(36) Stone & clay prod.		1			2	1	5	4
(37) Iron & steel		21					1	1
(38) Nonferrous metals		2			1		3	2
(39) Metal containers								
(40) Heating, etc.	1	30			3		4	3
(41) Stampings, etc.							1	1
(42) Hardware, etc.		10	40	44	1		2	1
(43) Engines & turbines		2					1	
(44) Farm equipment	1	1	18	21				
(45) Constr. & mining eq.	143	80	93	104				
(46) Materials hand. eq.	19	11	7	8	3	1	5	5
(47) Metalworking eq.	16	16					1	1
(48) Special ind. eq.		13			11	6	27	22
(49) General ind. eq.	22	21			2	1	6	5
(50) Machine shop prod.								
(51) Office & comp. mach.	1	1			1		2	1
(52) Service ind. mach.	1	1			1			2
(53) Electrical apparatus	49	30			1	1	3	3
(54) Household appliances	1	8					22	16
(55) Light. & wiring eq.							1	1
(56) Communication eq.		1						
(57) Electronic compon.								
(58) Batteries, etc.								
(59) Motor vehicles & eq.	28	28	3	3	56	4	17	6
(60) Aircraft								
(61) Trains, ships, etc.	13	6						
(62) Instruments, etc.	2	6	2	2	1		1	1
(63) Photo. apparatus								
(64) Misc. manufactures								
(65) Transportation	3	2	1	2	1		3	2
(66) Telephone								
(68) Utilities							1	
(70) Finance & insurance					1		2	2
(73) Business services		4			5	2	15	12
(77) Institutions		8						

Table D4 (*continued*)

Sector	Sector							
	(20)	(21)	(22)	(23)	(24)	(26)	(27)	(28)
(11) New construction	45	23	25	33	52	34	56	51
(16) Textiles								
(20) Wood & products	6	4	5	7	6	4	4	5
(22) Household furniture								
(23) Office furniture	1	1	1	1	1	1	1	1
(24) Paper & products								
(27) Basic chemicals							1	
(30) Paint	1	1	1	1	2	1	2	2
(32) Rubber prod., etc.								
(36) Stone & clay prod.	9	7	6	7	12	6	17	14
(37) Iron & steel	3	2	1	1	4	1	7	4
(38) Nonferrous metals	4	4	3	3	6	3	9	7
(39) Metal containers								
(40) Heating, etc.	10	8	4	4	15	4	29	24
(41) Stampings, etc.	1	1	1	1	1	1	2	2
(42) Hardware, etc.	3	3	2	2	5	2	10	6
(43) Engines & turbines	7	1	1	1	2		3	2
(44) Farm equipment								
(45) Constr. & mining eq.								
(46) Materials hand. eq.	15	4	8	11	12	5	6	8
(47) Metalworking eq.	2	2	13	43	1	1	1	2
(48) Special ind. eq.	53	56	25	8	64	47	41	77
(49) General ind. eq.	9	7	7	10	12	6	26	22
(50) Machine shop prod.								
(51) Office & comp. mach.	2	1	1	1	3	1	5	4
(52) Service ind. mach.					1		1	4
(53) Electrical apparatus	7	5	5	7	10	3	14	10
(54) Household appliances			6	3				
(55) Light. & wiring eq.	1	1	1	1	1	1	2	2
(56) Communication eq.							1	1
(57) Electronic compon.								
(58) Batteries, etc.					1		1	1
(59) Motor vehicles & eq.	37	28	12	18	19	18	28	7
(60) Aircraft								
(61) Trains, ships, etc.	2							
(62) Instruments, etc.	3	2	1	1	5	1	12	6
(63) Photo. apparatus					1		2	1
(64) Misc. manufactures							1	1
(65) Transportation	5	3	3	4	6	3	7	6
(66) Telephone								
(68) Utilities	1	1	1	1	1	1	1	1
(70) Finance & insurance	4	2	2	3	4	2	5	5
(73) Business services	28	18	16	21	30	15	37	35
(77) Institutions	1				1		1	1

Table D4 (*continued*)

Sector	Sector (29)	(30)	(31)	(32)	(33)	(34)	(35)	(36)
(11) New construction	19	23	39	35	17	14	51	66
(16) Textiles								
(20) Wood & products	3	3	1	4	4	3	5	5
(22) Household furniture								
(23) Office furniture		1	1	1	1	1	1	1
(24) Paper & products								
(27) Basic chemicals								
(30) Paint	1	1	1	1	1	1	2	2
(32) Rubber prod., etc.								
(36) Stone & clay prod.	4	5	9	9	5	4	11	12
(37) Iron & steel	1	1	4	2	1	1	3	3
(38) Nonferrous metals	2	2	5	5	2	2	5	6
(39) Metal containers								
(40) Heating, etc.	6	5	15	12	4	3	11	12
(41) Stampings, etc.		1	1	1	1		1	1
(42) Hardware, etc.	2	2	5	3	2	1	4	4
(43) Engines & turbines	1	1	1	2	1		2	2
(44) Farm equipment								
(45) Constr. & mining eq.								1
(46) Materials hand. eq.	6	6	2	22	6	10	18	13
(47) Metalworking eq.	1	1	1	5	1	4	3	3
(48) Special ind. eq.	18	25	17	49	33	20	51	79
(49) General ind. eq.	9	9	15	15	6	6	25	12
(50) Machine shop prod.								
(51) Office & comp. mach.	2	2	3	3	2	2	4	6
(52) Service ind. mach.	1		2	1				
(53) Electrical apparatus	4	4	6	9	4	3	10	9
(54) Household appliances						2		
(55) Light. & wiring eq.	1	1	1	1	1	1	1	1
(56) Communication eq.								
(57) Electronic compon.								
(58) Batteries, etc.								
(59) Motor vehicles & eq.	7	15	41	8	8	6	28	93
(60) Aircraft								
(61) Trains, ships, etc.								
(62) Instruments, etc.	2	2	7	3	1	1	3	3
(63) Photo. apparatus			1					
(64) Misc. manufactures								1
(65) Transportation	2	2	3	5	3	2	5	5
(66) Telephone								
(68) Utilities			1	1			1	1
(70) Finance & insurance	2	2	2	4	2	2	4	4
(73) Business services	11	13	18	26	14	12	27	29
(77) Institutions				1			1	1

Table D4 (*continued*)

Sector	Sector (37)	(38)	(39)	(40)	(41)	(42)	(43)	(44)
(11) New construction	93	40	46	26	43	38	30	33
(16) Textiles								
(20) Wood & products	9	3	11	4	6	7	6	5
(22) Household furniture								
(23) Office furniture	2	1	1	1	1	1	1	1
(24) Paper & products								
(27) Basic chemicals								
(30) Paint	3	1	2	1	2	2	1	1
(32) Rubber prod., etc.								
(36) Stone & clay prod.	23	8	13	5	11	10	7	8
(37) Iron & steel	5	2	2	1	1	1	1	1
(38) Nonferrous metals	9	4	6	2	5	4	3	3
(39) Metal containers								
(40) Heating, etc.	26	9	9	4	8	7	8	9
(41) Stampings, etc.	2	1	2	1	2	1	1	1
(42) Hardware, etc.	7	3	3	1	2	2	2	2
(43) Engines & turbines	4	2	1	1		1	1	1
(44) Farm equipment								
(45) Constr. & mining eq.	2							
(46) Materials hand. eq.	18	9	34	9	35	25	10	11
(47) Metalworking eq.	69	19	74	39	84	60	52	56
(48) Special ind. eq.	18	12	2	1	3	9	6	1
(49) General ind. eq.	45	26	19	7	15	12	8	10
(50) Machine shop prod.								
(51) Office & comp. mach.	6	2	2	1	4	2	1	1
(52) Service ind. mach.	1		1					
(53) Electrical apparatus	18	8	14	6	9	16	6	9
(54) Household appliances								
(55) Light. & wiring eq.	2	1	2	1	1	1	1	1
(56) Communication eq.	1		1					
(57) Electronic compon.								
(58) Batteries, etc.	1							
(59) Motor vehicles & eq.	8	7	11	31	17	18	6	29
(60) Aircraft								
(61) Trains, ships, etc.	5							
(62) Instruments, etc.	7	3	2	1	2	2	2	2
(63) Photo. apparatus	2							
(64) Misc. manufactures	1		1		1			
(65) Transportation	9	4	6	3	6	5	4	4
(66) Telephone								
(68) Utilities	2	1	1	1	1	1	1	1
(70) Finance & insurance	7	3	5	2	4	4	3	3
(73) Business services	50	21	35	16	33	28	22	23
(77) Institutions	1		1		1	1		1

Table D4 (*continued*)

Sector	(45)	(46)	(47)	(48)	(49)	(50)	(51)	(52)
(11) New construction	44	33	62	50	42	49	36	18
(16) Textiles								
(20) Wood & products	8	6	14	10	7	9	7	3
(22) Household furniture								
(23) Office furniture	1	1	2	1	1	2	1	1
(24) Paper & products								
(27) Basic chemicals								
(30) Paint	2	1	3	2	2	2	2	1
(32) Rubber prod., etc.								
(36) Stone & clay prod.	10	8	16	11	10	12	9	5
(37) Iron & steel	2	1	2	2	1	1	2	1
(38) Nonferrous metals	4	4	7	5	5	5	4	2
(39) Metal containers								
(40) Heating, etc.	9	6	10	8	8	7	9	4
(41) Stampings, etc.	1	1	2	2	1	2	1	1
(42) Hardware, etc.	2	2	4	3	2	2	2	1
(43) Engines & turbines	1		1	1	1	1	1	
(44) Farm equipment								
(45) Constr. & mining eq.								
(46) Materials hand. eq.	15	13	27	14	24	22	24	15
(47) Metalworking eq.	73	66	130	92	82	116	53	35
(48) Special ind. eq.	2	1	7	6	2	2	1	1
(49) General ind. eq.	13	11	22	14	13	13	15	7
(50) Machine shop prod.								
(51) Office & comp. mach.	2	1	3	2	2	2	2	1
(52) Service ind. mach.			1	1		1	1	
(53) Electrical apparatus	11	8	12	13	10	16	13	5
(54) Household appliances								
(55) Light. & wiring eq.	1	1	2	1	1	2	1	1
(56) Communication eq.			1			1		
(57) Electronic compon.								
(58) Batteries, etc.								
(59) Motor vehicles & eq.	28	21	22	29	21	15	3	9
(60) Aircraft								
(61) Trains, ships, etc.								
(62) Instruments, etc.	2	1	2	2	2	2	2	1
(63) Photo. apparatus								
(64) Misc. manufactures			1	1	1	1		
(65) Transportation	5	4	9	6	6	7	5	3
(66) Telephone								
(68) Utilities	1	1	2	1	1	1	1	
(70) Finance & insurance	4	3	6	5	4	5	4	2
(73) Business services	30	23	47	35	31	38	27	14
(77) Institutions	1	1	1	1	1	1	1	

Table D4 (*continued*)

Sector	Sector (53)	(54)	(55)	(56)	(57)	(58)	(59)	(60)
(11) New construction	33	20	23	18	21	27	24	24
(16) Textiles								
(20) Wood & products	6	4	5	4	5	4	4	3
(22) Household furniture								
(23) Office furniture	1	1	1	1	1	1	1	1
(24) Paper & products								
(27) Basic chemicals								
(30) Paint	2	1	1	1	1	2	1	1
(32) Rubber prod., etc.								
(36) Stone & clay prod.	11	5	9	6	7	9	6	4
(37) Iron & steel	2	1	1	1	1	1	1	1
(38) Nonferrous metals	5	2	3	3	3	4	3	2
(39) Metal containers								
(40) Heating, etc.	12	5	10	6	7	7	8	4
(41) Stampings, etc.	1	1	1	1	1	1	1	1
(42) Hardware, etc.	3	1	2	2	2	2	2	1
(43) Engines & turbines	1			1		1	1	
(44) Farm equipment								
(45) Constr. & mining eq.						1		
(46) Materials hand. eq.	21	12	19	14	16	25	17	10
(47) Metalworking eq.	42	25	31	5	6	30	38	26
(48) Special ind. eq.	17	1	12	19	23	18	1	1
(49) General ind. eq.	15	7	10	8	10	13	10	5
(50) Machine shop prod.								
(51) Office & comp. mach.	2	1	1	1	1	1	1	1
(52) Service ind. mach.	1		1		1			
(53) Electrical apparatus	25	6	9	18	17	23	8	5
(54) Household appliances							1	
(55) Light. & wiring eq.	1	1	1	1	1	1	1	1
(56) Communication eq.								
(57) Electronic compon.								
(58) Batteries, etc.								
(59) Motor vehicles & eq.	12	5	5	4	1	10	10	2
(60) Aircraft								
(61) Trains, ships, etc.								
(62) Instruments, etc.	2	1	2	2	2	1	1	1
(63) Photo. apparatus								
(64) Misc. manufactures	1							
(65) Transportation	6	3	4	3	4	5	3	2
(66) Telephone								
(68) Utilities	1		1	1	1	1	1	
(70) Finance & insurance	4	2	3	2	3	3	2	2
(73) Business services	31	14	22	16	19	25	18	13
(77) Institutions	1					1		

Table D4 (*continued*)

Sector	Sector (61)	(62)	(63)	(64)	(65)	(66)	(67)	(68)
(11) New construction	42	24	39	35	307	272	62	405
(16) Textiles								
(20) Wood & products	5	6	8	6		39		
(22) Household furniture					3	8	4	1
(23) Office furniture	1	1	1	1	1	6	1	
(24) Paper & products								
(27) Basic chemicals								
(30) Paint	1	1	2	2				
(32) Rubber prod., etc.								
(36) Stone & clay prod.	8	7	13	9				
(37) Iron & steel	1	2	3	2				101
(38) Nonferrous metals	3	3	6	4		301		
(39) Metal containers								
(40) Heating, etc.	8	7	11	7				3
(41) Stampings, etc.	1	1	2	1				
(42) Hardware, etc.	2	2	4	2	1			1
(43) Engines & turbines	1		2	1				49
(44) Farm equipment								
(45) Constr. & mining eq.								5
(46) Materials hand. eq.	11	9	25	25				2
(47) Metalworking eq.	49	15	19	33	6			
(48) Special ind. eq.	5	28	54	26				
(49) General ind. eq.	11	11	20	14	2			52
(50) Machine shop prod.								
(51) Office & comp. mach.	1	2	5	3	2	2	14	1
(52) Service ind. mach.		1	2	1	7	1	6	1
(53) Electrical apparatus	8	6	13	11	3	8		230
(54) Household appliances				1	6			1
(55) Light. & wiring eq.	1	1	2	1	1	48		
(56) Communication eq.			1		11	405	109	2
(57) Electronic compon.								
(58) Batteries, etc.			1			3		2
(59) Motor vehicles & eq.	15	4	13	27	344	92		3
(60) Aircraft					9			
(61) Trains, ships, etc.					568			
(62) Instruments, etc.	2	6	4	2		1		12
(63) Photo. apparatus		1	1					
(64) Misc. manufactures			1				2	
(65) Transportation	4	4	6	5	10	6		4
(66) Telephone								
(68) Utilities	1	1	1	1				
(70) Finance & insurance	3	3	5	4				
(73) Business services	22	20	34	26				
(77) Institutions			1	1				

Table D4 (*continued*)

Sector	Sector (69)	(70)	(71)	(72)	(73)	(75)	(76)	(77)
(11) New construction	133	62	9	203	74	378	281	428
(16) Textiles	3			10				
(20) Wood & products	1							
(22) Household furniture	25	13	1	84	28			41
(23) Office furniture	12	4		13	8	1		11
(24) Paper & products				3	2			1
(27) Basic chemicals								
(30) Paint								
(32) Rubber prod., etc.								
(36) Stone & clay prod.	2							
(37) Iron & steel				3				
(38) Nonferrous metals								
(39) Metal containers								
(40) Heating, etc.	15			43		142		
(41) Stampings, etc.	3			1				
(42) Hardware, etc.				8	6	16		3
(43) Engines & turbines								
(44) Farm equipment								
(45) Constr. & mining eq.								
(46) Materials hand. eq.				1		19		
(47) Metalworking eq.						24		
(48) Special ind. eq.	4				3	9		
(49) General ind. eq.	34			2		46		
(50) Machine shop prod.								
(51) Office & comp. mach.	7	21		129	17	9		1
(52) Service ind. mach.	10	9		55	8	58		
(53) Electrical apparatus						2		
(54) Household appliances	9			2		11		
(55) Light. & wiring eq.						1		
(56) Communication eq.	1					2		
(57) Electronic compon.								
(58) Batteries, etc.					4			2
(59) Motor vehicles & eq.	49			63		249	2	
(60) Aircraft								
(61) Trains, ships, etc.								
(62) Instruments, etc.	1	1		39	19	1		9
(63) Photo. apparatus						6	3	
(64) Misc. manufactures	8	3		16	2	6		
(65) Transportation	2	1		6	1			1
(66) Telephone								
(68) Utilities								
(70) Finance & insurance								6
(73) Business services								13
(77) Institutions								29

[a] Rows where all entries were zero in 1947 and 1958 are omitted from the table.

Table D5 Replacement Coefficients for 1947, 76 Order[a] (dollars per ten thousand dollars of output in 1947 prices)

Sector	Sector (1)	(2)	(3)	(4)	(5)	(6)	(7)	(8)
(11) New construction	62	66	55	36	239	285	120	142
(16) Textiles								
(20) Wood & products					1			
(22) Household furniture						1		
(23) Office furniture								
(24) Paper & products								
(27) Basic chemicals								
(30) Paint								
(32) Rubber prod., etc.					3	2		
(36) Stone & clay prod.								
(37) Iron & steel					3	12	4	
(38) Nonferrous metals								
(39) Metal containers								
(40) Heating, etc.					1	1		64
(41) Stampings, etc.								
(42) Hardware, etc.						1		80
(43) Engines & turbines								69
(44) Farm equipment	134	142	136	76	2	4	1	
(45) Constr. & mining eq.	11	11	11	6	195	261	287	242
(46) Materials hand. eq.	9	10	9	5	24	29	18	
(47) Metalworking eq.					20	21	3	
(48) Special ind. eq.								
(49) General ind. eq.					24	19	5	
(50) Machine shop prod.								
(51) Office & comp. mach.					1	1	1	1
(52) Service ind. mach.					1	1		
(53) Electrical apparatus					55	49	19	
(54) Household appliances					1	1		
(55) Light. & wiring eq.								
(56) Communication eq.						1		
(57) Electronic compon.								
(58) Batteries, etc.								
(59) Motor vehicles & eq.	26	28	23	15	41	29		
(60) Aircraft								
(61) Trains, ships, etc.			203		19	29	7	
(62) Instruments, etc.					2	1		15
(63) Photo. apparatus								
(64) Misc. manufactures								
(65) Transportation	4	4	4	2	4	5	4	5
(66) Telephone								
(68) Utilities								
(70) Finance & insurance								
(73) Business services								
(77) Institutions								

Table D5 (*continued*)

Sector	Sector (9)	(10)	(11)	(12)	(14)	(15)	(16)	(18)
(11) New construction	207	127	12	12	27	10	38	32
(16) Textiles								
(20) Wood & products	1	2						
(22) Household furniture	1	1			1			1
(23) Office furniture								
(24) Paper & products								
(27) Basic chemicals		1						
(30) Paint		1						
(32) Rubber prod., etc.	5	2	1	1				
(36) Stone & clay prod.		1						
(37) Iron & steel		25						
(38) Nonferrous metals		3						
(39) Metal containers								
(40) Heating, etc.	2	36			8		1	
(41) Stampings, etc.								
(42) Hardware, etc.		12	48	48	2			21
(43) Engines & turbines		3			3			
(44) Farm equipment	1	1	22	22				
(45) Constr. & mining eq.	219	97	112	112				
(46) Materials hand. eq.	29	14	8	8	5			
(47) Metalworking eq.	25	20						
(48) Special ind. eq.		15			26	20	83	88
(49) General ind. eq.	33	25			5		1	
(50) Machine shop prod.								
(51) Office & comp. mach.	2	1			1		1	2
(52) Service ind. mach.	2	2			3		1	1
(53) Electrical apparatus	74	36			2		1	
(54) Household appliances	1	10			2			10
(55) Light. & wiring eq.								
(56) Communication eq.		1						
(57) Electronic compon.								
(58) Batteries, etc.								
(59) Motor vehicles & eq.	43	33	3	3	25	10	3	1
(60) Aircraft								
(61) Trains, ships, etc.	20	8			1		2	
(62) Instruments, etc.	3	7	2	2				
(63) Photo. apparatus								
(64) Misc. manufactures								
(65) Transportation	5	3	2	2	1		1	1
(66) Telephone								
(68) Utilities								
(70) Finance & insurance								
(73) Business services		5						
(77) Institutions		10						

Table D5 (*continued*)

Sector	Sector							
	(20)	(21)	(22)	(23)	(24)	(26)	(27)	(28)
(11) New construction	87	26	31	29	55	45	52	69
(16) Textiles								
(20) Wood & products	1	1			1			
(22) Household furniture	1			1		1		1
(23) Office furniture				1				
(24) Paper & products								
(27) Basic chemicals								
(30) Paint								
(32) Rubber prod., etc.		1						
(36) Stone & clay prod.							1	1
(37) Iron & steel	1	1						1
(38) Nonferrous metals			1	1	3		1	2
(39) Metal containers								
(40) Heating, etc.	12	18	1	8	9		55	61
(41) Stampings, etc.								
(42) Hardware, etc.	2	3			2		10	9
(43) Engines & turbines	2	5						
(44) Farm equipment	1	1	11	4	2			1
(45) Constr. & mining eq.	7	1	3	2	1		1	
(46) Materials hand. eq.	19	4	6	3	5	5	3	8
(47) Metalworking eq.	3		45	60	1			
(48) Special ind. eq.	52	29	33	24	89	91	53	120
(49) General ind. eq.	10	14	6	17	8		24	27
(50) Machine shop prod.								
(51) Office & comp. mach.	1	1	1	1	1	2	1	1
(52) Service ind. mach.				1	1	1	4	6
(53) Electrical apparatus	3	5	3	3	11		12	11
(54) Household appliances			1	1	1		3	5
(55) Light. & wiring eq.			1	1			2	2
(56) Communication eq.								
(57) Electronic compon.								
(58) Batteries, etc.								
(59) Motor vehicles & eq.	63	26	10	9	10	5	8	2
(60) Aircraft								
(61) Trains, ships, etc.	31						1	
(62) Instruments, etc.				1	1	1	4	5
(63) Photo. apparatus						1		
(64) Misc. manufactures								1
(65) Transportation	2	1	1	1	1	1	2	3
(66) Telephone								
(68) Utilities								
(70) Finance & insurance								
(73) Business services								
(77) Institutions								

Table D5 (*continued*)

Sector	Sector							
	(29)	(30)	(31)	(32)	(33)	(34)	(35)	(36)
(11) New construction	20	15	34	39	34	49	107	127
(16) Textiles								
(20) Wood & products								
(22) Household furniture	1			3		1		1
(23) Office furniture				1				
(24) Paper & products								
(27) Basic chemicals								
(30) Paint								
(32) Rubber prod., etc.								
(36) Stone & clay prod.	1	2	1				1	8
(37) Iron & steel				7			1	1
(38) Nonferrous metals				1				1
(39) Metal containers								
(40) Heating, etc.	15	11	29	11			5	13
(41) Stampings, etc.								
(42) Hardware, etc.	19	4	22	1			1	1
(43) Engines & turbines	2		6		1			
(44) Farm equipment				1			1	2
(45) Constr. & mining eq.								32
(46) Materials hand. eq.	1	3	4	18			26	16
(47) Metalworking eq.	1		3			2	3	5
(48) Special ind. eq.	30	14	34	13	47	24	192	32
(49) General ind. eq.	10	9	21	69			4	13
(50) Machine shop prod.								
(51) Office & comp. mach.	1	1		1	1	1	1	1
(52) Service ind. mach.	3	1	1	2			1	1
(53) Electrical apparatus	5	6	12	2	1		29	15
(54) Household appliances	3	1	1	1				1
(55) Light. & wiring eq.	1	1					8	1
(56) Communication eq.								
(57) Electronic compon.								
(58) Batteries, etc.								
(59) Motor vehicles & eq.	5	5	8	23	2	2	6	45
(60) Aircraft								
(61) Trains, ships, etc.								1
(62) Instruments, etc.	7	1	7	1				2
(63) Photo. apparatus								
(64) Misc. manufactures								
(65) Transportation	1	1	2	2	1		3	2
(66) Telephone								
(68) Utilities								
(70) Finance & insurance								
(73) Business services								
(77) Institutions								

Table D5 (*continued*)

Sector	Sector (37)	(38)	(39)	(40)	(41)	(42)	(43)	(44)
(11) New construction	127	99	22	35	54	46	23	55
(16) Textiles								
(20) Wood & products								
(22) Household furniture				1	1	1	1	1
(23) Office furniture					1			
(24) Paper & products								
(27) Basic chemicals								
(30) Paint								
(32) Rubber prod., etc.								
(36) Stone & clay prod.		3						
(37) Iron & steel		1					1	
(38) Nonferrous metals		5	1	2		4		4
(39) Metal containers								
(40) Heating, etc.	2	15	2	3	2	3	2	2
(41) Stampings, etc.							1	
(42) Hardware, etc.		6					7	
(43) Engines & turbines	2							
(44) Farm equipment	2	1	1	1	2	3		4
(45) Constr. & mining eq.	5	34		1	1	1		3
(46) Materials hand. eq.	29	15	18	4	8	4		5
(47) Metalworking eq.	67	48	239	92	104	168	98	149
(48) Special ind. eq.	12	5	1	4	2	4		6
(49) General ind. eq.	11	18	13	9	5	19	1	14
(50) Machine shop prod.								
(51) Office & comp. mach.	1	1	1	1	2	1	2	2
(52) Service ind. mach.	1	1		1	1	1	1	1
(53) Electrical apparatus	38	20	2	9	7	13		11
(54) Household appliances		1	1	2		4		3
(55) Light. & wiring eq.		2		1		2		2
(56) Communication eq.					1			
(57) Electronic compon.								
(58) Batteries, etc.								
(59) Motor vehicles & eq.	3	4	7	12	4	4	2	5
(60) Aircraft								
(61) Trains, ships, etc.	10	1						
(62) Instruments, etc.	2	2				1		1
(63) Photo. apparatus								
(64) Misc. manufactures								
(65) Transportation	2	2	3	1	1	3	1	2
(66) Telephone								
(68) Utilities								
(70) Finance & insurance								
(73) Business services								
(77) Institutions								

Table D5 (*continued*)

Sector	Sector							
	(45)	(46)	(47)	(48)	(49)	(50)	(51)	(52)
(11) New construction	35	36	41	44	31	66	21	26
(16) Textiles								
(20) Wood & products								
(22) Household furniture	1	1	1	1	1	1	1	1
(23) Office furniture								
(24) Paper & products								
(27) Basic chemicals								
(30) Paint								
(32) Rubber prod., etc.								
(36) Stone & clay prod.								
(37) Iron & steel								
(38) Nonferrous metals	3	3	4	5	3	6	3	2
(39) Metal containers								
(40) Heating, etc.	2	1	2	2	2	3	2	1
(41) Stampings, etc.								
(42) Hardware, etc.								
(43) Engines & turbines								
(44) Farm equipment	2	3	2	2	2	2	1	2
(45) Constr. & mining eq.	2	2	1	1	1	1	2	1
(46) Materials hand. eq.	9	6	8	7	6	4	1	2
(47) Metalworking eq.	105	143	212	228	149	243	142	88
(48) Special ind. eq.	2	2	1	5	3	1	1	1
(49) General ind. eq.	11	11	14	16	11	23	14	9
(50) Machine shop prod.								
(51) Office & comp. mach.	1	2	2	2	1	2	1	1
(52) Service ind. mach.	1	1	1	1	1	1	1	1
(53) Electrical apparatus	8	11	14	17	10	18	11	7
(54) Household appliances	2	3	4	5	3	5	3	2
(55) Light. & wiring eq.	1	2	3	3	2	3	2	1
(56) Communication eq.								
(57) Electronic compon.								
(58) Batteries, etc.								
(59) Motor vehicles & eq.	4	4	3	5	4	8	2	3
(60) Aircraft								
(61) Trains, ships, etc.								
(62) Instruments, etc.	1	1	1	1	1	1	1	1
(63) Photo. apparatus								
(64) Misc. manufactures								
(65) Transportation	2	2	3	3	2	4	2	1
(66) Telephone								
(68) Utilities								
(70) Finance & insurance								
(73) Business services								
(77) Institutions								

Table D5 (*continued*)

Sector	Sector (53)	(54)	(55)	(56)	(57)	(58)	(59)	(60)
(11) New construction	33	28	33	33	36	28	36	24
(16) Textiles								
(20) Wood & products								
(22) Household furniture	3	1	1	1	2	1		1
(23) Office furniture	1			1	2			
(24) Paper & products								
(27) Basic chemicals								
(30) Paint								
(32) Rubber prod., etc.								
(36) Stone & clay prod.								
(37) Iron & steel								
(38) Nonferrous metals	2	2	3	1	1	3	2	2
(39) Metal containers								
(40) Heating, etc.	2	1	3	3	3	3	1	1
(41) Stampings, etc.								
(42) Hardware, etc.				1				
(43) Engines & turbines								
(44) Farm equipment	2	2	3	2	2	2	2	1
(45) Constr. & mining eq.	1	1	1	1	1	1	1	1
(46) Materials hand. eq.	4	2	5	5	5	5	3	2
(47) Metalworking eq.	97	70	142	104	107	107	101	91
(48) Special ind. eq.	2	2	3	3	2	2	1	
(49) General ind. eq.	11	8	18	10	9	15	9	8
(50) Machine shop prod.								
(51) Office & comp. mach.	3	1	1	3	3	1	1	2
(52) Service ind. mach.	2	1	1	1	1	1		1
(53) Electrical apparatus	8	6	11	8	9	8	8	7
(54) Household appliances	2	1	3	1	1	2	2	2
(55) Light. & wiring eq.	1	1	2	1		2	1	1
(56) Communication eq.								
(57) Electronic compon.								
(58) Batteries, etc.								
(59) Motor vehicles & eq.	2	2	2	2	2	3	2	4
(60) Aircraft								
(61) Trains, ships, etc.								
(62) Instruments, etc.	1		1	1	1	1	1	1
(63) Photo. apparatus								
(64) Misc. manufactures	1				1			
(65) Transportation	2	1	2	2	2	2	2	1
(66) Telephone								
(68) Utilities								
(70) Finance & insurance								
(73) Business services								
(77) Institutions								

Table D5 (*continued*)

Sector	Sector (61)	(62)	(63)	(64)	(65)	(66)	(67)	(68)
(11) New construction	50	36	33	28	448	388	60	710
(16) Textiles								
(20) Wood & products						50		
(22) Household furniture		1	1	1	1	11	4	1
(23) Office furniture						7	1	
(24) Paper & products								
(27) Basic chemicals								
(30) Paint								
(32) Rubber prod., etc.								
(36) Stone & clay prod.								
(37) Iron & steel								160
(38) Nonferrous metals	1	5	4			379		1
(39) Metal containers								
(40) Heating, etc.	1	3	3	1		1		5
(41) Stampings, etc.								
(42) Hardware, etc.	6			6	2			1
(43) Engines & turbines								79
(44) Farm equipment	1	2	1					
(45) Constr. & mining eq.	4	1	1					9
(46) Materials hand. eq.	3	1	1					3
(47) Metalworking eq.	70	195	180	73	7			1
(48) Special ind. eq.	1	2	2	17				
(49) General ind. eq.	7	21	19	10	2			82
(50) Machine shop prod.								
(51) Office & comp. mach.	1	2	2	2	1	2	14	2
(52) Service ind. mach.		1	1	1	4	1	6	1
(53) Electrical apparatus	3	15	14	1	3	11		368
(54) Household appliances	1	4	4		3			1
(55) Light. & wiring eq.		3	3		1	60		
(56) Communication eq.					9	511	105	3
(57) Electronic compon.								
(58) Batteries, etc.						4		3
(59) Motor vehicles & eq.	4	2	4	6	228	113		5
(60) Aircraft					5			
(61) Trains, ships, etc.				1	480			
(62) Instruments, etc.		1	1			1		19
(63) Photo. apparatus								
(64) Misc. manufactures							2	
(65) Transportation	1	3	2	1	9	7		6
(66) Telephone								
(68) Utilities								
(70) Finance & insurance								
(73) Business services	3							
(77) Institutions	6							

Table D5 (*continued*)

Sector	Sector (69)	(70)	(71)	(72)	(73)	(75)	(76)	(77)
(11) New construction	139	47	12	205	49	240	254	581
(16) Textiles	3			10				
(20) Wood & products	1							
(22) Household furniture	26	10	2	85	19			13
(23) Office furniture	13	3		13	5	1		4
(24) Paper & products				3	1			
(27) Basic chemicals								
(30) Paint								
(32) Rubber prod., etc.								
(36) Stone & clay prod.	2							
(37) Iron & steel				3				
(38) Nonferrous metals								
(39) Metal containers								
(40) Heating, etc.	16			44		83		
(41) Stampings, etc.	3			1				
(42) Hardware, etc.	1			8	4	11		1
(43) Engines & turbines								
(44) Farm equipment								
(45) Constr. & mining eq.								
(46) Materials hand. eq.				1		14		
(47) Metalworking eq.						18		
(48) Special ind. eq.	5				2	6		
(49) General ind. eq.	35			2		33		
(50) Machine shop prod.								
(51) Office & comp. mach.	7	16		130	11	5		
(52) Service ind. mach.	10	7		56	5	32		
(53) Electrical apparatus						2		
(54) Household appliances	9			2		6		
(55) Light. & wiring eq.						1		
(56) Communication eq.	1					1		
(57) Electronic compon.								
(58) Batteries, etc.					3			1
(59) Motor vehicles & eq.	51			64		161	2	
(60) Aircraft								
(61) Trains, ships, etc.								
(62) Instruments, etc.	1	1		39	12	1		3
(63) Photo. apparatus						3	3	
(64) Misc. manufactures	9	2		16	2	3		
(65) Transportation	2	1		6	1			
(66) Telephone								
(68) Utilities								
(70) Finance & insurance								2
(73) Business services								4
(77) Institutions								9

[a] Rows where all entries were zero in 1947 and 1958 are omitted from the table.

References

Alexander, W. O. "The Competition of Materials," *Scientific American*, 217(3), 1967, pp. 255–266.

Almon, Clopper, Jr. *Matrix Methods in Economics*. Reading, Mass.: Addison-Wesley Publishing Co., 1967.

——— "Investment in Input-Output Models and the Treatment of Secondary Products," in Anne P. Carter and Andrew Brody (eds.), *Applications of Input-Output Analysis*. Amsterdam: North-Holland Publishing Co., 1969, pp. 103–116.

Ames, Edward and Nathan Rosenberg. "The Progressive Division and Specialization of Industries," Institute Paper No. 39, Institute for Quantitative Research in Economics and Management, Purdue University, 1963.

Balderston, J. B. and T. M. Whitin. "Aggregation in the Input-Output Model," in Oskar Morgenstern (ed.), *Economic Activity Analysis*. New York: John Wiley & Sons, Inc., 1954, pp. 79–128.

Brody, Andrew. "A Simplified Growth Model," *The Quarterly Journal of Economics*, 80 (February 1966), pp. 137–146.

Brown, Murray (ed.). *The Theory and Empirical Analysis of Production*, Studies in Income and Wealth, vol. 31. New York: Columbia University Press, 1967.

Brubaker, Sterling. *Trends in the World Aluminum Industry*. Baltimore: The Johns Hopkins Press, 1967.

Bureau of Labor Statistics. *Science and Engineering in American Industry*, Report on a 1956 Survey, NSF 59–50. Washington, D.C.: U. S. Government Printing Office, 1960.

——— *Outlook for Numerical Control of Machine Tools*, BLS Bulletin No. 1437. Washington, D.C.: U. S. Government Printing Office, 1965.

——— *Projections 1970*, BLS Bulletin No. 1536. Washington, D.C.: U. S. Government Printing Office, 1966.

——— *1970 Input-Output Coefficients*, BLS Report No. 326. Washington, D.C.: U. S. Government Printing Office, 1967.

——— *Capital Flow Matrix, 1958*, BLS Bulletin No. 1601. Washington, D.C.: U. S. Government Printing Office, 1968.

Carter (Grosse), Anne P. "The Technological Structure of the Cotton Textile Industry," in Wassily Leontief et al., *Studies in the Structure of the American Economy*. New York: Oxford University Press, 1953, pp. 360–420.

Carter, Anne P. "Capital Coefficients as Economic Parameters: The Problem of Instability," in *Problems of Capital Formation*, Studies in Income and Wealth, vol. 19. Princeton: Princeton University Press, 1957, pp. 287–310.

——— "Investment, Capacity Utilization, and Change in Input Structures in the Tin Can Industry," *The Review of Economics and Statistics*, 42 (August 1960), pp. 283–291.

——— "Incremental Flow Coefficients for a Dynamic Input-Output Model with Changing Technology," in Tibor Barna (ed.), *Structural Interdependence and Economic Development*. New York: St. Martin's Press, 1963, pp. 277–302.

Carter, Anne P. "The State of the Arts in Projecting Input-Output Structures," Harvard Economic Research Project; presented at the Institute of Management Sciences Boston Meetings, April 7, 1967a.

——— "Changes in the Structure of the American Economy, 1947 to 1958 and 1962," *The Review of Economics and Statistics*, 49 (May 1967b), pp. 209–224.

——— "How to Handle Changing Technical Coefficients in Input-Output Tables," in Keith Cox and Ben M. Enis (eds.), *A New Measure of Responsibility for Marketing*, 1968 June Conference Proceedings, Series No. 27. Chicago: American Marketing Association, 1968, pp. 306–309.

——— "A Linear Programming System Analyzing Embodied Technological Change," in Anne P. Carter and Andrew Brody (eds.), *Contributions to Input-Output Analysis*. Amsterdam: North-Holland Publishing Co., 1969, pp. 77–98.

Chandler, Margaret K. and Leonard R. Sayles. *Contracting-Out*. New York: Graduate School of Business, Columbia University, 1959.

Chenery, H. B. "Process and Production Functions from Engineering Data," in Wassily Leontief et al., *Studies in the Structure of the American Economy*. New York: Oxford University Press, 1953, pp. 297–325.

——— and Paul G. Clark. *Interindustry Economics*. New York: John Wiley and Sons, Inc., 1959.

Conrad, Alfred H. "Labor Utilization and Income Generation," in "Report on Research for 1955," Harvard Economic Research Project, 1956, pp. 33–52, 139–151.

Cornfield, Jerome, W. Duane Evans, and Marvin Hoffenberg. "Full Employment Patterns, 1950," *Monthly Labor Review*, 64 (February–March 1947), pp. 163–190, 420–432.

Creamer, Daniel, Sergei P. Dobrovolsky, and Israel Borenstein. *Capital in Manufacturing and Mining*. Princeton: Princeton University Press, 1960.

David, Paul A. "The Deflation of Value-Added," *The Review of Economics and Statistics*, 44 (May 1962), pp. 148–155.

Denison, Edward F. *The Sources of Economic Growth in the United States and the Alternatives Before Us*, Supplementary Paper No. 13. New York: Committee for Economic Development, 1962.

——— "The Unimportance of the Embodied Question," *American Economic Review*, March 1964, pp. 90–94.

Denton, Norman. "Technology of Plastics," in Arthur Garrat (ed.), *Penguin Technology Survey, 1967*. Baltimore: Penguin Books, 1967, pp. 68–105.

Domar, Evsey D. "An Index Number Tournament," *The Quarterly Journal of Economics*, 81 (May 1967), pp. 169–188.

Eisner, Robert. "Capital and Labor in Production: Some Direct Estimates," with comments by Bert G. Hickman, Dale W. Jorgenson, and Evsey D. Domar; in Murray Brown (ed.), *The Theory and Empirical Analysis of Production*, Studies in Income and Wealth, vol. 31. New York: Columbia University Press, 1967, pp. 431–475.

Evans, W. Duane and Marvin Hoffenberg. "The Interindustry Relations Study for 1947, *The Review of Economics and Statistics*, 34 (May 1952), pp. 97–142.

Fellner, William. "Does the Market Direct the Relative Factor-Saving Effects of Technological Progress?," in *The Rate and Direction of Inventive Activity:*

Economic and Social Factors, Special Conference Series, No. 13. Princeton: Princeton University Press, 1962, pp. 171–188.

Fisher, Walter D. "Criteria for Aggregation in Input-Output Analysis," *The Review of Economics and Statistics*, 40 (August 1958), pp. 250–260.

Foote, Nelson N. and Paul K. Hatt. "Social Mobility and Economic Advancement," *American Economic Review*, May 1953, pp. 364–378.

Freeman, Richard. "The Labor Market for College Trained Manpower," Ph.D. dissertation, Harvard University, 1967.

Fuchs, Victor R. *The Growing Importance of the Service Industries*, Occasional Paper 96. New York: Columbia University Press, 1965.

——— and Jean A. Wilburn. *Productivity Differences within the Service Sector*, Occasional Paper 102. New York: Columbia University Press, 1967.

Gigantes, T. "The Representation of Technology in Input-Output Systems," in Anne P. Carter and Andrew Brody (eds.), *Contributions to Input-Output Analysis.* Amsterdam: North-Holland Publishing Co., 1969, pp. 270–290.

Goldman, Morris R., Martin L. Marimont, and Beatrice N. Vaccara. "The Interindustry Structure of the United States: A Report on the 1958 Input-Output Study," *Survey of Current Business*, November 1964, pp. 10–29.

Goldsmith, Raymond W. *The National Wealth of the United States in the Postwar Period.* Princeton: Princeton University Press, 1962.

Greenfield, Harry I. *Manpower and the Growth of Producer Services.* New York: Columbia University Press, 1966.

Grose, Lawrence, Irving Rottenberg, and Robert C. Wasson. "New Estimates of Fixed Business Capital in the United States, 1925–1965," *Survey of Current Business*, December 1966, pp. 34–40.

Grosse, Robert N. *Capital Requirements for the Expansion of Industrial Capacity*, vol. 1, Interindustry Economic Research Program. Washington, D.C.: Executive Office of the President, 1953.

Hadley, G. *Linear Programming*, Series in Industrial Management. Reading, Mass.: Addison-Wesley Publishing Co., Inc., 1962.

Harvard Economic Research Project. "Report to the Bureau of Labor Statistics, Interagency Growth Project," Harvard Economic Research Project, 1964.

Hickman, Bert G. *Investment Demand and U.S. Economic Growth.* Washington, D.C.: The Brookings Institution, 1965.

Holzman, Mathilda. "Problems of Classification and Aggregation," in Wassily Leontief et al., *Studies in the Structure of the American Economy.* New York: Oxford University Press, 1953, pp. 326–359.

Horowitz, Morris A. and Irwin L. Herrnstadt. "Changes in the Skill Requirements of Occupations in Selected Industries," in *The Employment Impact of Technological Change*, appendix vol. 2 of *Technology and the American Economy.* Washington, D.C.: U. S. Government Printing Office, 1966, pp 223–287.

Jack Faucett Associates, Inc. "Input-Output Transactions by Transportation Mode, 1947 and 1958"; prepared for Federal Aviation Administration, Silver Spring, Maryland, 1968a.

——— "Statistical Supplement to Input-Output Transactions by Transportation Mode, 1947 and 1958"; prepared for Federal Aviation Administration, Silver Spring, Maryland, 1968b.

Jaszi, George, Robert C. Wasson, and Lawrence Grose. "Expansion of Fixed Business Capital in the United States," *Survey of Current Business*, November 1962, pp. 9–18.

Jewkes, John, David Sawers, and Richard Stillerman. *The Sources of Invention.* New York: St. Martin's Press, 1958.

Kendrick, John W. *Productivity Trends in the United States.* Princeton: Princeton University Press, 1961.

Kennedy, Charles. "Induced Bias in Innovation and the Theory of Distribution," *Economic Journal*, 74 (September 1964), pp. 541–547.

Klein, L. R. "Some Theoretical Issues in the Measurement of Capacity," *Econometrica*, 28 (April 1960), pp. 272–286.

Koenig, Lou P. and Philip M. Ritz. "Investment Factors: National Economic Strength Model," National Resource Evaluation Center Technical Report No. 66. Washington, D.C.: National Planning Association, March 1967.

Komiya, Ryutaro. "Technological Progress and the Prediction of Input Coefficient in the United States Steam Power Industry," in "Report on Research for 1958–1959," Harvard Economic Research Project, 1960, pp. 70–93.

——— "Technological Progress and the Production Function in the United States Steam Power Industry," *The Review of Economics and Statistics*, 44 (May 1962), pp. 156–166.

Kossov, V. V. "The Theory of Aggregation in Interindustry Models," in Anne P. Carter and Andrew Brody (eds.), *Contributions to Input-Output Analysis.* Amsterdam: North-Holland Publishing Co., 1969, pp. 241–248.

Kuznets, Simon. *National Product since 1869.* Publications of National Bureau of Economic Research No. 46, New York: National Bureau of Economic Research, Inc., 1946.

Landsberg, Hans H., Leonard L. Fischman, and Joseph L. Fisher. *Resources in America's Future.* Baltimore: The Johns Hopkins Press, 1963.

——— and Sam H. Schurr. *Energy in the United States: Sources, Uses, and Policy Issues.* New York: Random House, 1968.

Lave, Lester B. *Technological Change: Its Conception and Measurement.* Englewood Cliffs, N.J.: Prentice-Hall, Inc., 1966.

Leontief, Wassily W. *The Structure of the American Economy, 1919–1939*, 2nd ed. New York: Oxford University Press, 1951.

——— et al. *Studies in the Structure of the American Economy.* New York: Oxford University Press, 1953.

——— "Structural Change," in Wassily Leontief et al., *Studies in the Structure of the American Economy.* New York: Oxford University Press, 1953, pp. 17–52.

——— "Technological Change, Consumption, Saving, and Employment in Economic Growth," in *Saving in Contemporary Economic Research*, Transactions of the Congress Organized by the General Savings Bank and Pension Fund of Belgium on the Occasion of its Centenary, Brussels, November 16–19, 1965, pp. 123–132.

——— and Anne P. Carter. "The Position of Metalworking Industries in the Structure of an Industrializing Economy," Harvard Economic Research Project, 1966.

Lynn, Frank, Thomas Roseberry, and Victor Babich. "A History of Recent

Technological Innovations," in *Employment Impact of Technological Change*, appendix vol. 2 of *Technology and the American Economy*. Washington, D.C.: U. S. Government Printing Office, 1966, pp. 46–91.

Manne, Alan S. and Harry M. Markowitz (eds.). *Studies in Process Analysis*, Monograph 18. New York: John Wiley and Sons, Inc., 1963.

Mansfield, Edwin. *Industrial Research and Technological Innovation*. New York: W. W. Norton & Company, Inc., 1968.

May, Kenneth O. "Technological Change and Aggregation," *Econometrica*, 15 (January 1947), pp. 51–63.

McGraw-Hill. *McGraw-Hill Encyclopedia of Science and Technology*, vol. 13. New York: McGraw-Hill, 1960.

Melman, Seymour. *Dynamic Factors in Industrial Productivity*. New York: John Wiley and Sons, Inc., 1956.

Middelhoek, A. J. "Tests of Marginal Stability of Input-Output Coefficients: 1970 Projections of a Coefficient Matrix," in Anne P. Carter and Andrew Brody (eds.), *Applications of Input-Output Analysis*. Amsterdam: North-Holland Publishing Co., 1969, pp. 261–279.

Miernyk, William H. *The Elements of Input-Output Analysis*. New York: Random House, 1965.

——— "Sampling Techniques in Making Regional Industry Forecasts," in Anne P. Carter and Andrew Brody (eds.), *Applications of Input-Output Analysis* Amsterdam: North-Holland Publishing Co., 1969, pp. 305–321.

Morishima, Michio. "Prices, Interest and Profits in a Dynamic Leontief System," *Econometrica*, 26 (July 1958), pp. 358–380.

——— *Equilibrium Stability and Growth*. Oxford, England: The Clarendon Press, 1964.

National Commission on Technology, Automation, and Economic Progress. *Technology and the American Economy*, vol. 1. Washington, D.C.: U. S. Government Printing Office, 1966.

National Science Foundation. *Funds for Research and Development in Industry, 1958*, Surveys of Science Research Series, NSF 61–32. Washington, D.C.: U. S. Government Printing Office, 1961.

Nelson, Richard R., Merton J. Peck, and Edward J. Kalachek. *Technology, Economic Growth, and Public Policy*. Washington, D.C.: The Brookings. Institution, 1967.

Nelson, Ralph L. "Market Growth, Company Diversification, and Product Concentration, 1947–1954," *Journal of the American Statistical Association*, 55 (December 1960), pp. 640–649.

Nerlove, Marc. "Recent Empirical Studies of the CES and Related Production Functions," in Murray Brown (ed.), *The Theory and Empirical Analysis of Production*, Studies in Income and Wealth, vol. 31. New York: Columbia University Press, 1967, pp. 55–122.

Office of Business Economics. *Business Statistics, 1963: Biennial Edition*, a Supplement to the *Survey of Current Business*. Washington, D.C.: U. S. Government Printing Office, 1963.

——— "The Transactions of the 1958 Input-Output Study and Revised Direct and Total Requirements Data," *Survey of Current Business*, September 1965, pp. 33–49, 56.

Office of Business Economics. "Additional Industry Detail for the 1958 Input-Output Study," *Survey of Current Business*, April 1966, pp. 14–17.

——— "Fixed Business Capital in the United States, 1925–1966," *Survey of Current Business*, December 1967, pp. 46–52.

Peck, Merton J. *Competition in the Aluminum Industry, 1945–1958*. Cambridge, Mass.: Harvard University Press, 1961.

Phelps, Edmund S. "The New View of Investment: A Neoclassical Analysis," *The Quarterly Journal of Economics*, 76 (November 1962), pp. 548–567.

——— "Substitution, Fixed Proportions, Growth and Distribution," *International Economic Review*, 4 (September 1963), pp. 265–288.

——— *Golden Rules of Economic Growth*. New York: W. W. Norton and Company, Inc., 1966.

Regan, William J. "Economic Growth and Services," *Journal of Business*, 36 (April 1963), pp. 200–209.

Ricardo, David. *The Principles of Political Economy and Taxation*. London: J. M. Dent and Sons, Ltd., 1911.

Salter, W. E. G. *Productivity and Technical Change*. Cambridge, England: Cambridge University Press, 1960.

Samuelson, Paul A. "Abstract of a Theorem Concerning Substitutability in Open Leontief Models," in Tjalling C. Koopmans (ed.), *Activity Analysis of Production and Allocation*, Monograph No. 13. New York: John Wiley and Sons, Inc., 1951, pp. 142–146.

——— "A Theory of Induced Innovation along Kennedy-Weisäcker Lines," *The Review of Economics and Statistics*, 47 (November 1965), pp. 343–356.

Schon, Donald A. *Technology and Change: The New Heraclitus*. New York: Delacorte Press, 1967.

Schmookler, Jacob. *Invention and Economic Growth*. Cambridge, Mass.: Harvard University Press, 1966.

Schumpeter, Joseph A. *History of Economic Analysis*. New York: Oxford University Press, 1954.

Schurr, Sam H. and Bruce C. Netschert. *Energy in the American Economy, 1850–1975*. Baltimore: The Johns Hopkins Press, 1960.

Schwartz, Eugene S. and Theodore O. Prenting. "Automation in the Fabricating Industries," in *The Outlook for Technological Change and Employment*, appendix vol. 1 of *Technology and the American Economy*. Washington, D.C.: U. S. Government Printing Office, 1966, pp. 291–361.

Scientific American, Inc. *Competition of Materials*. New York: Scientific American, Inc., 1968.

Searle, Allan D. "Capital Goods Pricing," in *Measuring the Nation's Wealth*, Studies in Income and Wealth, vol. 29. New York: Columbia University Press, 1964, pp. 355–366.

Sevaldson, Per. "Changes in Input-Output Coefficients," in Tibor Barna (ed.), *Structural Interdependence and Economic Development*, New York: St. Martin's Press, 1963, pp. 303–328.

——— "The Stability of Input-Output Coefficients," in Anne P. Carter and Andrew Brody (eds.), *Applications of Input-Output Analysis*. Amsterdam: North-Holland Publishing Co., 1969, pp. 207–237.

Simon, Herbert. "Effects of Technological Change in a Linear Model," in Tjalling C. Koopmans (ed.), *Activity Analysis of Production and Allocation*, Monograph No. 13. New York: John Wiley and Sons, Inc., 1951, pp. 260–277.

Simpson, David and Jinkichi Tsukui. "The Fundamental Structure of Input-Output Tables, An International Comparison," *The Review of Economics and Statistics*, 47 (November 1965), pp. 434–446.

Solow, Robert M. "Technical Change and the Aggregate Production Function," *The Review of Economics and Statistics*, 39 (August 1957), pp. 312–320.

——— "Competitive Valuation in a Dynamic Input-Output System," *Econometrica*, 27 (January 1959), pp. 30–53.

——— "Investment and Technical Progress," in Kenneth J. Arrow, Samuel Karlin, and Patrick Suppes (eds.), *Mathematical Methods in the Social Sciences, 1959*, Stanford Mathematical Studies in the Social Sciences IV. Stanford: Stanford University Press, 1960, pp. 89–104.

Stigler, George. "The Division of Labor is Limited by the Extent of the Market," *The Journal of Political Economy*, 59 (June 1951), pp. 185–193.

Strout, Alan M. "Technological Change and the United States Energy Consumption, 1939–1954," Ph.D. dissertation, University of Chicago, 1967.

Taskier, Charlotte E. *Input-Output Bibliography, 1955–1960*, ST/STAT 7. New York: United Nations, 1961.

Theil, Henri. *Applied Economic Forecasting*, vol. 4. *Studies in Mathematical and Managerial Economics*. Edited by Henri Theil. Chicago: Rand McNally Company, 1966.

Tilanus, Christiaan B. *Input-Output Experiments; the Netherlands, 1948–1961*. Rotterdam: Rotterdam University Press, 1966.

Tostlebe, Alvin S. *Capital in Agriculture*. Princeton: Princeton University Press, 1957.

Ulmer, Melville J. *Capital in Transportation, Communications, and Public Utilities*. Princeton: Princeton University Press, 1960.

United Nations. *Input-Output Bibliography, 1960–1963*, Statistical Papers, Series M, No. 39. New York: United Nations, 1964.

——— *Input-Output Bibliography, 1963–1966*, Statistical Papers, Series M, No. 46. New York: United Nations, 1967.

U. S. Bureau of the Census. *Historical Statistics of the United States: Colonial Times to 1957*. Washington, D.C.: U. S. Government Printing Office, 1960.

——— *Historical Statistics of the United States: Colonial Times to 1957: Continuation to 1962 and Revisions*. Washington, D.C.: U. S. Government Printing Office, 1965.

——— Industry Division. "Evaluation of Proposed New (Hypothetical) SIC Industry," U. S. Bureau of the Census, June 1968.

U. S. Department of Commerce. *See* Office of Business Economics.

U. S. Department of Labor. *See* Bureau of Labor Statistics.

U. S. Steel Corporation, Marketing Department. "Highlights of U. S. Steel's Work with Input-Output for the Fastener Industry," U. S. Steel Corporation, 1967.

Vaccara, Beatrice N. and Nancy W. Simon. "Factors Affecting the Postwar Industrial Composition of Real Product," in *The Industrial Composition of Income and Products*, Studies in Income and Wealth, vol. 32. New York: Columbia University Press, 1968, pp. 19–66.

——— "Changes Over Time in Input-Output Coefficients for the United States," in Anne P. Carter and Andrew Brody (eds.), *Applications of Input-Output Analysis*. Amsterdam: North-Holland Publishing Co., 1969, pp. 238–260.

Waddell, Robert M., Philip M. Ritz, John DeWitt Norton, and Marshall K. Wood. *Capacity Expansion Planning Factors*. Washington, D.C.: National Planning Association, 1966.

von Weizsäcker, C. C. "Tentative Notes on a Two Sector Model with Induced Technical Progress," *The Review of Economic Studies*, 33 (July 1966), pp. 245–251.

Yan, Chiou-shuang and Edward Ames. "Economic Interrelatedness," *The Review of Economic Studies*, 32 (October 1965), pp. 299–310.

Index